P9-CDB-902

Student Solutions Manual

Daniel Miller • Laurel Tech

Thinking Mathematically

Second Edition

Robert Blitzer

EXPANDED VERSION

Thinking Mathematically

Second Edition

Robert Blitzer

Prentice Hall

Upper Saddle River, NJ 07458

Editor in Chief: Sally Yagan
Supplement Editor: Joanne Wendelken
Assistant Managing Editor: John Matthews
Production Editor: Wendy A. Perez
Supplement Cover Manager: Paul Gourhan
Supplement Cover Designer: Joanne Alexandris
Manufacturing Buyer: Ilene Kahn

© 2003 by Pearson Education, Inc.
Pearson Education, Inc.
Upper Saddle River, NJ 07458

All rights reserved. No part of this book may be reproduced in any form or
by any means, without permission in writing from the publisher.

The author and publisher of this book have used their best efforts in
preparing this book. These efforts include the development, research, and
testing of the theories and programs to determine their effectiveness. The
author and publisher make no warranty of any kind, expressed or implied,
with regard to these programs or the documentation contained in this
book. The author and publisher shall not be liable in any event for
incidental or consequential damages in connection with, or arising out of,
the furnishing, performance, or use of these programs.

Printed in the United States of America

10 9 8 7 6 5 4 3 2
ISBN 0-13-044971-7

Pearson Education Ltd., *London*
Pearson Education Australia Pty. Ltd., *Sydney*
Pearson Education Singapore, Pte. Ltd.
Pearson Education North Asia Ltd., *Hong Kong*
Pearson Education Canada, Inc., *Toronto*
Pearson Educacíon de Mexico, S.A. de C.V.
Pearson Education—Japan, *Tokyo*
Pearson Education Malaysia, Pte. Ltd.
Pearson Education, *Upper Saddle River, New Jersey*

TABLE OF CONTENTS

PROPERTY OF
TUTORING LAB
INTERNATIONAL COLLEGE

Chapter 1
Problem Solving and Critical Thinking

Check Points 1.1

1. Inductive; the general conclusion for all people who increase time spent exercising was based on specific observations.

2. There are many counterexamples including $50 \times 60 = 3000$.

3. **a.** $3 + 6 = 9$
   $9 + 6 = 15$
   $15 + 6 = 21$
   $21 + 6 = 27$
   $27 + 6 = 33$
   $3, 9, 15, 21, 27, \underline{33}$

   b. $2 \times 5 = 10$
   $10 \times 5 = 50$
   $50 \times 5 = 250$
   $250 \times 5 = 1250$
   $2, 10, 50, 250, \underline{1250}$

4. The shapes alternate between square and triangle.
   The number of little legs cycles from 1 to 2 to 3 and then back to 1.
   Therefore the next figure will be a square with 2 little legs.

5. **a.** Conjecture based on results: The original number is doubled.

Select a number	4	10	0	3
Multiply the number by 4.	$4 \times 4 = 16$	$10 \times 4 = 40$	$0 \times 4 = 0$	$3 \times 4 = 12$
Add 6 to the product.	$16 + 6 = 22$	$40 + 6 = 46$	$0 + 6 = 6$	$12 + 6 = 18$
Divide this sum by 2.	$22 \div 2 = 11$	$46 \div 2 = 23$	$6 \div 2 = 3$	$18 \div 2 = 9$
Subtract 3 from the quotient.	$11 - 3 = 8$	$23 - 3 = 20$	$3 - 3 = 0$	$9 - 3 = 6$
Summary of results:	$4 \rightarrow 8$	$10 \rightarrow 20$	$0 \rightarrow 0$	$3 \rightarrow 6$

   b. Select a number: n
   Multiply the number by 4: $4n$
   Add 6 to the product: $4n + 6$

   Divide this sum by 2: $\dfrac{4n+6}{2} = \dfrac{4n}{2} + \dfrac{6}{2} = 2n + 3$

   Subtract 3 from the quotient: $2n + 3 - 3 = 2n$

Exercise Set 1.1

1. Inductive; the general conclusion for all people's fingerprints was based on specific observations.

3. Deductive; the specific conclusion about my snake was based on general statements.

5. Counterexample: George W. Bush was younger than 65 at the time of his inauguration.

7. Counterexample: *War and Peace* by Leo Tolstoy has more than 600 pages.

9. Counterexample: $7 \times 7 = 49$

11. Counterexample: $\dfrac{1}{2} \neq \dfrac{1+1}{2+1} = \dfrac{2}{3}$

13. $8 + 4 = 12$
    $12 + 4 = 16$
    $16 + 4 = 20$
    $20 + 4 = 24$
    $24 + 4 = 28$
    $8, 12, 16, 20, 24, \underline{28}$

15. $37 - 5 = 32$
$32 - 5 = 27$
$27 - 5 = 22$
$22 - 5 = 17$
$17 - 5 = 12$
37, 32, 27, 22, 17, <u>12</u>

17. $3 \times 3 = 9$
$9 \times 3 = 27$
$27 \times 3 = 81$
$81 \times 3 = 243$
$243 \times 3 = 729$
3, 9, 27, 81, 243, <u>729</u>

19. $1 \times 2 = 2$
$2 \times 2 = 4$
$4 \times 2 = 8$
$8 \times 2 = 16$
$16 \times 2 = 32$
1, 2, 4, 8, 16, <u>32</u>

21. $4 \times 2 = 8$
$8 \times 2 = 16$
$16 \times 2 = 32$
1, 4, 1, 8, 1, 16, 1, <u>32</u>

23. The figure cycles from square to triangle to circle and then repeats. So the next figure is

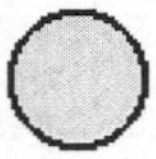

25. The pattern is to add one more letter to the previous figure and use the next consecutive letter in the alphabet. The next figure shown at right.

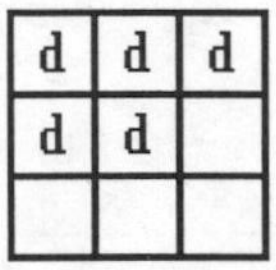

d	d	d
d	d	

27. a. Conjecture based on results: The original number is doubled.

Select a number	4	10	0	3
Multiply the number by 4.	$4 \times 4 = 16$	$10 \times 4 = 40$	$0 \times 4 = 0$	$3 \times 4 = 12$
Add 8 to the product.	$16 + 8 = 24$	$40 + 8 = 48$	$0 + 8 = 8$	$12 + 8 = 20$
Divide this sum by 2.	$24 \div 2 = 12$	$48 \div 2 = 24$	$8 \div 2 = 4$	$20 \div 2 = 10$
Subtract 4 from the quotient.	$12 - 4 = 8$	$24 - 4 = 20$	$4 - 4 = 0$	$10 - 4 = 6$
Summary of results:	$4 \to 8$	$10 \to 20$	$0 \to 0$	$3 \to 6$

b. $4n$
$4n + 8$
$$\frac{4n + 8}{2} = \frac{4n}{2} + \frac{8}{2} = 2n + 4$$
$2n + 4 - 4 = 2n$

29. a. Conjecture based on results: The result is always 3.

Select a number.	4	10	0	3
Add 5 to the number.	$4 + 5 = 9$	$10 + 5 = 15$	$0 + 5 = 5$	$3 + 5 = 8$
Double the result.	$9 \times 2 = 18$	$15 \times 2 = 30$	$5 \times 2 = 10$	$8 \times 2 = 16$
Subtract 4.	$18 - 4 = 14$	$30 - 4 = 26$	$10 - 4 = 6$	$16 - 4 = 12$
Divide the result by 2.	$14 \div 2 = 7$	$26 \div 2 = 13$	$6 \div 2 = 3$	$12 \div 2 = 6$
Subtract the original number.	$7 - 4 = 3$	$13 - 10 = 3$	$3 - 0 = 3$	$6 - 3 = 3$
Summary of results:	$4 \to 3$	$10 \to 3$	$0 \to 3$	$3 \to 3$

b. $n + 5$
$2(n + 5) = 2n + 10$
$2n + 10 - 4 = 2n + 6$
$$\frac{2n + 6}{2} = \frac{2n}{2} + \frac{6}{2} = n + 3$$
$n + 3 - n = 3$

31. Using inductive reasoning we predict $1+2+3+4+5+6 = \dfrac{6 \times 7}{2}$. Arithmetic verifies this result: $21 = 21$

33. Using inductive reasoning we predict $1+3+5+7+9+11 = 6 \times 6$. Arithmetic verifies this result: $36 = 36$

35. a.
$3 - 1 = 2$
$6 - 3 = 3$
$10 - 6 = 4$
$15 - 10 = 5$
$21 - 15 = 6$
The successive differences increase by 1.

    b. 1, 3, 6, 10, 15, and 21 are followed by
$21 + 7 = 28$
$28 + 8 = 36$
$36 + 9 = 45$
$45 + 10 = 55$
$55 + 11 = 66$
1, 3, 6, 10, 15, 21, 28, 36, 45, 55, and 66.

37. a. Each additional inch increases the shoe size by 3 units.

Foot length	9	10	11	12	13	14	15	16	17
Shoe size	5	8	11	14	17	20	23	26	29

    b. Continuing the pattern suggests Matthew McGrory's feet are 17 inches.

39–45. Answers will vary.

47.
$1 + 1 = 2$
$1 + 2 = 3$
$2 + 3 = 5$
$3 + 5 = 8$
$5 + 8 = 13$
$8 + 13 = 21$
$13 + 21 = 34$
$21 + 34 = 55$
1, 1, 2, 3, 5, 8, 13, 21, 34, <u>55</u>

49. The first multiplier increases by 33. $132 + 33 = 165$
The second multiplier is 3367. The product increases by 111,111. $165 \times 3367 = 555,555$ is correct.

51. Answers will vary. Possible answer:

$5 \times 1 = 5$ $5 \times 2^0 = 5$
$5 \times 2 = 10$ $5 \times 2^1 = 10$
$5 \times 3 = 15$ $5 \times 2^2 = 20$

53. a.
$3367 \times 3 = 10101$
$3367 \times 6 = 20202$
$3367 \times 9 = 30303$
$3367 \times 12 = 40404$

    b. The first multiplier is always 3367. The second multipliers are successive multiples of 3. The product increases by 10101.

    c.
$3367 \times 15 = 50505$
$3367 \times 18 = 60606$

    d. Inductive reasoning; it uses an observed pattern and draws conclusions from that pattern.

Check Points 1.2

1. 58 rounded to the nearest ten is 60.

2. a. $\$2.40+\$1.25+\$4.60+\$4.40+\$1.40+\$1.85+2.95 \approx \$2+\$1+\$5+\$4+\$1+\$2+3 \approx \$18$

   b. The bill of $21.85 is not reasonable. It is too high.

3. $\dfrac{0.2489 \times 48}{0.5103} = \dfrac{\frac{1}{4} \times 48}{\frac{1}{2}} = \dfrac{12}{\frac{1}{2}} = 12 \times \dfrac{2}{1} = 24$

4. a. Round $52 per hour to $50 per hour and assume 40 hours per week.

   $\dfrac{40 \text{ hours}}{\text{week}} \times \dfrac{\$50}{\text{hour}} = \dfrac{\$2000}{\text{week}}$ Therefore, the architect's salary is $\approx \$2000$ per week.

   b. Round 52 weeks per year to 50 weeks per year.

   $\dfrac{\$2000}{\text{week}} \times \dfrac{50 \text{ weeks}}{\text{year}} = \dfrac{\$100,000}{\text{year}}$ Therefore, the architect's salary is $\approx \$100,000$ per year.

5. First approximate minutes: $\dfrac{10,000 \text{ numbers}}{\frac{60 \text{ numbers}}{\text{minute}}} = 10,000 \text{ numbers} \times \dfrac{\text{minute}}{60 \text{ numbers}} = \dfrac{10,000 \text{ minutes}}{60} = 166\frac{2}{3} \text{ minutes.}$

$166\frac{2}{3}$ minutes equals about 2 hours and 45 minutes. Therefore 3 hours would be a reasonable estimate.

6. a. The sector labeled "African American" shows that 13.6% of the U.S. population in 2050 will be African American.

   b. Round 393,931,000 to 400,000,000 and round 8.2% to 8%. $400,000,000 \times \dfrac{8}{100} = 32,000,000$ Asian Americans.

7. a. $\approx 14\%$

   b. Tommy Hilfiger, Calvin Klein, and Nike

8. Between 1900 and 1950 the graph's maximum is about 21½. This occurs in 1900.

Exercise Set 1.2

1. Round $0.19 to $0.20. $12 \times \$0.19 \approx 12 \times \$0.20 \approx \$2.40$

3. Round 48 mph to 50 mph. Round 7 hours and 8 minutes to 7 hours.
48 mph $\times (7 \text{ hours}+8 \text{ minutes}) \approx 50 \text{ mph} \times 7 \text{ hours} \approx 350 \text{ miles}$

5. $\$3.47+\$5.89+\$19.98+\$2.03+\$11.85+\$0.23 \approx \$3+\$6+\$20+\$2+\$12+\$0 \approx \$43$

7. $\$2037 \times 0.05 \approx \dfrac{\$2000 \times 0.10}{2} \approx \dfrac{\$200}{2} \approx \$100$

9. Round $19.50 to $20 per hour.
40 hours per week
$(40 \times \$20)$ per week = $800/week
Round 52 weeks to 50 weeks per year.
50 weeks per year
$(50 \times \$800)$ per year = $40,000
$19.50 per hour $\approx$ $40,000 per year

11. $\dfrac{0.57 \times 68}{0.493} \approx \dfrac{0.5 \times 68}{0.5} \approx 68$

13. Hours in a year:
    Round 365 days/year to 400 days/year. Round 24 hours/day to 25 hours/day
    Hours in a year ≈ 400 days/year × 25 hours/day ≈ (400 × 25) hours ≈ 10,000 hours

15. Round 78 years to 80 years. Round 365 days/year to 400 days/year. Round 24 hours/day to 25 hours/day.
    ≈ (80 years)(400 days/year)(25 hours/day) ≈ (80 × 400 × 25) hours ≈ 800,000 hours

17. Round $61,500 to $60,000 per year. Round 52 weeks per year to 50 weeks per year.
    50 weeks × 40 hours per week = 2000 hours
    $60,000 ÷ 2000 hours = $30 per hour
    $61,500 per year ≈ $30 per hour

19. Round the raise of $310,000 to $300,000. Round the 294 professors to 300.
    $300,000 ÷ 300 professors = $1000 per professor.
    $310,000 raise ≈ $1000 per professor.

21. Round the $605 monthly payment to $600. 3 years is 36 months. Round the 36 months to 40 months.
    $600 × 40 months = $24,000 total cost.
    $605 monthly payment for 3 years ≈ $24,000 total cost.

23. Round the $21.36 to $20. 15% of $20 is a $3.00 tip.

25. **a.** The distance on map is about 3 inches. So, at 80 miles/inch this means the total distance
    ≈ 3 × 80 miles ≈ 240 miles .

    b. $\dfrac{240 \text{ miles}}{40 \text{ miles / hour}} = 6 \text{ hours}$, therefore travel time ≈ 6 hours

27. Round US population to 300,000,000 people. Round percentage of Protestants to 60%.
    Protestant population ≈ $\dfrac{6}{10}$(300,000,000 people) ≈ 180,000,000 people.

29. Round the 9976 hate crimes to 10,000. Round percentage motivated by race to 60%.
    Hate crimes motivated by race ≈ $\dfrac{6}{10}$(10,000 crimes motivated by race) ≈ 6000 crimes motivated by race.

31. The graph indicates that approximately 33% of vacations include shopping.

33. historical places and museums, outdoor recreation, and shopping

35. The life expectancy for men born in 1900 was approximately 48 years.

37. Approximately 14 years more. The life expectancy for women born in 1996 was approximately 80 years. The life expectancy for men born in 1950 was approximately 66 years.

39. Approximately $470

41. Approximately $20

43. Approximately 5%

45. It reached a maximum in 1982. It was at about 9.9% for that year.

47. 1832; approximately 20%

49. Approximately 52%

Chapter 1: Problem Solving and Critical Thinking

51-59. Answers will vary.

61. c is false

63. Round days in a year to 400. $\dfrac{\$1,000,000,000}{\$1000/\text{day}} = 1,000,000 \text{ days} \approx \dfrac{1,000,000 \text{ days}}{400 \text{ days/year}} \approx 2500 \text{ years}$

Check Points 1.3

1. The amount of money given to the cashier is unknown.

2. It was unnecessary to state that the $30 included a $20 bill and a $10 bill. The change received should be $30.00 - $20.36 = 9.64 . To receive as few bills and coins as possible you should request one $5 bill, two $2 bills, one half-dollar, one dime, and 4 pennies. Due to the scarcity of $2 bills and half-dollars the next best answer would be one $5 bill, four $1 bills, two quarters, one dime, and 4 pennies.

3. Step 1: Understand the problem. Given the cost of the computer, the amount of cash paid up front, and the amount paid each month, we must determine the number of months it will take to finish paying for the computer.
   Step 2: Devise a plan. Subtract the amount paid in cash from the cost of the computer. This results in the amount still to be paid. Because the monthly payments are $45, divide the amount still to be paid by 45. This will give the number of months required to pay for the computer.
   Step 3: Carry out the plan and solve the problem.
   The balance is $980 - $350 = 630. Now divide the $630 balance by $45, the monthly payment.
   $$\$630 \div \frac{\$45}{\text{month}} = \$630 \times \frac{\text{month}}{\$45} = \frac{630 \text{ months}}{45} = 14 \text{ months}.$$
   Step 4: Look back and check the answer. This answer satisfies the conditions of the problem. 14 monthly payments at $45 each gives $14 \times \$45 = \$630$. Adding in the up front cash payment of $350 gives us $\$630 + \$350 = \$980$. $980 is the cost of the computer.

4. Step 1: Understand the problem.
   Step 2: Devise a plan. Make a list of all possible coin combinations. Begin with the coins of larger value and work toward the coins of smaller value.
   Step 3: Carry out the plan and solve the problem.

Quarters	Dimes	Nickels
1	0	1
0	3	0
0	2	2
0	1	4
0	0	6

   There are 5 ways.
   Step 4: Look back and check the answer. Check to see that no combinations are omitted, and that those given total 30 cents. Also double-check the count.

5. Step 1: Understand the problem. We must determine the number of jeans/T-shirt combinations that we can make. For example, one such combination would be to wear the blue jeans with the beige shirt.
    Step 2: Devise a plan. Each pair of jeans could be matched with any of the three shirts.
    We will make a tree diagram to show all combinations.
    Step 3: Carry out the plan and solve the problem.

JEANS	T-SHIRT	COMBINATIONS
	Beige shirt	Blue jeans-Beige shirt
Blue jeans	Yellow shirt	Blue jeans-Yellow shirt
	Blue shirt	Blue jeans-Blue shirt
	Beige shirt	Black jeans-Beige shirt
Black jeans	Yellow shirt	Black jeans-Yellow shirt
	Blue shirt	Black jeans-Blue shirt

There are 6 different outfits possible.
    Step 4: Look back and check the answer. Check to see that no combinations are omitted, and double-check the count.

6. Step 1: Understand the problem. We are in a log cabin with the materials stated and we must decide what to light first.
    Step 2: Devise a plan. I will consider the ramifications of my various options
    Step 3: Carry out the plan and solve the problem.
    Candle: I could strike the match to light the candle.
    Fireplace: I could strike the match and start the fireplace.
    Stove: I could strike the match to light the woodburning stove.
    Solution: I must light the *match* before I light the candle, fireplace, or the woodburning stove.
    Step 4: Check the answer. The solution is correct, indicating the need to think carefully when problem solving.

7. Step 1: Understand the problem. Despite this person's apparent irresponsibility, she/he is your friend.
    You would like to see your friend's attendance and academic achievement improve.
    Step 2: Devise a plan. List as many ideas as possible.
    Step 3: Carry out the plan and solve the problem. Your list might include…
    - Offer to drive your friend to school.
    - Suggest the friend take later classes in future semesters. (or see if instructor has later sections of the course)
    - Offer a reward to your friend based on good attendance or grades.
    - Offer help with assignments.
    - Try to address the reasons your friend is absent. (ex. offer to baby-sit)
    Step 4: Look back and check the answer. Real-life situations often have more than one solution.
    Also, some of the ideas in our list may be inappropriate or make matters worse. Choose carefully.

Trick Questions 1.3

1. The farmer has 12 sheep left since all but 12 sheep died.

2. All months have [at least] 28 days.

3. The doctor and brother are brother and sister.

Exercise Set 1.3

1. The price of the computer is needed.

3. The number of words per page is needed.

5. The weekly salary is unnecessary information.
    $212 - 200 = 12$ items sold in excess of 200
    $12 \times \$15 = \$180$ extra is received.

7. How much the attendant was given is not necessary.
   There were 5 hours of parking. 1st hour is $2.50 4 hours at $0.50/hr
   $2.50 + (4 \times \$0.50) = \$2.50 + \$2.00 = \$4.50$, therefore $4.50 was charged.

9. Step 1: Comparing two yearly salaries
   Step 2: We need to convert the second person's wages to yearly salary.
   Step 3: The person that earns $3750/month earns
   $\quad$ $12 \times \$3750 = \$45,000$/year. The person that earns $48,000/year gets $3000 more per year.
   Step 4: It appears to satisfy the conditions of the problem.

11. Step 1: We are trying the difference between two methods of payment.
    Step 2: Compute total costs and compare two figures.
    Step 3: By spreading purchase out, the total comes to: $100 + 14($50) = $100 + $700 = $800
    $\quad\quad\quad\quad\quad\quad\quad\quad\quad\quad$ $800 – $750 = $50 saved by paying all at once
    Step 4: It satisfies the conditions of problem.

13. Step 1: Trying to determine profit on goods sold.
    Step 2: Find total cost of buying product and comparing with gross sales.
    Step 3: Purchased: ($65 per dozen)(6 dozen) = $390
    $\quad\quad$ Sold: 6 dozen = 72 calculators
    $\quad\quad$ $\dfrac{72}{3} = 24$ groups of 3 at $20 per group.
    $\quad\quad$ $24 \times \$20 = \$480$
    $\quad\quad$ $480 – $390 = $90 profit
    Step 4: It satisfies the conditions of the problem.

15. Step 1: Determine profit for ten-day period.
    Step 2: Compare totals
    Step 3: (200 slices)($1.50) = $300 for pizza
    $\quad\quad$ (85 sandwiches)($2.50) = $212.50 for sandwiches
    $\quad\quad$ For 10 day period:
    $\quad\quad$ Gross: $10(\$300) + 10(\$212.50) = \$3000 + \$2125.00$
    $\quad\quad\quad\quad\quad\quad\quad\quad\quad\quad$ $= \$5125.00$
    $\quad\quad$ Expenses: 10($60) = $600
    $\quad\quad$ Profit: $5125.00 – $600 = $4525
    Step 4: It satisfies the conditions of the problem.

17. Step 1: We are trying to compute total rental cost.
    Step 2: Add rental cost and mileage cost to get total cost.
    Step 3: Rental costs:
    $\quad\quad$ (2 weeks)($220 per week) = $440
    $\quad\quad$ Mileage: (500 miles)($0.25) = $125
    $\quad\quad$ Total: $440 + $125 = $565
    Step 4: It satisfies the conditions of problem.

19. Step 1: A round trip was made; we need to determine how much was walked or ridden.
    Step 2: Add up the totals walked and ridden and compare.
    Step 3: It is 5 miles between the homes or a 10 mile round trip.
    $\quad\quad$ The first 3 were covered with the bicycle, leaving 7 miles covered by walking.
    $\quad\quad$ 7 miles – 3 miles = 4 miles more that was walked.
    Step 4: It satisfies the conditions of the problem.

21. Step 1: To determine profit by comparing expenses with gross sales
    Step 2: Calculate expenses and gross sales and compare.
    Step 3: Expense: (25 calculators)($30) = $750
    $\quad\quad$ Gross Sales: (22 calculators)($35.00) = $770

The storeowner receives $30 – $2 = $28 for each returned calculator.
(3 calculators)($28) = $84
Total Income: $770 + $84 = $854
Profit = Income – Expenses
$$= \$854 - \$750$$
$$= \$104$$

Step 4: It satisfies the conditions of the problem.

23. Use a list.

2 Quarters	3 Dimes	5 Nickels
1	2	0
1	1	2
1	0	4
0	3	3
0	2	5

There are 5 ways.

25. Make a list of all possible orders: GBBB, BGBB, BBGB, BBBG
The girl could be born 1st, 2nd, 3rd, or 4th. Therefore, there are 4 ways.

27.
```
       156
   28)4368
       28
      156
      140
      168
      168
        0
```

29. The completed diagonal has a sum of 75. Use this sum to complete rows or columns that have two known values. For example, the middle number of the right column must be $75 - 30 - 40 = 5$. Continuing this process results in the magic square shown at right.

10	35	30
45	25	5
20	15	40

31. The figure has 9 one by one's ☐, 4 two by two's ⊞, and 1 three by three ⊞. Giving a total of 9 + 4 + 1 = 14 squares.

33. Move the 3 shaded toothpicks as shown. Notice that there are 4 little squares and 1 big square.

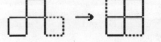

35. The message is: ESCAPE TONIGHT

37-43. Answers will vary.

45. You should choose the dentist whose teeth show the effects of poor dental work because he took good care of the other dentist's teeth.

47. The farmer can use the following strategy: Take the goat to the other side of the stream and return to get either the wolf or cabbage. The farmer should take that across the stream and bring the goat back to the original side. He then takes across the cabbage or wolf, whichever remains, and leaves it on the other side while he returns to get the goat.

49.

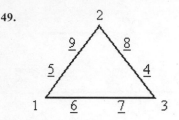

Chapter 1 Review Exercises

1. Vermont, Virginia, Washington, West Virginia, Wisconsin and Wyoming are all counterexamples.

2. Any two consecutive odd numbers, like 5 and 7, are a counterexample.

3. $4 + 5 = 9$
   $9 + 5 = 14$
   $14 + 5 = 19$
   $19 + 5 = 24$
   4, 9, 14, 19, <u>24</u>

4. $7 \times 2 = 14$
   $14 \times 2 = 28$
   $28 \times 2 = 56$
   $56 \times 2 = 112$
   7, 14, 28, 56, <u>112</u>

5. $1 + 2 = 3$
   $3 + 3 = 6$
   $6 + 4 = 10$
   $10 + 5 = 15$
   $15 + 6 = 21$
   1, 3, 6, 10, 15, <u>21</u>

6. The pattern is alternating between square and circle while the line rotates $90°$ clockwise. The next figure is shown at right.

7. **a.** Conjecture based on results: The result is the original number.

Select a number	4	10	0	3
Double the number.	$4 \times 2 = 8$	$10 \times 2 = 20$	$0 \times 2 = 0$	$3 \times 2 = 6$
Add 4 to the product.	$8 + 4 = 12$	$20 + 4 = 24$	$0 + 4 = 4$	$6 + 4 = 10$
Divide this sum by 2.	$12 \div 2 = 6$	$24 \div 2 = 12$	$4 \div 2 = 2$	$10 \div 2 = 5$
Subtract 2 from the quotient.	$6 - 2 = 4$	$12 - 2 = 10$	$2 - 2 = 0$	$5 - 2 = 3$
Summary of results:	4 → 4	10 → 10	0 → 0	3 → 3

   b. $2n$
   $2n + 4$
   $$\frac{2n + 4}{2} = \frac{2n}{2} + \frac{4}{2} = n + 2$$
   $n + 2 - 2 = n$

8. $\$8.47 + \$0.89 + \$2.79 + \$0.14 + \$1.19 + \$4.76 \approx \$8 + \$1 + \$3 + \$0 + \$1 + \$5 \approx \$18$

9. Round 78 hours to 80, round $6.85 to $7.00. $78 \times \$6.85 \approx 80 \times \$7.00 \approx \$560$

10. $60\,\dfrac{\text{seconds}}{\text{minute}} \times 60\,\dfrac{\text{minutes}}{\text{hour}} = 3600\,\dfrac{\text{seconds}}{\text{hour}}$

    Round $3600\,\dfrac{\text{seconds}}{\text{hour}}$ to $4000\,\dfrac{\text{seconds}}{\text{hour}}$. Round $24\,\dfrac{\text{hours}}{\text{day}}$ to $20\,\dfrac{\text{hours}}{\text{day}}$.

    $20\,\dfrac{\text{hours}}{\text{day}} \times 4000\,\dfrac{\text{seconds}}{\text{hour}} = 80,000\,\dfrac{\text{seconds}}{\text{day}}$

    $3600\,\dfrac{\text{seconds}}{\text{hour}} \times 24\,\dfrac{\text{hours}}{\text{day}} \approx 80,000\,\dfrac{\text{seconds}}{\text{day}}$

11. Round $27\dfrac{19}{20}$ to 28. Round 6.823 to 7. $\quad 27\dfrac{19}{20} \div 6.823 \approx 28 \div 7 \approx 4$

12. Round book price to $1.00 each. Round chair price to $12.00 each. Round plate price to $15.00.
    $(21 \times \$0.85) + (2 \times \$11.95) + \$14.65 \approx (21 \times \$1) + (2 \times \$12) + \$15 \approx \$21 + \$24 + \$15 \approx \$60$

13. Round 66.6 million households to 70 million. 25% of 70 million $\approx \left(\dfrac{70}{4}\right)$ million ≈ 17.5 million

14. Possible answer: Successful students are in class more often than unsuccessful students.

15. The veterinary costs for cats in 1983 was approximately $1 billion.

16. The veterinary costs for dogs in 1996 was approximately $8 billion.

17. The veterinary costs for cats in 2000 was approximately $6 billion. The veterinary costs for dogs in 2000 was approximately $14 billion. The difference between the two was approximately $8 billion.

18. This occurs for dogs in 1991, for cats in 1996, and for cats in 2000. Note that the wording in the question *excludes* dogs in 1987 but *includes* cats in 2000.

19. The population was a minimum in 1980, when there were approximately 54 million people in the U.S. under 16.

20. The weight of the child is needed.

21. The unnecessary information is the customer giving the driver a $20 bill. For a 6 mile trip, the first mile is $3.00, and the next 5 miles are $0.50/half-mile or $1.00/mile. The cost is $\$3.00 + (5 \times \$1.00) = \$3.00 + \$5.00 = \$8.00$.

22. Total of $28 \times 2 = 56$ frankfurters would be needed. $\dfrac{56}{7} = 8$. Therefore, 8 pounds would be needed.

23. Rental for 3 weeks at $175 per week is $3 \times \$175 = \$525$. Mileage for 1200 miles at $0.30 per mile is $1200 \times \$0.30 = \$360$. Total cost is $\$525 + \$360 = \$885$.

24. Healthy Bodies is a better deal by $40 for the year.
    Superfit: $\$500 + 80(\$1.00) = \$580$; Healthy Bodies: $\$400 + 80(\$1.75) = \$400 + \$140 = \$540$

25. The flight leaves Miami at 7:00 A.M. Pacific Standard Time. With a lay-over of 45 minutes, it arrives in San Francisco at 1:30 P.M. Pacific Standard Time, 6 hrs 30 min. − 45 min = 5 hours 45 minutes.

26. At steady decrease in value: $\dfrac{\$37,000 - \$2600}{8 \text{ years}} = \dfrac{\$34,400}{8 \text{ years}} = \$4300 / \text{year}$

   After 5 years: $\$4300 \times 5 = \$21,500$ decrease in value

   Value of car: $\$37,000 - \$21,500 = \$15,500$

27. The machine will accept nickels, dimes, quarters.

nickels	dimes	quarters
7	0	0
5	1	0
3	2	0
2	0	1
1	3	0
0	1	1

   There are 6 combinations.

Chapter 1 Test

1. One possible answer is: $50 \times 2 = 100$ which is a three-digit number.

2. $0 + 5 = 5$
   $5 + 5 = 10$
   $10 + 5 = 15$
   $15 + 5 = 20$
   $0, 5, 10, 15, \underline{20}$

3. $\dfrac{1}{6 \times 2} = \dfrac{1}{12}$

   $\dfrac{1}{12 \times 2} = \dfrac{1}{24}$

   $\dfrac{1}{24 \times 2} = \dfrac{1}{48}$

   $\dfrac{1}{48 \times 2} = \dfrac{1}{96}$

   $\dfrac{1}{6}, \dfrac{1}{12}, \dfrac{1}{24}, \dfrac{1}{48}, \dfrac{1}{\underline{96}}$

4. The outer figure is always a square. The inner figure appears to cycle from triangle to circle to square. The line segments at the bottom alternate from two to one. The next shape is shown at right.

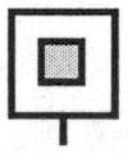

5. **a.** Conjecture based on results: The original number is doubled.

Select a number	4	10	3
Multiply the number by 4.	$4 \times 4 = 16$	$10 \times 4 = 40$	$3 \times 4 = 12$
Add 8 to the product.	$16 + 8 = 24$	$40 + 8 = 48$	$12 + 8 = 20$
Divide this sum by 2.	$24 \div 2 = 12$	$48 \div 2 = 24$	$20 \div 2 = 10$
Subtract 4 from the quotient.	$12 - 4 = 8$	$24 - 4 = 20$	$10 - 4 = 6$
Summary of results:	$4 \rightarrow 8$	$10 \rightarrow 20$	$3 \rightarrow 6$

    b. $4n$

$$4n + 8$$

$$\frac{4n + 8}{2} = \frac{4n}{2} + \frac{8}{2} = 2n + 4$$

$$2n + 4 - 4 = 2n$$

6. Round $47.00 to $50.00. Round $311.00 to $300.00. Round $405.00 to $400.00. Round $681.79 to $680.00.
   Total needed for expenses: $47.00 + $311.00 + $405.00 \approx $50.00 + $300.00 + $400.00 \approx 750.00
   Additional money needed: $750.00 - $681.79 \approx $750.00 - $680.00 \approx 70

7. Round $485,000 to $500,000. Round the number of people to 20. $\dfrac{\$485,000}{19 \text{ people}} \approx \dfrac{\$500,000}{20 \text{ people}} \approx \$25,000$ per person

8. $0.48992 \times 120 \approx 0.5 \times 120 \approx 60$

9. Round percentage of impaired lakes to 40%. Round number of acres to 17 million.
   37% of $17.1 \approx 0.40 \times 17 \approx 6.8$ Approximately 6.8 million acres of lakes are impaired.

10. Approximately 90% of U.S. households have VCR's.

11. fax machine, personal computer, and cable TV

12. Approximately 7 murders.

13. The murder rate for this time period hit a maximum in 1980 of 10 murders per 100,000 people.

14. For 3 hours:
    Estes: $9 per $\dfrac{1}{4}$ hour
    $3 \times 4 = 12$ quarter-hours $\rightarrow 12 \times \$9 = \$108$
    Ship and Shore: $20 per $\dfrac{1}{2}$ hour
    $3 \times 2 = 6$ half-hours $\rightarrow 6 \times \$20 = \$120$
    Estes is a better deal by $120 - $108 = $12.00.

15. 20 round trips mean 40 one-way trips at $11/trip. (40 trips)(32 passengers)($11) = $14,080 in one day

16. $960 - $50 = 910 remaining to pay, which gives $\dfrac{\$910}{\$35 \text{ per week}} = 26$ weeks

Chapter 2
Set Theory

Check Points 2.1

1. Set *L* is the set of the first six lowercase letters in the English alphabet.

2. $D = \{1, 3, 4, 7, 8\}$

3. L = {m, a, r, c, h}

4. **a.** True; 8 is an element of the given set.

   b. True; r is not an element of the given set.

5. **a.** $A = \{1, 2, 3\}$

   b. $B = \{15, 16, 17, \ldots\}$

   c. $O = \{1, 3, 5, \ldots\}$

6. **a.** $n(A) = 5$; the set has 5 elements

   b. $n(B) = 1$; the set has only 1 element

   c. $n(C) = 8$; Though this set lists only five elements, the three dots indicate 12, 13, and 14 are also elements.

   d. $n(D) = 0$ because the set has no elements.

7. **a.** True, {O, L, D} = {D, O, L} because the sets contain exactly the same elements.

   b. False, the two sets do not contain exactly the same elements.

8. **a.** No they are not equal because they have different elements.

   b. Yes they are equivalent because they have the same number of elements.

Exercise Set 2.1

1. The known planets in the Solar System.

3. The signs of the Zodiac.

5. Odd natural numbers less than 100.

7. {Winter, Spring, Summer, Autumn}

9. {1, 2, 3}

11. {January, June, July}

13. {*Titanic*}

15. True; 3 is a member of the set.

17. True; 12 is a member of the set.

19. False; 5 is *not* a member of the set.

21. True; 11 is *not* a member of the set.

23. False; 37 is a member of the set.

25. $\notin$
    Mark McGwire is not, nor has ever been, President.

27. $\in$
    13,791 is an odd natural number.

29. $\notin$
    $\varnothing$ is not a member of { }

31. {1, 3, 5, 7, 9, 11}

33. {101, 102, 103, ...}

35. {11, 12, 13, 14, 15}

37. 5; There are 5 elements in the set.

39. 25; There are 25 elements in the set.

41. 0; There are *no* days of the week beginning with A.

43. True; These two sets contain exactly the same elements.

45. False; These two sets do not contain the exact same elements.

47. True; These two sets contain exactly the same elements.

49. **a.** Not equal
    The two sets contain different elements.

   b. Not equivalent
    The number of elements is not the same.

51. **a.** Not equal
    The elements are not exactly the same.

   b. Equivalent
    The number of elements is the same.

53. a. Equal. The elements are exactly the same.

    b. Equivalent
       The number of elements is the same.

55. {amusement parks, gardening, movies, exercise}

57. {home improvement, amusement parks, gardening}

59. {Chicago, Newark, LaGuardia}

61. {Atlanta, Philadelphia, Boston}

63. {75}

65-73. Answers will vary.

75. a

77. Yes; $968,924 \div 2684 = 361$

79. No; $140,542 \div 244 \approx 575.99$

Check Points 2.2

  1. U is the set of TV game shows.

  2. $A' = \{b, c, e\}$; those are the elements in U but not in A.

  3. $U = \{a, b, c, d, e\}$; $A = \{a, d\}$; and $A' = \{b, c, e\}$

  4. a. $\not\subseteq$; because 7 is not in set B.

     b. $\subseteq$, because all elements in set A are also in set B.

     c. $\subseteq$; because all elements in set A are also in set B.

  5. a. $\subseteq$

     b. Both $\subseteq$ and $\subset$ are correct.

  6. Yes, the empty set is a subset of any set.

7. a. This set has 3 elements. Therefore there are $2^3 = 2 \times 2 \times 2 = 8$ subsets. This number means that there are 8 ways to make a selection, including the option to bring no books.

b.
No books packed:	{ }
One book packed:	{*The Color Purple*}
	{*Hannibal*}
	{*The Royals*}
Two books packed:	{*The Color Purple, Hannibal*}
	{*The Color Purple, The Royals*}
	{*Hannibal, The Royals*}
Three books packed:	{*The Color Purple, Hannibal, The Royals*}

c. 7 of the 8 subsets are proper subsets. {*The Color Purple, Hannibal, The Royals*} is not a proper subset.

Exercise Set 2.2

1. U is the set of composers.

3. U is the set of soft drinks.

5. $A' = \{c, d, e\}$

7. $C' = \{b, c, d, e, f\}$

9. $A' = \{6, 7, 8, \ldots, 20\}$

11. $C' = \{2, 4, 6, 8, \ldots, 20\}$

13. $A' = \{21, 22, 23, 24, \ldots\}$

15. $C' = \{1, 3, 5, 7, \ldots\}$

17. $A' = \{3, 4, 6, 7\}$

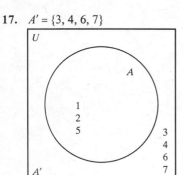

19. $A' = \{c, d, e\}$

21. $A' = \{$February, March, April, May, August, September, October, November, December$\}$
A' is the set of all months not beginning with the letter J.

23. $\subseteq$

25. $\not\subseteq$

27. $\not\subseteq$

29. $\not\subseteq$
Subset cannot be larger than the set.

31. $\subseteq$

33. $\not\subseteq$

35. $\subseteq$ or $\subset$

37. $\subseteq$

39. Neither symbol will make the statement true. The second set is smaller than the first.

41. True

43. False
{Ralph} is a subset, not Ralph.

45. True

47. False
The symbol "$\varnothing$" is not a member of the set.

49. { } {Border Collie} {Poodle} {Border Collie, Poodle}

51. { } {t} {a} {b} {t, a} {t, b} {a, b} {t, a, b}

53. { } {0}

55. 16 subsets, 15 proper subsets
There are 4 elements, which means there are 2^4 or 16 subsets. There are $2^4 - 1$ proper subsets or 15.

57. 64 subsets, 63 proper subsets
There are 6 elements, which means there are 2^6 or 64 subsets. There are $2^6 - 1$ proper subsets or 63.

59. 128 subsets, 127 proper subsets
There are 7 elements, which means there are 2^7 or 128 subsets. There are $2^7 - 1$ proper subsets or 127.

61. 8 subsets, 7 proper subsets
There are 3 elements, which means there are 2^3 or 8 subsets. There are $2^3 - 1$ proper subsets or 7.

63. $2^5 = 32$ option combinations

65. $2^6 = 64$ viewing combinations

67. $2^8 = 256$ city combinations

69-79. Answers will vary.

81. 0, 5¢, 10¢, 25¢, 40¢, 15¢, 30¢, 35¢
Since there are 3 elements or coins, there are 2^3 or 8 different coin combinations.

Check Points 2.3

1. a. $\{1, 3, 5, \underline{7}, \underline{10}\} \cap \{6, 7, 10, 11\} = \{7, 10\}$

b. $\{1, 2, 3\} \cap \{4, 5, 6, 7\} = \varnothing$

c. $\{1, 2, 3\} \cap \varnothing = \varnothing$

2. a. $\{1, 3, 5, 7, 10\} \cup \{6, 7, 10, 11\}$
$= \{1, 3, 5, 6, 7, 10, 11\}$

b. $\{1, 2, 3\} \cup \{4, 5, 6, 7\} = \{1, 2, 3, 4, 5, 6, 7\}$

c. $\{1, 2, 3\} \cup \varnothing = \{1, 2, 3\}$

3. a. $A \cup B = \{b, c, e\}$
$(A \cup B)' = \{a, d\}$

b. $A' = \{a, d, e\}$
$B' = \{a, d\}$
$A' \cap B' = \{a, d\}$

Exercise Set 2.3

1. $A = \{1, 3, 5, 7\}$
$B = \{1, 2, 3\}$
$A \cap B = \{1, 3\}$

3. $A = \{1, 3, 5, 7\}$
$B = \{1, 2, 3\}$
$A \cup B = \{1, 2, 3, 5, 7\}$

5. $A = \{1, 3, 5, 7\}$
$U = \{1, 2, 3, 4, 5, 6, 7\}$
$A' = \{2, 4, 6\}$

7. $A' = \{2, 4, 6\}$
$B' = \{4, 5, 6, 7\}$
$A' \cap B' = \{4, 6\}$

9. $A = \{1, 3, 5, 7\}$
$C' = \{1, 7\}$
$A \cup C' = \{1, 3, 5, 7\}$

11. $A - \{1, 3, 5, 7\}$
$C = \{2, 3, 4, 5, 6\}$
$A \cap C = \{3, 5\}$
$(A \cap C)' = \{1, 2, 4, 6, 7\}$

13. $A = \{1, 3, 5, 7\}$ $C = \{2, 3, 4, 5, 6\}$
$A' = \{2, 4, 6\}$ $C' = \{1, 7\}$
$A' \cup C' = \{1, 2, 4, 6, 7\}$

15. $A = \{1, 3, 5, 7\}$ $B = \{1, 2, 3\}$
$(A \cup B) - \{1, 2, 3, 5, 7\}$
$(A \cup B)' = \{4, 6\}$

17. $A = \{1, 3, 5, 7\}$
$A \cup \varnothing = \{1, 3, 5, 7\}$

19. $A \cap \varnothing = \varnothing$

21. $A \cup U = U$
$U = \{1, 2, 3, 4, 5, 6, 7\}$

23. $A \cap U = A$
$A = \{1, 3, 5, 7\}$

25. $A = \{a, g, h\}$
$B = \{b, g, h\}$
$A \cap B = \{g, h\}$

27. $A = \{a, g, h\}$
$B = \{b, g, h\}$
$A \cup B = \{a, b, g, h\}$

29. $A = \{a, g, h\}$
$U = \{a, b, c, d, e, f, g, h\}$
$A' = \{b, c, d, e, f\}$

31. $A' = \{b, c, d, e, f\}$
$B' = \{a, c, d, e, f\}$
$A' \cap B' = \{c, d, e, f\}$

33. $A = \{a, g, h\}$
$C' = \{a, g, h\}$
$A \cup C' = \{a, g, h\}$

35. $A = \{a, g, h\}$
$C = \{b, c, d, e, f\}$
$A \cap C = \varnothing$
$(A \cap C)' = \{a, b, c, d, e, f, g, h\}$

37. $A' = \{b, c, d, e, f\}$
$C' = \{a, g, h\}$
$A' \cup C' = \{a, b, c, d, e, f, g, h\}$

39. $A = \{a, g, h\}$
$B = \{b, g, h\}$
$A \cup B = \{a, b, g, h\}$
$(A \cup B)' = \{c, d, e, f\}$

41. $A \cup \varnothing = A$
$A = \{a, g, h\}$

43. $A \cap \varnothing = \varnothing$

45. $A = \{a, g, h\}$
$U = \{a, b, c, d, e, f, g, h\}$
$A \cup U = \{a, b, c, d, e, f, g, h\}$

47. $A = \{a, g, h\}$
$U = \{a, b, c, d, e, f, g, h\}$
$A \cap U = \{a, g, h\}$

49. $A = \{1, 3, 4, 7\}$

51. $U = \{1, 2, 3, 4, 5, 6, 7, 8, 9\}$

53. $A \cap B = \{3, 7\}$

55. $B' = \{1, 4, 8, 9\}$

57. $(A \cup B)' = \{8, 9\}$

59. {<u>spatial-temporal</u>, <u>sports equipment</u>, <u>toy cars and trucks</u>} $\cap$ {dollhouses, <u>spatial-temporal</u>, <u>sports equipment</u>, <u>toy cars and trucks</u>}
= {spatial-temporal, sports equipment, toy cars and trucks}

61. {spatial-temporal, sports equipment, toy cars and trucks} $\cup$ {dollhouses, spatial-temporal, sports equipment, toy cars and trucks}
= {dollhouses, spatial-temporal, sports equipment, toy cars and trucks}

63. {toy cars and trucks } $\cap$ {dollhouses, domestic accessories, dolls, spatial-temporal, sports equipment}
= $\varnothing$

65-69. Answers will vary.

71. b is true.

73. Example of a possible answer: Suppose there is an element in set A that is not also in set B. This element would be part of $A \cup B$ *and* B' This element would then be part of $A' \cup B'$. However, since this element is part of $A \cup B$ it is *not* part of $(A \cup B)'$. Since this element is a part of $A' \cup B'$ but *not* $(A \cup B)'$, this serves as a counterexample to the proposition that $(A \cup B)' = A' \cup B'$.

Check Points 2.4

1. a. $B \cap C = \{b, f\}$
$A \cup (B \cap C) = \{a, b, c, d, f\}$

b. $A \cup B = \{a, b, c, d, f\}$
$A \cup C = \{a, b, c, d, f\}$
$(A \cup B) \cap (A \cup C) = \{a, b, c, d, f\}$

c. $C' = \{a, d, e\}$
$B \cup C' = \{a, b, d, e, f\}$
$A \cap (B \cup C') = \{a, b, d\}$

2. a. Regions IV, V, VI, and VII

b. Regions II, III, IV, V, VI, and VII

c. Regions IV and V

d. Regions I, II, III, and VIII

e. Regions I, II, III, IV, V, VI, VII

3. a. $A \cup B$ is represented by regions I, II, and III. Therefore $(A \cup B)'$ is represented by region IV.

b. A' is represented by regions III and IV.
B' is represented by regions I and IV.
Therefore $A' \cap B'$ is represented by region IV.

c. $(A \cup B)' = A' \cap B'$ because they both represent region IV.

4. a. $B \cup C$ is represented by regions II, III, IV, V, VI, and VII.
Therefore $A \cap (B \cup C)$ is represented by regions II, IV, and V.

b. $A \cap B$ is represented by regions II and V.
$A \cap C$ is represented by regions IV and V.
Therefore $(A \cap B) \cup (A \cap C)$ is represented by regions II, IV, and V.

c. $A \cap (B \cup C) = (A \cap B) \cup (A \cap C)$ because they both represent region IV.

Exercise Set 2.4

1. $B \cap C = \{2, 3\}$
$A \cup (B \cap C) = \{1, 2, 3, 5, 7\}$

3. $A \cup B = \{1, 2, 3, 5, 7\}$
$A \cup C = \{1, 2, 3, 4, 5, 6, 7\}$
$(A \cup B) \cap (A \cup C) = \{1, 2, 3, 5, 7\}$

5. $A' = \{2, 4, 6\}$ $C' = \{1, 7\}$
$B \cup C' = \{1, 2, 3, 7\}$
$A' \cap (B \cup C') = \{2\}$

7. $A' = \{2, 4, 6\}$ $C' = \{1, 7\}$
$A' \cap B = \{2\}$
$A' \cap C' = \emptyset$
$(A' \cap B) \cup (A' \cap C') = \{2\}$

9. $A = \{1, 3, 5, 7\}$
$B = \{1, 2, 3\}$
$C = \{2, 3, 4, 5, 6\}$
$A \cup B \cup C = \{1, 2, 3, 4, 5, 6, 7\}$
$(A \cup B \cup C)' = \emptyset$

11. $A = \{1, 3, 5, 7\}$
$B = \{1, 2, 3\}$
$A \cup B = \{1, 2, 3, 5, 7\}$
$(A \cup B)' = \{4, 6\}$
$C = \{2, 3, 4, 5, 6\}$
$(A \cup B)' \cap C = \{4, 6\}$

13. $B \cap C = \{b\}$
$A \cup (B \cap C) = \{a, b, g, h\}$

15. $A \cup B = \{a, b, g, h\}$
$A \cup C = \{a, b, c, d, e, f, g, h\}$
$(A \cup B) \cap (A \cup C) = \{a, b, g, h\}$

17. $A' = \{b, c, d, e, f\}$
$C' = \{a, g, h\}$
$B \cup C' = \{a, b, g, h\}$
$A' \cap (B \cup C') = \{b\}$

19. $A' = \{b, c, d, e, f\}$
    $A' \cap B = \{b\}$
    $C' = \{a, g, h\}$
    $A' \cap C' = \varnothing$
    $(A' \cap B) \cup (A' \cap C') = \{b\}$

21. $A \cup B \cup C = \{a, b, c, d, e, f, g, h\}$
    $(A \cup B \cup C)' = \varnothing$

23. $A \cup B = \{a, b, g, h\}$
    $(A \cup B)' = \{c, d, e, f\}$
    $(A \cup B)' \cap C = \{c, d, e, f\}$

25. II, III, V, VI

27. I, II, IV, V, VI, VII

29. II, V

31. I, IV, VII, VIII

33. a. II

    b. II

    c. $A \cap B = B \cap A$

35. a. I, III, IV

    b. IV

    c. No, $(A \cap B)' \neq A' \cap B'$

37. a. II, III, IV

    b. I

    c. No, $A' \cup B \neq A \cap B'$

39. The regions represented by $(A' \cap B)'$ are I, II, IV.
    The regions represented by $A \cup B'$ are I, II, IV.
    Thus, $(A' \cap B)' = A \cup B'$.

41. a. II, IV, V, VI, VII

    b. II, IV, V, VI, VII

    c. $(A \cap B) \cup C = (A \cup C) \cap (B \cup C)$

43. a. II, IV, V

    b. I, II, IV, V, VI

    c. No. The results in **a** and **b** show
       $A \cap (B \cup C) \neq A \cup (B \cap C)$ because of the
       different regions represented.

45. B consists of regions II, III, V, and VI.
    $A \cap C$ consists of regions IV and V.
    Therefore $B \cup (A \cap C)$ consists of regions II, III,
    IV, V, and VI.
    $A \cup B$ consists of regions I, II, III, IV, V, and VI.
    $B \cup C$ consists of regions II, III, IV, V, VI and VII.
    Therefore $(A \cup B) \cap (B \cup C)$ consists of regions II,
    III, IV, V, and VI.
    Thus $B \cup (A \cap C) = (A \cup B) \cap (B \cup C)$.

47. I

49. II

51. V

53. V

55. IV

57. Answers will vary

59. AB^+

61. Yes

Check Points 2.5

1. a. $55 + 20 = 75$

   b. $20 + 70 = 90$

   c. 20

   d. $55 + 20 + 70 = 145$

   e. 55

   f. 70

   g. 30

   h. $55 + 20 + 70 + 30 = 175$

2. $n(A \cup B) = n(A) + n(B) - n(A \cap B)$
   $= 45 + 15 - 5 = 55$

3. The 50 Hispanic Americans that agreed with the statement belong in region II. Since region II accounts for 50 of the 550 people who agreed with the statement, the remaining 500 people belong in region III. Since region II also accounts for 50 of the 250 Hispanic Americans, the remaining 200 Hispanic Americans belong in region I. Regions I, II, and III account for a total of 750 of the 1000 people surveyed. This leaves 250 people in region IV.

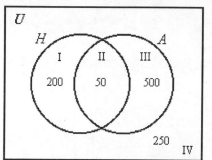

a. 500 non-Hispanics agreed with the statement. (Region III)

b. 250 non-Hispanics disagreed with the statement. (Region IV)

4. Parts a, b, and c are labeled.

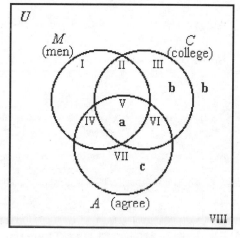

Exercise Set 2.5

1. 26

3. 17

5. 37

7. 7

9. $n(A \cup B) = n(A) + n(B) - n(A \cap B)$
$= 17 + 20 - 6$
$= 31$

11. $n(A \cup B) = n(A) + n(B) - n(A \cap B)$
$= 17 + 17 - 7$
$= 27$

13. Parts b, c, and d are labeled.

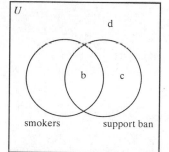

15. Parts b, c, and d are labeled. Answers for part e will vary.

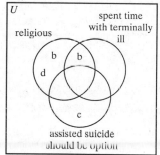

17. a. Venn diagram:

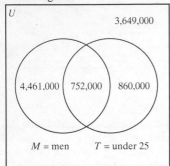

b. U.S. workers who lost job and are under 25 minus male workers who lost job and are under 25
= 1,612,000 − 752,000 = 860,000 female workers who lost job and are under 25

c. Number of female workers 25 and over who lost job = Total number of workers who lost job minus male workers who lost job minus female workers who lost job and under 25.
9,722,000 − 5,213,000 − 860,000 = 3,649,000

19. Answers will vary.

21. a. 0; This would assume none of the psychology students were taking mathematics.

b. 30; This would assume all 30 students taking psychology were taking mathematics.

c. 60; $U = 150$ so with 90 taking mathematics, if we assume all the psychology students are taking mathematics courses, $U − 90 = 60$.

Chapter 2 Review Exercises

1. Possible description: Natural numbers less than 11.

2. {Tuesday, Thursday}

3. $\in$
93 is an element of the set.

4. $\notin$
{d} is a subset, not a member; "d" would be a member.

5. 12 months in the year.

6. 15

7. $\neq$
The two sets do not contain exactly the same elements.

8. $\neq$
One set is infinite. The other is finite.

9. Equivalent
Same number of elements, but different elements.

10. Equal and equivalent
The two sets have exactly the same elements.

11. $A' = \{1, 2, 8, 9\}$

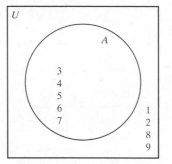

12. {January, February, March, May, July, August, October, December}

13. $\subseteq$

14. $\not\subseteq$

15. $\subseteq$

16. $\subseteq$

17. Both $\subseteq$ and $\subset$ are correct.

18. False. Texas is not a member of the set.

19. False. 4 is not a subset. {4} is a subset.

20. True

21. False. It is a subset, but not a proper subset.

22. $\varnothing$ {1} {5} {1, 5}
{1, 5} is not a proper subset.

23. There are 5 elements.
This means there are $2^5 = 32$ subsets.
There are $2^5 − 1 = 31$ proper subsets.

24. {January, June, July}
    There are 3 elements.
    This means there are $2^3 = 8$ subsets.
    There are $2^3 - 1 = 7$ proper subsets.

25. $A \cap B = \{1, 2, 4\}$

26. $A \cup B' = \{1, 2, 3, 4, 6, 7, 8\}$

27. $A' \cap B = \{5\}$

28. $(A \cup B)' = \{6, 7, 8\}$

29. $A' \cap B' = \{6, 7, 8\}$

30. $\{4, 5, 6\}$

31. $\{2, 3, 6, 7\}$

32. $\{1, 4, 5, 6, 8, 9\}$

33. $\{4, 5\}$

34. $\{1, 2, 3, 6, 7, 8, 9\}$

35. $\{2, 3, 7\}$

36. $\{6\}$

37. $\{1, 2, 3, 4, 5, 6, 7, 8, 9\}$

38. $B \cap C = \{1, 5\}$
    $A \cup (B \cap C) = \{1, 2, 3, 4, 5\}$

39. $A \cap C = \{1\}$
    $(A \cap C)' = (2, 3, 4, 5, 6, 7, 8)$
    $(A \cap C)' \cup B = \{1, 2, 3, 4, 5, 6, 7, 8\}$

40. The shaded regions are the same for $(A \cup B)'$ and
    $A' \cap B'$. Therefore $(A \cup B)' = A' \cap B'$

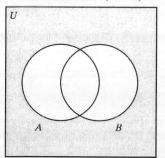

41. The statement is false because the shaded regions
    are different.

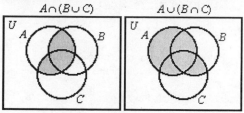

42. $n(A \cup B) = n(A) + n(B) - n(A \cap B)$
    $\qquad\qquad = 25 + 17 - 9$
    $\qquad\qquad = 33$

43. a.
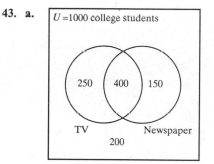

   b. $250 + 400 + 150 = 800$

   c.   250

   d.   200

44. a.   Parts b and c are labeled.
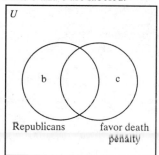

Chapter 2 Test

1. {18, 19, 20, 21, 22, 23, 24, 25}

2. 12

3. False, {6} is not an element of the set,
   but 6 is an element.

4. True, both sets have seven elements.

5. True

6. False, g is not an element in the larger set.

7. True

8. False, 14 is an element of the set.

9. False, Number of subsets: 2^N where N is the number of elements.
   There are 5 elements. $2^5 = 32$ subsets

10. False, $\varnothing$ is *not* a proper subset of itself.

11. $\varnothing$ {6} {9} {6, 9}
    {6, 9} is not a proper subset.

12. {a, b, c, d, e, f}

13. $B \cap C = \{e\}$
    $(B \cap C)' = \{a, b, c, d, f, g\}$

14. $C' = \{b, c, d, f\}$
    $A \cap C' = \{b, c, d\}$

15. $A \cup B = \{a, b, c, d, e, f\}$
    $(A \cup B) \cap C = \{a, e\}$

16. I, II, IV, V, VI

17. **a.**

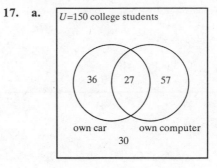

U=150 college students

36   27   57

own car   own computer

30

b. Car Only: $63 - 27 = 36$
Computer Only: $84 - 27 = 57$
Both: 27
$150 - 36 - 57 - 27 = 30$

c. 57

d. 36

e. 120

Chapter 3
Logic

Check Points 3.1

1. **a.** Paris is not the capital of Spain.

   b. July is a month.

2. **a.** $\sim p$

   b. $\sim q$

3. Tom Hanks is not a jazz singer.

4. Some new tax dollars will not be used to improve education.
   At least one new tax dollar will not be used to improve education.

Exercise Set 3.1

1. Statement

3. Statement

5. Not a statement

7. Not a statement

9. Statement

11. Statement

13. It is not raining.

15. Macbeth is a comedy show on television.

17. $\sim p$

19. $\sim r$

21. *I Love Lucy* is not a Broadway musical.

23. Benjamin Franklin invented the rocking chair.

25. The United States is not the country with the most Internet users.

27. Michael is the most common of boys' names.

29. **a.** There are no whales that are not mammals.

   b. Some whales are not mammals.

31. **a.** There exists at least one student that is a business major.

   b. No students are business majors.

33. **a.** Not all thieves are criminals.

   b. All thieves are criminals.

35. **a.** All Democratic presidents have not been impeached.

   b. Some Democratic presidents have been impeached.

37. b

39-45. Answers will vary.

47. Answers will vary. Possible answer: Everything I say or write is false.

49. The statement could mean that she *is* dating him, but that his muscularity is *not* the reason. The statement could also mean that she *is not* dating him because she does not like the fact that he is muscular.

Check Points 3.2

1. **a.** $p \wedge q$ **b.** $\sim p \wedge q$

2. **a.** $p \vee \sim q$ **b.** $q \vee p$

3. **a.** $p \to q$ **b.** $\sim p \to \sim q$
   c. $\sim q \to \sim p$

4. **a.** $q \leftrightarrow p$ **b.** $\sim p \leftrightarrow \sim q$

5. **a.** It is not true that Steven Spielberg is an actor and a director.

   b. Steven Spielberg is not an actor and Steven Spielberg is a director.

   c. It is not true that Steven Spielberg is an actor or a director.

6. **a.** If the plant is fertilized and the plant is watered, then the plant does not wilt.

   b. The plant is fertilized, and if the plant is watered then the plant does not wilt.

Exercise Set 3.2

1. $p \wedge q$;

$$\underline{\textit{I Love Lucy} \text{ is a television show}} \quad \underline{\text{and}}$$
$$\qquad\qquad\qquad p \qquad\qquad\qquad\quad \wedge$$
$$\underline{\textit{Macbeth} \text{ is a television show.}}$$
$$\qquad\qquad q$$

3. $p \wedge \sim q$;

$$\underline{\textit{I Love Lucy} \text{ is a television show}} \quad \underline{\text{and}}$$
$$\qquad\qquad\qquad p \qquad\qquad\qquad\quad \wedge$$
$$\underline{\textit{Macbeth} \text{ is not a television show.}}$$
$$\qquad\qquad \sim q$$

5. $p \vee q$;

$$\underline{\text{I study}} \quad \underline{\text{or}} \quad \underline{\text{I pass the course.}}$$
$$\quad p \qquad \vee \qquad\qquad q$$

7. $p \vee \sim q$;

$$\underline{\text{I study}} \quad \underline{\text{or}} \quad \underline{\text{I do not pass the course.}}$$
$$\quad p \qquad \vee \qquad\qquad \sim q$$

9. $p \rightarrow q$;

$$\underline{\text{If}} \quad \underline{\text{this is an alligator,}} \quad \underline{\text{then}} \quad \underline{\text{this is a reptile.}}$$
$$\qquad\qquad p \qquad\qquad \rightarrow \qquad\qquad q$$

11. $\sim p \rightarrow \sim q$;

$$\underline{\text{If}} \quad \underline{\text{this is not an alligator,}} \quad \underline{\text{then}}$$
$$\qquad\qquad \sim p \qquad\qquad\qquad \rightarrow$$
$$\underline{\text{this is not a reptile.}}$$
$$\qquad \sim q$$

13. $p \leftrightarrow q$;

$$\underline{\text{The campus is closed}} \quad \underline{\text{if and only if}} \quad \underline{\text{it is Sunday.}}$$
$$\qquad\quad p \qquad\qquad\qquad \leftrightarrow \qquad\qquad\quad q$$

15. $\sim q \leftrightarrow \sim p$;

$$\underline{\text{It is not Sunday}} \quad \underline{\text{if and only if}}$$
$$\qquad \sim q \qquad\qquad\qquad \leftrightarrow$$
$$\underline{\text{the campus is not closed.}}$$
$$\qquad\qquad \sim p$$

17. The heater is not working and the house is cold.

19. The heater is working or the house is not cold.

21. If the heater is working then the house is not cold.

23. The heater is working if and only if the house is not cold.

25. It is July 4th and we are not having a barbeque.

27. It is not July 4th or we are having a barbeque.

29. If we are having a barbeque, then it is not July 4th.

31. It is not July 4th if and only if we are having a barbeque.

33. It is not true that Romeo loves Juliet and Juliet loves Romeo.

35. Romeo does not love Juliet and Juliet loves Romeo.

37. It is not true that Juliet loves Romeo or Romeo loves Juliet.

39. Juliet does not love Romeo or Romeo loves Juliet.

41. Romeo does not love Juliet and Juliet does not love Romeo.

43. $(p \wedge q) \vee r$;

$$\left(\underline{\text{The temperature outside is freezing}} \quad \underline{\text{and}} \quad \underline{\text{the heater is working,}} \right) \underline{\text{or}} \quad \underline{\text{the house is cold.}}$$
$$\qquad\qquad\qquad p \qquad\qquad\qquad\qquad \wedge \qquad\qquad q \qquad\qquad\qquad \vee \qquad\qquad r$$

45. $(p \vee \sim q) \rightarrow r$;

$$\left(\underline{\text{If the temperature outside is freezing}} \quad \underline{\text{or}} \quad \underline{\text{the heater is not working,}} \right) \underline{\text{then}} \quad \underline{\text{the house is cold.}}$$
$$\qquad\qquad\qquad\quad p \qquad\qquad\qquad\qquad \vee \qquad\qquad \sim q \qquad\qquad\qquad \rightarrow \qquad\qquad r$$

47. $r \leftrightarrow (p \wedge \sim q)$;

$$\underline{\text{The house is cold}} \quad \underline{\text{if and only if}} \left(\underline{\text{the temperature outside is freezing}} \quad \underline{\text{and}} \quad \underline{\text{the heater isn't working.}} \right)$$
$$\qquad r \qquad\qquad\qquad \leftrightarrow \qquad\qquad\qquad\qquad p \qquad\qquad\qquad\qquad \wedge \qquad\qquad \sim q$$

49. If the temperature is above 85° and we finished studying, then we go to the beach.

51. The temperature is above 85°, and if we finished studying then we go to the beach.

53. If we do not go to the beach, then the temperature is not above 85° or we did not finish studying.

55. If we do not go to the beach then we didn't finish studying, or the temperature is above 85°.

57. We go to the beach if and only if the temperature is above 85° and we finished studying.

59. The temperature is above 85° if and only if we finished studying, and we go to the beach.

61. If we don't go to the beach, then it is not true that the temperature is above 85° and we finished studying.

63. $p \rightarrow q$; $\underbrace{\text{If you give other people permission to do so,}}_{p}$ $\underset{\rightarrow}{\text{then}}$ $\underbrace{\text{you can be made to feel less than you are.}}_{q}$

65. $(p \wedge q) \rightarrow (r \wedge s)$;

$\left(\underbrace{\text{If Ed Muskie is ahead on the first ballot}}_{p} \text{ and } \underbrace{\text{Humphrey is close behind,}}_{q} \right) \underset{\rightarrow}{\text{then}}$

$\left(\underbrace{\text{we'll switch to Ted Kennedy on the second ballot}}_{r} \text{ and } \underbrace{\text{start a draft.}}_{s} \right)$

67-73. Answers will vary.

75. $\left(\underbrace{\text{Shooting unarmed civilians is morally justifiable}}_{p} \underset{\leftrightarrow}{\text{ if and only if }} \right.$

$\underbrace{\text{bombing unarmed civilians is morally justifiable,}}_{q} \left) \underset{\wedge}{\text{ and }} \left(\underbrace{\text{as the former is not morally justifiable,}}_{\sim p} \underbrace{\text{ neither is the latter.}}_{\wedge \sim q} \right)$

77. $p \wedge \sim q$; $\underbrace{\text{I will pass math}}_{p} \underset{\wedge}{\text{ but }} \underbrace{\text{[I will] not [pass] history.}}_{\sim q}$

79. $(\sim p \vee \sim q) \rightarrow \sim r$; $\left(\underbrace{\text{If I do not pass math}}_{\sim p} \underset{\vee}{\text{ or }} \underbrace{\text{[I do not pass] history,}}_{\sim q} \right) \underset{\rightarrow}{\text{then}} \underbrace{\text{I will not graduate.}}_{\sim r}$

Check Points 3.3

1. The statement is <u>true</u> because it is of the form: [true] *and* [true].

2. The statement is <u>false</u> because it is of the form: [false] *or* [false].

3. $\sim(p \vee q)$

p	q	$p \vee q$	$\sim(p \vee q)$
T	T	T	F
T	F	T	F
F	T	T	F
F	F	F	T

4. $\sim p \wedge \sim q$

p	q	$\sim p$	$\sim q$	$\sim p \wedge \sim q$
T	T	F	F	F
T	F	F	T	F
F	T	T	F	F
F	F	T	T	T

5. Will Smith is not an actor or Bob Dylan is not an actor.

6. $(p \wedge \sim q) \vee \sim p$

p	q	$\sim p$	$\sim q$	$p \wedge \sim q$	$(p \wedge \sim q) \vee \sim p$
T	T	F	F	F	F
T	F	F	T	T	T
F	T	T	F	F	T
F	F	T	T	F	T

Chapter 3: Logic

Exercise Set 3.3

1. p: The Beatles were a rock group – true
q: Ernest Hemingway was a writer – true
A conjunction is true when both statements are true.
Hence, $p \wedge q$ is true.

3. p: $3 + 2 = 5$ – true
q: 3 is an even number – false
A conjunction is true when both statements are true.
Hence, $p \wedge q$ is false.

5. p: Fir trees produce fur coats – false
q: Some first basemen are right-handed– true
A conjunction is true when both statements are true.
Hence, $p \wedge q$ is false.

7. p: $7 \times 3 = 21$ – true
q: Martin Luther King fought for racial equality – true
A disjunction is false when both statements are false.
Hence, $p \vee q$ is true.

9. p: Some athletes are not college professors – true
q: Sammy Sosa was a U.S. President – false
A disjunction is false when both statements are false.
Hence, $p \vee q$ is true.

11. p: All politicians have law degrees – false
q: No students are business majors – false
A disjunction is false when both statements are false.
Hence, $p \vee q$ is false.

13. $p \vee \sim p$

p	$\sim p$	$p \vee \sim p$
T	F	T
F	T	T

15. $\sim p \wedge q$

p	q	$\sim p$	$\sim p \wedge q$
T	T	F	F
T	F	F	F
F	T	T	T
F	F	T	F

17. $\sim(p \vee q)$

p	q	$p \vee q$	$\sim(p \vee q)$
T	T	T	F
T	F	T	F
F	T	T	F
F	F	F	T

19. $\sim p \wedge \sim q$

p	q	$\sim p$	$\sim q$	$\sim p \wedge \sim q$
T	T	F	F	F
T	F	F	T	F
F	T	T	F	F
F	F	T	T	T

21. $p \vee \sim q$

p	q	$\sim q$	$p \vee \sim q$
T	T	F	T
T	F	T	T
F	T	F	F
F	F	T	T

23. $\sim(\sim p \vee q)$

p	q	$\sim p$	$\sim p \vee q$	$\sim(\sim p \vee q)$
T	T	F	T	F
T	F	F	F	T
F	T	T	T	F
F	F	T	T	F

25. $(p \vee q) \wedge \sim p$

p	q	$\sim p$	$p \vee q$	$(p \vee q) \wedge \sim p$
T	T	F	T	F
T	F	F	T	F
F	T	T	T	T
F	F	T	F	F

27. $\sim p \vee (p \wedge \sim q)$

p	q	$\sim p$	$\sim q$	$p \wedge \sim q$	$\sim p \vee (p \wedge \sim q)$
T	T	F	F	F	F
T	F	F	T	T	T
F	T	T	F	F	T
F	F	T	T	F	T

29. $(p \vee q) \wedge (\sim p \vee \sim q)$

p	q	$\sim p$	$\sim q$	$p \vee q$	$\sim p \vee \sim q$	$(p \vee q) \wedge (\sim p \vee \sim q)$
T	T	F	F	T	F	F
T	F	F	T	T	T	T
F	T	T	F	T	T	T
F	F	T	T	F	T	F

31. $(p \wedge \sim q) \vee (p \wedge q)$

p	q	$\sim p$	$\sim q$	$p \wedge \sim q$	$p \wedge q$	$(p \wedge \sim q) \vee (p \wedge q)$
T	T	F	F	F	T	T
T	F	F	T	T	F	T
F	T	T	F	F	F	F
F	F	T	T	F	F	F

33. Texas is not a state or Paris is not a state.

35. *Romeo and Juliet* was not written by Shakespeare or *Jurassic Park* was not written by Shakespeare.

37. Two statements linked by "and."
p: A motorist in Los Angeles averages more than 80 hours in traffic per year. – true
q: motorists in Atlanta average less hours in traffic per year than in Seattle. – true
Both statements are true, so $p \wedge q$ is true.

39. Two statements linked by "or."
p: Men born in 2000 will live longer than women born in 2000. – false
q: People in the U.S. are living longer. – true
First statement is false, second is true; so $p \vee q$ is true.

41. The chance of divorce peaks during the fourth year of marriage and the chance of divorce does not decrease after 4 years of marriage. This statement is false.

43. The chance of divorce does not peak during the fourth year of marriage and more than 2% of all divorces occur during the 25th year of marriage. This statement is false.

45. The chance of divorce peaks during the fourth year of marriage or the chance of divorce does not decrease after four years of marriage. This statement is true.

47. The chance of divorce does not peak during the fourth year of marriage or more than 2% of divorces occur during the 25th year of marriage. This statement is false.

49-57. Answers will vary.

59. $p \veebar q$

p	q	$p \veebar q$
T	T	F
T	F	T
F	T	T
F	F	F

61. $(p \wedge \sim q) \vee q$

p	q	$\sim q$	$p \wedge \sim q$	$(p \wedge \sim q) \vee q$
T	T	F	F	T
T	F	T	T	T
F	T	F	F	T
F	F	T	F	F

Check Points 3.4

1. $\sim p \to \sim q$

p	q	$\sim p$	$\sim q$	$\sim p \to \sim q$
T	T	F	F	T
T	F	F	T	T
F	T	T	F	F
F	F	T	T	T

2. If Shakespeare is not the author, then the play is not *Macbeth*.

3. $[(p \to q) \land \sim q] \to \sim p$ is a tautology because the final column is always true.

p	q	$\sim p$	$\sim q$	$p \to q$	$(p \to q) \land \sim q$	$[(p \to q) \land \sim q] \to \sim p$
T	T	F	F	T	F	T
T	F	F	T	F	F	T
F	T	T	F	T	F	T
F	F	T	T	T	T	T

4. $(p \to q) \land (\sim p \to \sim q)$

p	q	$\sim p$	$\sim q$	$p \to q$	$\sim p \to \sim q$	$(p \to q) \land (\sim p \to \sim q)$
T	T	F	F	T	T	T
T	F	F	T	F	T	F
F	T	T	F	T	F	F
F	F	T	T	T	T	T

5. $(p \lor q) \leftrightarrow (\sim p \to q)$ is a tautology.

p	q	$\sim p$	$p \lor q$	$\sim p \to q$	$(p \lor q) \leftrightarrow (\sim p \to q)$
T	T	F	T	T	T
T	F	F	T	T	T
F	T	T	T	T	T
F	F	T	F	F	T

6. $(p \to q) \land r$ is not a tautology.

p	q	r	$p \to q$	$(p \to q) \land r$
T	T	T	T	T
T	T	F	T	F
T	F	T	F	F
T	F	F	F	F
F	T	T	T	T
F	T	F	T	F
F	F	T	T	T
F	F	F	T	F

Exercise Set 3.4

1. $p \to \sim q$

p	q	$\sim q$	$p \to \sim q$
T	T	F	F
T	F	T	T
F	T	F	T
F	F	T	T

3. $\sim(q \to p)$

p	q	$q \to p$	$\sim(q \to p)$
T	T	T	F
T	F	T	F
F	T	F	T
F	F	T	F

5. $(p \land q) \to (p \lor q)$

p	q	$p \land q$	$p \lor q$	$(p \land q) \to (p \lor q)$
T	T	T	T	T
T	F	F	T	T
F	T	F	T	T
F	F	F	F	T

7. $(p \to q) \land \sim q$

p	q	$p \to q$	$\sim q$	$(p \to q) \land \sim q$
T	T	T	F	F
T	F	F	T	F
F	T	T	F	F
F	F	T	T	T

9. If I do not live in Washington, then I do not live in Seattle.

11. If a plant does not wilt, then it is watered.

13. $[(p \to q) \land q] \to p$ is not a tautology.

p	q	$p \to q$	$(p \to q) \land q$	$[(p \to q) \land q] \to p$
T	T	T	T	T
T	F	F	F	T
F	T	T	T	F
F	F	T	F	T

15. $\left[(p \to q) \land \sim q\right] \to \sim p$ is a tautology.

p	q	$\sim p$	$\sim q$	$p \to q$	$(p \to q) \land \sim q$	$\left[(p \to q) \land \sim q\right] \to \sim p$
T	T	F	F	T	F	T
T	F	F	T	F	F	T
F	T	T	F	T	F	T
F	F	T	T	T	T	T

17. $\left[(p \lor q) \land p\right] \to \sim q$ is not a tautology.

p	q	$\sim q$	$p \lor q$	$(p \lor q) \land p$	$\left[(p \lor q) \land p\right] \to \sim q$
T	T	F	T	T	F
T	F	T	T	T	T
F	T	F	T	F	T
F	F	T	F	F	T

19. $(p \rightarrow q) \rightarrow (\sim p \vee q)$ is a tautology.

p	q	$\sim p$	$p \rightarrow q$	$\sim p \vee q$	$(p \rightarrow q) \rightarrow (\sim p \vee q)$
T	T	F	T	T	T
T	F	F	F	F	T
F	T	T	T	T	T
F	F	T	T	T	T

21. $(p \wedge q) \wedge (\sim p \vee \sim q)$ is not a tautology.

p	q	$\sim p$	$\sim q$	$p \wedge q$	$\sim p \vee \sim q$	$(p \wedge q) \wedge (\sim p \vee \sim q)$
T	T	F	F	T	F	F
T	F	F	T	F	T	F
F	T	T	F	F	T	F
F	F	T	T	F	T	F

23. $p \leftrightarrow \sim q$

p	q	$\sim q$	$p \leftrightarrow \sim q$
T	T	F	F
T	F	T	T
F	T	F	T
F	F	T	F

25. $\sim (p \leftrightarrow q)$

p	q	$p \leftrightarrow q$	$\sim (p \leftrightarrow q)$
T	T	T	F
T	F	F	T
F	T	F	T
F	F	T	F

27. $(p \leftrightarrow q) \rightarrow p$

p	q	$p \leftrightarrow q$	$(p \leftrightarrow q) \rightarrow p$
T	T	T	T
T	F	F	T
F	T	F	T
F	F	T	F

29. $(\sim p \leftrightarrow q) \rightarrow (\sim p \rightarrow q)$

p	q	$\sim p$	$\sim p \leftrightarrow q$	$\sim p \rightarrow q$	$(\sim p \leftrightarrow q) \rightarrow (\sim p \rightarrow q)$
T	T	F	F	T	T
T	F	F	T	T	T
F	T	T	T	T	T
F	F	T	F	F	T

31. $[(p \wedge q) \wedge (q \rightarrow p)] \leftrightarrow (p \wedge q)$

p	q	$p \wedge q$	$q \rightarrow p$	$(p \wedge q) \wedge (q \rightarrow p)$	$[(p \wedge q) \wedge (q \rightarrow p)] \leftrightarrow (p \wedge q)$
T	T	T	T	T	T
T	F	F	T	F	T
F	T	F	F	F	T
F	F	F	T	F	T

33. $\sim(p \wedge q) \leftrightarrow (\sim p \wedge \sim q)$ is not a tautology.

p	q	$\sim p$	$\sim q$	$p \wedge q$	$\sim(p \wedge q)$	$\sim p \wedge \sim q$	$\sim(p \wedge q) \leftrightarrow (\sim p \wedge \sim q)$
T	T	F	F	T	F	F	T
T	F	F	T	F	T	F	F
F	T	T	F	F	T	F	F
F	F	T	T	F	T	T	T

35. $(p \to q) \leftrightarrow (q \to p)$ is not a tautology.

p	q	$p \to q$	$q \to p$	$(p \to q) \leftrightarrow (q \to p)$
T	T	T	T	T
T	F	F	T	F
F	T	T	F	F
F	F	T	T	T

37. $(p \to q) \leftrightarrow (\sim p \vee q)$ is a tautology.

p	q	$\sim p$	$p \to q$	$\sim p \vee q$	$(p \to q) \leftrightarrow (\sim p \vee q)$
T	T	F	T	T	T
T	F	F	F	F	T
F	T	T	T	T	T
F	F	T	T	T	T

39. $(p \leftrightarrow q) \leftrightarrow [(q \to p) \wedge (p \to q)]$ is a tautology.

p	q	$q \to p$	$p \to q$	$p \leftrightarrow q$	$(q \to p) \wedge (p \to q)$	$(p \leftrightarrow q) \leftrightarrow [(q \to p) \wedge (p \to q)]$
T	T	T	T	T	T	T
T	F	T	F	F	F	T
F	T	F	T	F	F	T
F	F	T	T	T	T	T

41. $(p \wedge \sim q) \vee r$

p	q	r	$\sim q$	$p \wedge \sim q$	$(p \wedge \sim q) \vee r$
T	T	T	F	F	T
T	T	F	F	F	F
T	F	T	T	T	T
T	F	F	T	T	T
F	T	T	F	F	T
F	T	F	F	F	F
F	F	T	T	F	T
F	F	F	T	F	F

43. $(p \vee q) \to r$ is not a tautology.

p	q	r	$p \vee q$	$(p \vee q) \to r$
T	T	T	T	T
T	T	F	T	F
T	F	T	T	T
T	F	F	T	F
F	T	T	T	T
F	T	F	T	F
F	F	T	F	T
F	F	F	F	T

45. $(p \wedge q) \rightarrow (p \vee r)$ is a tautology.

p	q	r	$p \wedge q$	$p \vee r$	$(p \wedge q) \rightarrow (p \vee r)$
T	T	T	T	T	T
T	T	F	T	T	T
T	F	T	F	T	T
T	F	F	F	T	T
F	T	T	F	T	T
F	T	F	F	F	T
F	F	T	F	T	T
F	F	F	F	F	T

47. $r \rightarrow (p \wedge q)$ is not a tautology.

p	q	r	$p \wedge q$	$r \rightarrow (p \wedge q)$
T	T	T	T	T
T	T	F	T	T
T	F	T	F	F
T	F	F	F	T
F	T	T	F	F
F	T	F	F	T
F	F	T	F	F
F	F	F	F	T

49. $[(p \rightarrow q) \wedge (q \rightarrow r)] \rightarrow (p \rightarrow r)$ is a tautology.

p	q	r	$p \rightarrow q$	$q \rightarrow r$	$[(p \rightarrow q) \wedge (q \rightarrow r)]$	$p \rightarrow r$	$[(p \rightarrow q) \wedge (q \rightarrow r)] \rightarrow (p \rightarrow r)$
T	T	T	T	T	T	T	T
T	T	F	T	F	F	F	T
T	F	T	F	T	F	T	T
T	F	F	F	T	F	F	T
F	T	T	T	T	T	T	T
F	T	F	T	F	F	T	T
F	F	T	T	T	T	T	T
F	F	F	T	T	T	T	T

51. The statement is of the form "If p then q" with p true and q false. Therefore the statement is false.

53. If the *Lion King* sold 3 million more albums than *Purple Rain*, then *Titanic* sold 10 million albums. The statement is <u>true</u> because it is of the form: If [false], then [true].

55. If *Titanic* sold 10 million albums, then the *Lion King* sold 3 million more albums than *Purple Rain*. The statement is <u>false</u> because it is of the form: If [true], then [false].

57. If *Dirty Dancing* is not the best-selling album, then *Titanic* did not sell 10 million albums. The statement is <u>false</u> because it is of the form: If [true], then [false].

59. a. *p unless q* <u>means</u> $\sim q \rightarrow p$.

p	q	$\sim q$	$\sim q \rightarrow p$	p unless q
T	T	F	T	T
T	F	T	T	T
F	T	F	T	T
F	F	T	F	F

b.

$$\underline{\text{The curfew will not be lifted}}\ \underline{\text{unless}}\ \underline{\text{the violence ends.}}$$
$$p \qquad\qquad unless \qquad q$$

Because $p\ unless\ q$ is equivalent to $\sim q \to p$, the sentence can be written as:

$$\underline{\text{If the violence does not end,}}\ \underline{\text{then}}\ \underline{\text{the curfew will not be lifted.}}$$
$$\sim q \qquad\qquad \to \qquad\qquad p$$

This statement is false when the curfew is lifted while the violence continues.

61-65. Answers will vary.

67. No, you cannot conclude you got an A. The person could still take you out to dinner if you received a different grade. That would be an example of an *if-then* statement with a false antecedent and a true consequent.

69. Answers will vary. Possible column headings:

p	q	$p \vee q$	$p \wedge q$	$\sim(p \wedge q)$	$(p \vee q) \to \sim(p \wedge q)$

71. $\sim p \to q$

p	q	$\sim p$	$\sim p \to q$
T	T	F	T
T	F	F	T
F	T	T	T
F	F	T	F

Check Points 3.5

1. a. $p \vee q$ and $\sim q \to p$ are equivalent.

p	q	$\sim q$	$p \vee q$	$\sim q \to p$
T	T	F	T	T
T	F	T	T	T
F	T	F	T	T
F	F	T	F	F

The statements are equivalent since their truth vales are the same.

b.
$$p \qquad\quad \vee \qquad\quad q$$
I attend classes or I lose my scholarship.

…is equivalent to…
$$\sim q \qquad\quad \to \qquad\quad p$$
If I do not lose my scholarship, then I attend classes.

2. $\sim p$ and $\sim[\sim(\sim p)]$ are equivalent.

p	$\sim p$	$\sim(\sim p)$	$\sim[\sim(\sim p)]$
T	F	T	F
F	T	F	T

The statements are equivalent since their truth vales are the same.

3. Given: If it's raining, then I need a jacket.
   p: It's raining.
   q: I need a jacket.
   a: It's not raining or I need a jacket.
   b: I need a jacket or it's not raining.
   c: If I need a jacket, then it's raining.
   d: If I do not need a jacket, then it's not raining.

   The given is *not* equivalent to statement **c**.

p	q	$\sim p$	$\sim q$	Given $p \rightarrow q$	a $\sim p \vee q$	b $q \vee \sim p$	c $q \rightarrow p$	d $\sim q \rightarrow \sim p$
T	T	F	F	T	T	T	T	T
T	F	F	T	F	F	F	T	F
F	T	T	F	T	T	T	F	T
F	F	T	T	T	T	T	T	T

4. **a.** If you do not pay a fine, then the book is not overdue.

   b. If we use the pool, then it is not cold.

   c. If supervision is needed, then some students do not take exams honestly.

5. Converse: If it can fly, then it's a bird.
   Inverse: If it's not a bird, then it can not fly.

6. The triangle is isosceles and it does not have two equal sides.

7. Kelsey Grammer is not an actor or Katie Couric is not an actor.

8. Oprah Winfrey is not a jazz musician and Oprah Winfrey is not a presidential candidate.
   Alternatively: Oprah Winfrey is neither a jazz musician nor a presidential candidate.

9. It does not fit and you must acquit.

Exercise Set 3.5

1. **a.** $p \rightarrow q$ and $\sim p \vee q$ are equivalent.

p	q	$\sim p$	$p \rightarrow q$	$\sim p \vee q$
T	T	F	T	T
T	F	F	F	F
F	T	T	T	T
F	F	T	T	T

   b. $\underline{\text{If}}$ a major dam on the upper Nile River is not in place, $\underline{\text{then}}$ the lower Nile overflows its banks each year.
   $ p \rightarrow q$

   …is equivalent to…

   $ \sim p \vee q$
   A major dam on the upper Nile River is in place or the lower Nile overflows its banks each year.

3. Given: I saw *Rent* or *Ragtime*.
   p: I saw *Rent*.
   q: I saw *Ragtime*.
   a: If I did not see *Rent*, I saw *Ragtime*.
   b: I saw both *Rent* and *Ragtime*.
   c: If I saw *Rent*, I did not see *Ragtime*.
   d: If I saw *Ragtime*, I did not see *Rent*.

The given is equivalent to statement **a**.

p	q	**Given** $p \vee q$	**a** $\sim p \rightarrow q$	**b** $p \wedge q$	**c** $p \rightarrow \sim q$	**d** $q \rightarrow \sim p$
T	T	T	T	T	F	F
T	F	T	T	F	T	T
F	T	T	T	F	T	T
F	F	F	F	F	T	T

5. Given: It is not true that Sondheim and Picasso are both musicians.
p: Sondheim is a musician.
q: Picasso is a musician.
a: Sondheim is not a musician or Picasso is not a musician.
b: If Sondheim is a musician, then Picasso is not a musician.
c: Sondheim is not a musician and Picasso is not a musician.
d: If Picasso is a musician, then Sondheim is not a musician.

The given is *not* equivalent to statement **c**.

p	q	**Given** $\sim(p \wedge q)$	**a** $\sim p \vee \sim q$	**b** $p \rightarrow \sim q$	**c** $\sim p \wedge \sim q$	**d** $q \rightarrow \sim p$
T	T	F	F	F	F	F
T	F	T	T	T	F	T
F	T	T	T	T	F	T
F	F	T	T	T	T	T

7. If I am not in Illinois, I am not in Chicago.

9. If I can hear you, then the stereo is not playing.

11. If you don't die, you laugh.

13. If some troops were not withdrawn, then the president was not telling the truth.

15. If no people suffer, then some institutions do not profit above human need.

17. Converse: If you get a skin rash, then you have touched poison oak.
Inverse: If you have not touched poison oak, then you will not get a skin rash.

19. Converse: If Shakespeare is the author, then the play is *Macbeth*.
Inverse: If the play is not *Macbeth*, then Shakespeare is not the author.

21. Converse: If you are not sleeping, then you are driving the car.
Inverse: If you are not driving the car, then you are sleeping.

23. Converse: If I am not in the West, then I am in Charleston.
Inverse: If I am not in Charleston, then I am in the West.

25. Converse: If some people wear green, then it is St. Patrick's Day.
Inverse: If it's not St. Patrick's Day, then no people wear green.

27. The negation of $p \rightarrow q$ is $p \wedge \sim q$: I am in Los Angeles and not in California.

29. The negation of $p \rightarrow q$ is $p \wedge \sim q$: It is purple and it is a carrot.

31. The negation of $p \rightarrow q$ is $p \wedge \sim q$: He doesn't, and I won't.

33. The negation of $p \rightarrow q$ is $p \wedge \sim q$: There is a blizzard, and yet some schools are not closed.

35. Australia is not an island or China is not an island.

37. My high school did not encourage creativity or did not encourage diversity.

39. Babe Ruth was not a writer and was not a lawyer.

41. The United States has eradicated neither poverty nor racism.

43. I'm going to neither Seattle nor San Francisco.

45. I do not study and I pass.

47. I am going or he is not going.

49. A bill does not become law or it receives majority approval.

51. **a.** The given statement is true.

    b. If it is not the case that 96% of the patients are women, then the procedure is not a chemical peel.
True

    c. If 96% of the patients are women, then the procedure is a chemical peel.
False

    d. If the procedure is not a chemical peel, then it is not the case that 96% of the patients are women.
False

    e. The procedure is a chemical peel and it is not the case that 96% of the patients are women.
False

53. **a.** The given statement is true.

    b. There are not 32,283 face lifts per year and it is not the case that 8% of the facelift patients are men.

    c. The negation is false.

55-61. Answers will vary.

63. b is true.

65. Rewrite the given statement: If I observe the speed limit, then I do not get a ticket.
Contrapositive: If I get a ticket, then I did not observe the speed limit.
Alternative form of Contrapositive: I did not observe the speed limit if I got a ticket.

67. Given: $p \rightarrow (\sim q \vee r)$

    Converse: $(\sim q \vee r) \rightarrow p$

    Inverse: $\sim p \rightarrow \sim(\sim q \vee r)$ can be rewritten as: $\sim p \rightarrow (q \wedge \sim r)$

    Contrapositive : $\sim(\sim q \vee r) \rightarrow \sim p$ can be rewritten as: $(q \wedge \sim r) \rightarrow \sim p$

    Negation: $\sim\left[p \rightarrow (\sim q \vee r)\right] \equiv p \wedge \sim(\sim q \vee r)$ can be rewritten as: $p \wedge (q \wedge \sim r)$

69. $p \rightarrow (q \rightarrow r) \equiv (p \wedge q) \rightarrow r$

p	q	r	$p \wedge q$	$q \rightarrow r$	$p \rightarrow (q \rightarrow r)$	$(p \wedge q) \rightarrow r$
T	T	T	T	T	T	T
T	T	F	T	F	F	F
T	F	T	F	T	T	T
T	F	F	F	T	T	T
F	T	T	F	T	T	T
F	T	F	F	F	T	T
F	F	T	F	T	T	T
F	F	F	F	T	T	T

Check Points 3.6

1. The argument is valid. *p*: I study for 5 hours. *q*: I fail.

$p \vee q$

$\dfrac{\sim p}{\therefore q}$

p	q	$\sim p$	$p \vee q$	$(p \vee q) \wedge \sim p$	$[(p \vee q) \wedge \sim p] \to q$
T	T	F	T	F	T
T	F	F	T	F	T
F	T	T	T	T	T
F	F	T	F	F	T

2. The argument is invalid. *p*: I study for 5 hours. *q*: I fail.

$p \vee q$

$\dfrac{p}{\therefore \sim q}$

p	q	$p \vee q$	$(p \vee q) \wedge p$	$[(p \vee q) \wedge p] \to \sim q$
T	T	T	T	T
T	F	T	T	F
F	T	T	F	T
F	F	F	F	T

3. The argument is invalid.
   p: You do not know how to read. *q*: You cannot read "War and Peace." *r*: Leo Tolstoy will hate you.

$p \to q$

$\dfrac{q \to r}{\therefore \sim p \to \sim r}$

p	q	r	$\sim p$	$\sim r$	$p \to q$	$q \to r$	$\sim p \to \sim r$	$(p \to q) \wedge (q \to r)$	$[(p \to q) \wedge (q \to r)] \to (\sim p \to \sim r)$
T	T	T	F	F	T	T	T	T	T
T	T	F	F	T	T	F	T	F	T
T	F	T	F	F	F	T	T	F	T
T	F	F	F	T	F	T	T	F	T
F	T	T	T	F	T	T	F	T	F
F	T	F	T	T	T	F	T	F	T
F	F	T	T	F	T	T	F	T	F
F	F	F	T	T	T	T	T	T	T

4. Using the contrapositive reasoning form of a valid argument, the following can be concluded: It is not midnight.

Exercise Set 3.6

1. This is an invalid argument.

p	q	$\sim p$	$\sim q$	$p \to q$	$(p \to q) \wedge \sim p$	$[(p \to q) \wedge \sim p] \to \sim q$
T	T	F	F	T	F	T
T	F	F	T	F	F	T
F	T	T	F	T	T	F
F	F	T	T	T	T	T

3. This is a valid argument.

p	q	$\sim p$	$\sim q$	$p \to \sim q$	$(p \to \sim q) \wedge q$	$[(p \to \sim q) \wedge q] \to \sim p$
T	T	F	F	F	F	T
T	F	F	T	T	F	T
F	T	T	F	T	T	T
F	F	T	T	T	F	T

5. This is a valid argument.

p	q	$\sim q$	$p \wedge \sim q$	$(p \wedge \sim q) \wedge p$	$[(p \wedge \sim q) \wedge p] \to \sim q$
T	T	F	F	F	T
T	F	T	T	T	T
F	T	F	F	F	T
F	F	T	F	F	T

7. This is an invalid argument.

p	q	$p \to q$	$q \to p$	$p \wedge q$	$[(p \to q) \wedge (q \to p)]$	$[(p \to q) \wedge (q \to p)] \to (p \wedge q)$
T	T	T	T	T	T	T
T	F	F	T	F	F	T
F	T	T	F	F	F	T
F	F	T	T	F	T	F

$[[(p \to q) \wedge (q \to p)] \wedge p] \to (p \vee q)$
T
T
T
T

9. This is a valid argument. p: It is cold. q: Motorcycle started.

$p \to \sim q$

q

$\therefore \sim p$

p	q	$\sim p$	$\sim q$	$p \to \sim q$	$(p \to \sim q) \wedge q$	$[(p \to \sim q) \wedge q \to \sim p]$
T	T	F	F	F	F	T
T	F	F	T	T	F	T
F	T	T	F	T	T	T
F	F	T	T	T	F	T

11. This an invalid argument. p: There is a dam. q: There is flooding.

$p \vee q$

q

$\therefore \sim p$

p	q	$\sim p$	$p \vee q$	$(p \vee q) \wedge q$	$[(p \vee q) \wedge q] \to \sim p$
T	T	F	T	T	F
T	F	F	T	F	T
F	T	T	T	T	T
F	F	T	F	F	T

13. This is an invalid argument. **p:** All people obey the law. **q:** No jails are needed.

$p \to q$

$\sim p$

$\therefore \sim q$

p	q	$\sim p$	$\sim q$	$p \to q$	$(p \to q) \wedge \sim p$	$[(p \to q) \wedge \sim p] \to \sim q$
T	T	F	F	T	F	T
T	F	F	T	F	F	T
F	T	T	F	T	T	F
F	F	T	T	T	T	T

15. This is an invalid argument.

p	q	r	$p \to q$	$q \to r$	$r \to p$	$(p \to q) \wedge (q \to r)$	$[(p \to q) \wedge (q \to r)] \to (r \to p)$
T	T	T	T	T	T	T	T
T	T	F	T	F	T	F	T
T	F	T	F	T	T	F	T
T	F	F	F	T	T	F	T
F	T	T	T	T	F	T	F
F	T	F	T	F	T	F	T
F	F	T	T	T	F	T	F
F	F	F	T	T	T	T	T

17. This is a valid argument.

p	q	r	$p \to q$	$q \wedge r$	$p \vee r$	$(p \to q) \wedge (q \wedge r)$	$[(p \to q) \wedge (q \wedge r)] \to (p \vee r)$
T	T	T	T	T	T	T	T
T	T	F	T	F	T	F	T
T	F	T	F	F	T	F	T
T	F	F	F	F	T	F	T
F	T	T	T	T	T	T	T
F	T	F	T	F	F	F	T
F	F	T	T	F	T	F	T
F	F	F	T	F	F	F	T

19. This is a valid argument.

p	q	r	$\sim p$	$\sim r$	$p \leftrightarrow q$	$q \to r$	$\sim r \to \sim p$
T	T	T	F	F	T	T	T
T	T	F	F	T	T	F	F
T	F	T	F	F	F	T	T
T	F	F	F	T	F	T	F
F	T	T	T	F	F	T	T
F	T	F	T	T	F	F	T
F	F	T	T	F	T	T	T
F	F	F	T	T	T	T	T

$(p \leftrightarrow q) \wedge (q \to r)$	$[(p \leftrightarrow q) \wedge (q \to r)] \to (\sim r \to \sim p)$
T	T
F	T
F	T
F	T
F	T
F	T
T	T
T	T

41

21. This is a valid argument. *p*: Tim plays   *q*: Janet plays   *r*: Team wins

$(p \wedge q) \rightarrow r$

$p \wedge \sim r$

$\therefore \sim q$

p	q	r	$\sim q$	$\sim r$	$p \wedge q$	$p \wedge \sim r$	$(p \wedge q) \rightarrow r$
T	T	T	F	F	T	F	T
T	T	F	F	T	T	T	F
T	F	T	T	F	F	F	T
T	F	F	T	T	F	T	T
F	T	T	F	F	F	F	T
F	T	F	F	T	F	F	T
F	F	T	T	F	F	F	T
F	F	F	T	T	F	F	T

$[(p \wedge q) \rightarrow r] \wedge (p \wedge \sim r)$	$\big[[(p \wedge q) \rightarrow r] \wedge (p \wedge \sim r)\big] \rightarrow \sim q$
F	T
F	T
F	T
T	T
F	T
F	T
F	T
F	T

23. This is a valid argument. *p*: It rains   *q*: It snows   *r*: I read

$(p \vee q) \rightarrow r$

$\sim r$

$\therefore \sim (p \vee q)$

p	q	r	$\sim r$	$p \vee q$	$\sim (p \vee q)$	$(p \vee q) \rightarrow r$
T	T	T	F	T	F	T
T	T	F	T	T	F	F
T	F	T	F	T	F	T
T	F	F	T	T	F	F
F	T	T	F	T	F	T
F	T	F	T	T	F	F
F	F	T	F	F	T	T
F	F	F	T	F	T	T

$[(p \vee q) \rightarrow r] \wedge \sim r$	$\big[[(p \vee q) \rightarrow r] \wedge \sim r\big] \rightarrow \sim (p \vee q)$
F	T
F	T
F	T
F	T
F	T
F	T
F	T
T	T

25. This is an invalid argument. *p*: It rains   *q*: It snows   *r*: I read

$(p \vee q) \rightarrow r$

r

$\therefore p \vee q$

p	q	r	$p \vee q$	$(p \vee q) \rightarrow r$	$[(p \vee q) \rightarrow r] \wedge r$	$[[(p \vee q) \rightarrow r] \wedge r] \rightarrow (p \vee q)$
T	T	T	T	T	T	T
T	T	F	T	F	F	T
T	F	T	T	T	T	T
T	F	F	T	F	F	T
F	T	T	T	T	T	T
F	T	F	T	F	F	T
F	F	T	F	T	T	F
F	F	F	F	T	F	T

27. This is an invalid argument. p: It's hot. q: It's humid. r: I complain.

$(p \wedge q) \rightarrow r$

$\underline{\sim p \vee \sim q}$

$\therefore \sim r$

p	q	r	$\sim p$	$\sim q$	$\sim r$	$p \wedge q$	$\sim p \vee \sim q$	$(p \wedge q) \rightarrow r$
T	T	T	F	F	F	T	F	T
T	T	F	F	F	T	T	F	F
T	F	T	F	T	F	F	T	T
T	F	F	F	T	T	F	T	T
F	T	T	T	F	F	F	T	T
F	T	F	T	F	T	F	T	T
F	F	T	T	T	F	F	T	T
F	F	F	T	T	T	F	T	T

$[(p \wedge q) \rightarrow r] \wedge (\sim p \vee \sim q)$	$[[(p \wedge q) \rightarrow r] \wedge (\sim p \vee \sim q)] \rightarrow \sim r$
F	T
F	T
T	F
T	T
T	F
T	T
T	F
T	T

29. p: We close the door.
q: There is less noise.

$p \rightarrow q$
$\underline{q}$
$\therefore p$
Invalid, by fallacy of the converse.

31. p: We criminalize drugs.
q: We damage the future of young people.

$p \vee q$
$\underline{\sim q}$
$\therefore p$
Valid, by disjunctive reasoning.

33. p: I am at the beach.
q: I swim in the ocean.
r: I feel refreshed.

$p \rightarrow q$
$\underline{q \rightarrow r}$
$\therefore p \rightarrow r$
Valid, by transitive reasoning.

35. p: I'm at the beach.
q: I swim in the ocean.
r: I feel refreshed.

$p \rightarrow q$
$q \rightarrow r$
$\therefore \sim p \rightarrow \sim r$
Invalid, by misuse of transitive reasoning.

43

37. p: A person is a chemist.
q: A person has a college degree.

$p \rightarrow q$

$\underline{\sim q}$

$\therefore \sim p$

My best friend is not a chemist. By contrapositive reasoning.

39. p: Writers improve.
q: "My Mother the Car" dropped from primetime.

$p \vee q$

$\underline{\sim p}$

$\therefore q$

"My Mother the Car" was dropped from primetime. By disjunctive reasoning.

41. p: All electricity off.
q: No lights work.

$p \rightarrow q$

$\underline{\sim q}$

$\therefore \sim p$

Some electricity is not off. By contrapositive reasoning.

43. p: I vacation in Paris.
q: I eat French pastries.
r: I gain weight.

$p \rightarrow q$

$\underline{q \rightarrow r}$

$\therefore p \rightarrow r$

If I vacation in Paris I gain weight. By transitive reasoning.

45. p: Poverty causes crime.
q: Crime sweeps American cities during the Great Depression.

$p \rightarrow q$

$\underline{\sim q}$

$\therefore \sim p$

Valid. By contrapositive reasoning.

47-53. Answers will vary.

55. This is a valid argument.
p: Secondary cigarette smoke is a health threat.
q: It's wrong to smoke in public.
r: The ALA says that secondary cigarette smoke is a health threat.

$p \rightarrow q$

$\sim p \rightarrow \sim r$

$\underline{r}$

$\therefore q$

p	q	r	$\sim p$	$\sim r$	$p \rightarrow q$	$\sim p \rightarrow \sim r$	$[(p \rightarrow q) \wedge (\sim p \rightarrow \sim r)] \wedge r$	$[[(p \rightarrow q) \wedge (\sim p \rightarrow \sim r)] \wedge r] \rightarrow q$
T	T	T	F	F	T	T	T	T
T	T	F	F	T	T	T	F	T
T	F	T	F	F	F	T	F	T
T	F	F	F	T	F	T	F	T
F	T	T	T	F	T	F	F	T
F	T	F	T	T	T	T	F	T
F	F	T	T	F	T	F	F	T
F	F	F	T	T	T	T	F	T

57. The statement is a tautology.

p	q	r	$\sim r$	$q \to r$	$p \to (q \to r)$	$p \wedge q$	$(p \wedge q) \wedge \sim r$	$[p \to (q \to r)] \vee [(p \wedge q) \wedge \sim r]$
T	T	T	F	T	T	T	F	T
T	T	F	T	F	F	T	T	T
T	F	T	F	T	T	F	F	T
T	F	F	T	T	T	F	F	T
F	T	T	F	T	T	F	F	T
F	T	F	T	F	T	F	F	T
F	F	T	F	T	T	F	F	T
F	F	F	T	T	T	F	F	T

Check Points 3.7

1. The argument is valid.

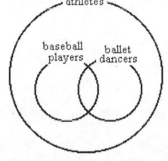

2. The argument is invalid.

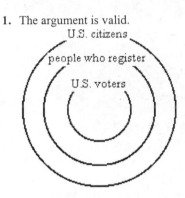

3. The argument is valid.

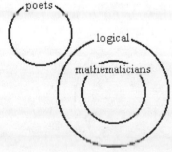

4. The argument is invalid.

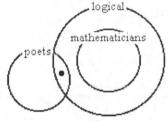

5. The argument is invalid.

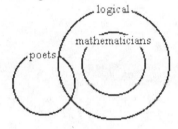

6. The argument is invalid.
The ● is Euclid.

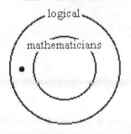

Exercise Set 3.7

1. Valid

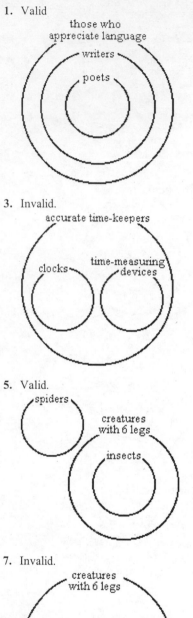

3. Invalid.

5. Valid.

7. Invalid.

9. Invalid.

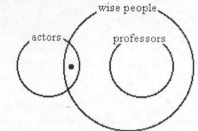

11. Valid.

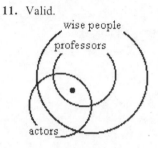

13. Valid. The ● is Savion Glover.

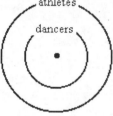

15. Invalid. The ● is Savion Glover.

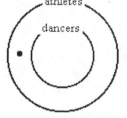

17. Invalid.

19. Valid.

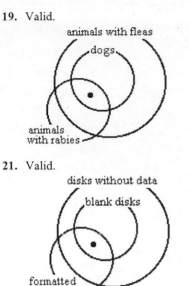

21. Valid.

disks without data

blank disks

formatted
disks

23. Valid. The ● is 8.

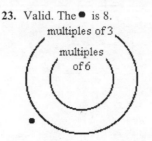

multiples of 3

multiples
of 6

25-27. Answers will vary.

29. All opera singers take voice lessons.

31. Invalid.

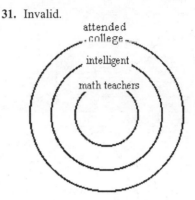

attended
college

intelligent

math teachers

Chapter 3 Review Exercises

1. If the temperature is below 32° and we finished studying, then we go to the movies.

2. If we don't go to the movies, then the temperature is not below 32° or we did not finish studying.

3. The temperature is below 32°, and if we finished studying we go to the movies.

4. We go to the movies if and only if the temperature is below 32° and we finished studying.

5. It is not true that the temperature is below 32° and we finished studying.

6. We won't go to the movies if and only if the temperature is not below 32° or we have not finished studying.

7. $(p \wedge q) \vee r$

8. $(p \vee \sim q) \to r$

9. $q \to (p \leftrightarrow r)$

10. $r \leftrightarrow (p \wedge \sim q)$

11. Some houses are not made of wood.

12. Some students major in business.

13. No crimes are motivated by passion.

14. All Democrats are registered voters.

15. Some new taxes are not for education.

16. The statement is false because it is of the form: [true] *and* [false].

17. The statement is true because it is of the form: [true] *or* [false].

18. The statement is true because it is of the form: If [false], then [false].

19. $p \vee (\sim p \wedge q)$

p	q	$\sim p$	$\sim p \wedge q$	$p \vee (\sim p \wedge q)$
T	T	F	F	T
T	F	F	F	T
F	T	T	T	T
F	F	T	F	F

20. $\sim p \vee \sim q$

p	q	$\sim p$	$\sim q$	$\sim p \vee \sim q$
T	T	F	F	F
T	F	F	T	T
F	T	T	F	T
F	F	T	T	T

21. $p \rightarrow (\sim p \vee q)$

p	q	$\sim p$	$\sim p \vee q$	$p \rightarrow (\sim p \vee q)$
T	T	F	T	T
T	F	F	F	F
F	T	T	T	T
F	F	T	T	T

22. $p \leftrightarrow \sim q$

p	q	$\sim q$	$p \leftrightarrow \sim q$
T	T	F	F
T	F	T	T
F	T	F	T
F	F	T	F

23. $\sim (p \vee q) \rightarrow (\sim p \wedge \sim q)$

p	q	$\sim p$	$\sim q$	$p \vee q$	$\sim (p \vee q)$	$\sim p \wedge \sim q$	$\sim (p \vee q) \rightarrow (\sim p \wedge \sim q)$
T	T	F	F	T	F	F	T
T	F	F	T	T	F	F	T
F	T	T	F	T	F	F	T
F	F	T	T	F	T	T	T

24. $(p \vee q) \rightarrow \sim r$

p	q	r	$\sim r$	$p \vee q$	$(p \vee q) \rightarrow \sim r$
T	T	T	F	T	F
T	T	F	T	T	T
T	F	T	F	T	F
T	F	F	T	T	T
F	T	T	F	T	F
F	T	F	T	T	T
F	F	T	F	F	T
F	F	F	T	F	T

25. $(p \wedge q) \leftrightarrow (p \wedge r)$

p	q	r	$p \wedge q$	$p \wedge r$	$(p \wedge q) \leftrightarrow (p \wedge r)$
T	T	T	T	T	T
T	T	F	T	F	F
T	F	T	F	T	F
T	F	F	F	F	T
F	T	T	F	F	T
F	T	F	F	F	T
F	F	T	F	F	T
F	F	F	F	F	T

26. a. $p \vee q \equiv \sim q \rightarrow p$

p	q	$\sim q$	$p \vee q$	$\sim q \rightarrow p$
T	T	F	T	T
T	F	T	T	T
F	T	F	T	T
F	F	T	F	F

b. If I get the part, then I learned the lines.

27. c

28. If I am not in the South, then I am not in Atlanta.

29. If today is a holiday, then I'm not in class.

30. If I do not pass some course, then I did not work hard.

31. Converse: If classes are cancelled, then there is a storm.
Inverse: If there is no storm, then classes are not cancelled.

32. Converse: If we do not talk, then the television is on.
Inverse: If the television is not on, then we talk.

33. I am in Bogota and I am not in Columbia.

34. I do not work hard and I succeed.

35. Chicago is not a city or Maine is not a city.

36. Ernest Hemingway was neither a musician nor an actor.

37. I don't work hard and I succeed.

38. She is using her car or she is not taking a bus.

39. The argument is invalid.

p	q	$\sim q$	$p \rightarrow q$	$(p \rightarrow q) \wedge \sim q$	$[(p \rightarrow q) \wedge \sim q] \rightarrow p$
T	T	F	T	F	T
T	F	T	F	F	T
F	T	F	T	F	T
F	F	T	T	T	F

40. The argument is valid.

p	q	r	$p \wedge q$	$q \to r$	$p \to r$	$(p \wedge q) \wedge (q \to r)$	$[(p \wedge q) \wedge (q \to r)] \to (p \to r)$
T	T	T	T	T	T	T	T
T	T	F	T	F	F	F	T
T	F	T	F	T	T	F	T
T	F	F	F	T	F	F	T
F	T	T	F	T	T	F	T
F	T	F	F	F	T	F	T
F	F	T	F	T	T	F	T
F	F	F	F	T	T	F	T

41. The argument is valid. *p*: Good baseball player. *q*: Good hand–eye coordination.

$$p \to q$$
$$\underline{\sim q}$$
$$\sim p$$

p	q	$\sim p$	$\sim q$	$p \to q$	$(p \to q) \wedge \sim q$	$[(p \to q) \wedge \sim q] \to \sim p$
T	T	F	F	T	F	T
T	F	F	T	F	F	T
F	T	T	F	T	F	T
F	F	T	T	T	T	T

42. The argument is invalid. *p*: Tony plays. *q*: Team wins.

$$p \to q$$
$$\underline{q}$$
$$\therefore p$$

p	q	$p \to q$	$(p \to q) \wedge q$	$[(p \to q) \wedge q] \to p$
T	T	T	T	T
T	F	F	F	T
F	T	T	T	F
F	F	T	F	T

43. The argument is invalid. *p*: Plant is fertilized. *q*: Plant turns yellow.

$$p \vee q$$
$$\underline{q}$$
$$\therefore \sim p$$

p	q	$\sim p$	$p \vee q$	$(p \vee q) \wedge q$	$[(p \vee q) \wedge q] \to \sim p$
T	T	F	T	T	F
T	F	F	T	F	T
F	T	T	T	T	T
F	F	T	F	F	T

44. The argument is valid. *p*: A majority of legislators vote for a bill. *q*: Bill does not become law.

$$p \vee q$$
$$\underline{\sim p}$$
$$\therefore q$$

p	q	$\sim p$	$p \vee q$	$(p \vee q) \wedge \sim p$	$[(p \vee q) \wedge \sim p] \to q$
T	T	F	T	F	T
T	F	F	T	F	T
F	T	T	T	T	T
F	F	T	F	F	T

45. The argument is invalid.
p: I purchase season tickets to the football games.
q: I do not attend all the lectures.
r: I do not do well in school.

$$p \to q$$
$$\underline{q \to r}$$
$$\therefore r \to p$$

p	q	r	$p \to q$	$q \to r$	$r \to p$	$(p \to q) \land (q \to r)$	$[(p \to q) \land (q \to r)] \to (r \to p)$
T	T	T	T	T	T	T	T
T	T	F	T	F	T	F	T
T	F	T	F	T	T	F	T
T	F	F	F	T	T	F	T
F	T	T	T	T	F	T	F
F	T	F	T	F	T	F	T
F	F	T	T	T	F	T	F
F	F	F	T	T	T	T	T

46. Invalid.

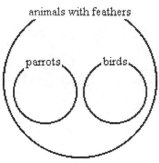

47. Valid.

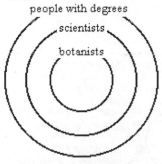

48. Valid.

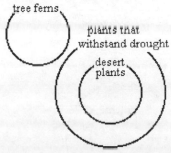

49. Invalid.

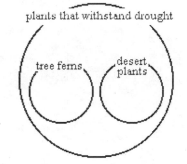

50. Invalid.

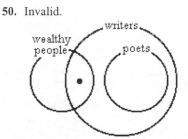

Chapter 3 Test

1. If I'm registered and I'm a citizen, then I vote.

2. I don't vote if and only if I'm not registered or I'm not a citizen.

3. I'm neither registered nor a citizen.

4. $(p \land q) \lor \sim r$

5. $(\sim p \lor \sim q) \to \sim r$

6. Some numbers are not divisible by 5.

7. No people wear glasses.

8. $p \wedge (\sim p \vee q)$

p	q	$\sim p$	$\sim p \vee q$	$p \wedge (\sim p \vee q)$
T	T	F	T	T
T	F	F	F	F
F	T	T	T	F
F	F	T	T	F

9. $\sim (p \wedge q) \leftrightarrow (\sim p \vee \sim q)$

p	q	$\sim p$	$\sim q$	$p \wedge q$	$\sim (p \wedge q)$	$(\sim p \vee \sim q)$	$\sim (p \wedge q) \leftrightarrow (\sim p \vee \sim q)$
T	T	F	F	T	F	F	T
T	F	F	T	F	T	T	T
F	T	T	F	F	T	T	T
F	F	T	T	F	T	T	T

10. $p \leftrightarrow (q \vee r)$

p	q	r	$q \vee r$	$p \leftrightarrow (q \vee r)$
T	T	T	T	T
T	T	F	T	T
T	F	T	T	T
T	F	F	F	F
F	T	T	T	F
F	T	F	T	F
F	F	T	T	F
F	F	F	F	T

11. b

12. If it snows, then it is not August.

13. Converse: If I cannot concentrate, then the radio is playing.

Inverse: If the radio is not playing, then I can concentrate.

14. It is cold and we use the pool.

15. The test is not today and the party is not tonight.

16. The banana is not green or it is ready to eat.

17. The argument is invalid. *p*: Parrot talks. *q*: It is intelligent.

$p \rightarrow q$
$\underline{q}$
$\therefore p$

18. The argument is valid. *p*: I am sick. *q*: I am tired.

$p \vee q$
$\underline{\sim q}$
$\therefore p$

19. Invalid.

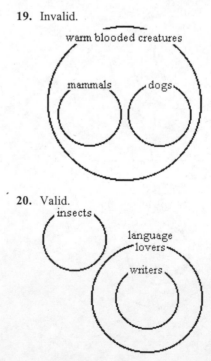

20. Valid.

Chapter 4
Number Representation and Calculation

Check Points 4.1

1. **a.** $10^5 = 10 \times 10 \times 10 \times 10 \times 10 = 100,000$ **b.** $10^6 = 10 \times 10 \times 10 \times 10 \times 10 \times 10 = 1,000,000$

2. **a.** $8^2 = 8 \times 8 = 64$ **b.** $6^3 = 6 \times 6 \times 6 = 216$

    c. $2^5 = 2 \times 2 \times 2 \times 2 \times 2 = 32$ **d.** $18^1 = 18$

3. **a.** $4026 = (4 \times 10^3) + (0 \times 10^2) + (2 \times 10^1) + (6 \times 1) = (4 \times 1000) + (0 \times 100) + (2 \times 10) + (6 \times 1)$

    b. $24,232 = (2 \times 10^4) + (4 \times 10^3) + (2 \times 10^2) + (3 \times 10^1) + (2 \times 1)$
    $$= (2 \times 10,000) + (4 \times 1000) + (2 \times 100) + (3 \times 10) + (2 \times 1)$$

4. **a.** $6000 + 70 + 3 = 6073$ **b.** $80,000 + 900 = 80,900$

5.
$$\begin{array}{ccc} \vee\vee\vee & << & <<< \vee \\ \downarrow & \downarrow & \downarrow \end{array}$$
$$(3 \times 60^2) + (2 \times 60) + (31 \times 1)$$
$$= (3 \times 3600) + (2 \times 60) + (31 \times 1)$$
$$= 10,800 + 1200 + 31$$
$$= 12,031$$

6.
$$\begin{array}{rclr} 11 & \times & 7200 = & 79,200 \\ 3 & \times & 360 = & 1080 \\ 0 & \times & 20 = & 0 \\ 13 & \times & 1 = & \underline{13} \\ & & & 80,293 \end{array}$$

Exercise Set 4.1

1. $5^2 = 5 \times 5 = 25$

3. $2^3 = 2 \times 2 \times 2 = 8$

5. $3^4 = 3 \times 3 \times 3 \times 3 = 81$

7. $10^5 = 10 \times 10 \times 10 \times 10 \times 10 = 100,000$

9. $36 = (3 \times 10^1) + (6 \times 1) = (3 \times 10) + (6 \times 1)$

11. $249 = (2 \times 10^2) + (4 \times 10^1) + (9 \times 1) = (2 \times 100) + (4 \times 10) + (9 \times 1)$

13. $703 = (7 \times 10^2) + (0 \times 10^1) + (3 \times 1) = (7 \times 100) + (0 \times 10) + (3 \times 1)$

15. $4856 = (4 \times 10^3) + (8 \times 10^2) + (5 \times 10^1) + (6 \times 1) = (4 \times 1000) + (8 \times 100) + (5 \times 10) + (6 \times 1)$

17. $3070 = (3 \times 10^3) + (0 \times 10^2) + (7 \times 10^1) + (0 \times 1) = (3 \times 1000) + (0 \times 100) + (7 \times 10) + (0 \times 1)$

19. $34,569 = (3 \times 10^4) + (4 \times 10^3) + (5 \times 10^2) + (6 \times 10^1) + (9 \times 1)$
$= (3 \times 10,000) + (4 \times 1000) + (5 \times 100) + (6 \times 10) + (9 \times 1)$

21. $230,007,004 = (2 \times 10^8) + (3 \times 10^7) + (0 \times 10^6) + (0 \times 10^5) + (0 \times 10^4) + (7 \times 10^3) + (0 \times 10^2) + (0 \times 10^1) + (4 \times 1)$
$= (2 \times 100,000,000) + (3 \times 10,000,000) + (0 \times 1,000,000) + (0 \times 100,000)$
$+ (0 \times 10,000) + (7 \times 1000) + (0 \times 100) + (0 \times 10) + (4 \times 1)$

23. $70 + 3 = 73$

25. $300 + 80 + 5 = 385$

27. $500,000 + 20,000 + 8000 + 700 + 40 + 3 = 528,743$

29. $7000 + 2 = 7002$

31. $600,000,000 + 2000 + 7 = 600,002,007$

33. $(10 + 10 + 1 + 1 + 1) \times 1 = 23 \times 1 = 23$

35. $(10 + 10 + 1) \times 60^1 + (1 + 1) \times 1 = (21 \times 60) + 2 = 1260 + 2 = 1262$

37. $(1 + 1 + 1) \times 60^2 + (10 + 1 + 1) \times 60^1 + (1 + 1 + 1) \times 1 = (3 \times 3600) + (12 \times 60) + 3 = 10,800 + 720 + 3 = 11,523$

39. $(10 + 1) \times 60^3 + (10 + 1) \times 60^2 + (10 + 1) \times 60^1 + (10 + 1) \times 1$
$= (11 \times 60^3) + (11 \times 60^2) + (11 \times 60^1) + (11 \times 1)$
$= (11 \times 216,000) + (11 \times 3600) + (11 \times 60) + 11$
$= 2,376,000 + 39,600 + 660 + 11$
$= 2,416,271$

41. $14 \times 1 = 14$

43.
$$\begin{array}{rcrcr} 19 & \times & 360 & = & 6840 \\ 0 & \times & 20 & = & 0 \\ 6 & \times & 1 & = & \underline{6} \\ & & & & 6846 \end{array}$$

45.
$$\begin{array}{rcrcr} 8 & \times & 360 & = & 2880 \\ 8 & \times & 20 & = & 160 \\ 8 & \times & 1 & = & \underline{8} \\ & & & & 3048 \end{array}$$

47.
$$\begin{array}{rcrcr} 2 & \times & 7200 & = & 14,400 \\ 0 & \times & 360 & = & 0 \\ 0 & \times & 20 & = & 0 \\ 11 & \times & 1 & = & \underline{11} \\ & & & & 14,411 \end{array}$$

49.
$$\begin{array}{rcrcr} 10 & \times & 7200 & = & 72,000 \\ 10 & \times & 360 & = & 3,600 \\ 0 & \times & 20 & = & 0 \\ 10 & \times & 1 & = & \underline{10} \\ & & & & 75,610 \end{array}$$

51.
$$\begin{array}{rcrcr} 4 & \times & 360 & = & 1440 \\ 16 & \times & 20 & = & 320 \\ 16 & \times & 1 & = & \underline{16} \\ & & & & 1776 \end{array}$$
Declaration of Independence

53-59. Answers will vary.

61. Change to Hindu-Arabic:

$(1 \times 60^2) + (10 + 1 + 1) \times 60^1 + (10 + 1) \times 1$

$= 3600 + (12 \times 60) + 11$

$= 3600 + 720 + 11 = 4331$

Change to Mayan:

$12 \times 360 = 4320$

$0 \times 20 = \quad 0$

$\underline{11 \times 1 = \quad 11}$

$\qquad\qquad 4331$

63. *Writing the numeral that precedes:*
Subtract 1 or $\vee$ from the numeral on the right:

$< \vee \quad <<<<< \vee \vee \vee \vee \vee \vee \vee \vee$

Writing the numeral that follows:
First, add 1, or $\vee$, to the numeral on the right:

$< \vee \quad <<<<< \vee \vee \vee \vee \vee \vee \vee \vee \vee \vee$

Next, change the 10 $\vee$'s to a $<$:

$< \vee \quad <<<<<<$

Since $<<<<<<$ has a value of 60 it
can be carried as a $\vee$ in the next higher place:

$< \vee \vee$ (note the missing place value)

The missing place value is indicated by $\overset{<}{_<}$

giving a final answer of: $< \vee \vee \quad \overset{<}{_<}$

65. $5^4 = 625$

67. $84^3 = 592,704$

Check Points 4.2

1. $3422_{\text{five}} = (3 \times 5^3) + (4 \times 5^2) + (2 \times 5^1) + (2 \times 1) = (3 \times 5 \times 5 \times 5) + (4 \times 5 \times 5) + (2 \times 5) + (2 \times 1) = 375 + 100 + 10 + 2 = 487$

2. $110011_{\text{two}} = (1 \times 2^5) + (1 \times 2^4) + (0 \times 2^3) + (0 \times 2^2) + (1 \times 2^1) + (1 \times 1)$
$\qquad\qquad = (1 \times 32) + (1 \times 16) + (0 \times 8) + (0 \times 4) + (1 \times 2) + (1 \times 1) = 32 + 16 + 2 + 1 = 51$

3. $AD4_{\text{sixteen}} = (10 \times 16^2) + (13 \times 16^1) + (4 \times 1) = (10 \times 16 \times 16) + (13 \times 16) + (4 \times 1) = 2560 + 208 + 4 = 2772$

4. $6_{\text{ten}} = (1 \times 5) + (1 \times 1) = 11_{\text{five}}$

5. The place values in base 7 are $...7^4, \ 7^3, \ 7^2, \ 7^1, \ 1$ or $...2401, \ 343, \ 49, \ 7, \ 1$

$$\begin{array}{ccc}
1 & 0 & 3 \\
343\overline{)365} & 49\overline{)22} & 7\overline{)22} \\
\underline{343} & \underline{0} & \underline{21} \\
22 & 22 & 1
\end{array}$$

$365_{\text{ten}} = (1 \times 343) + (0 \times 49) + (3 \times 7) + (1 \times 1) = (1 \times 7^3) + (0 \times 7^2) + (3 \times 7^1) + (1 \times 1) = 1031_{\text{seven}}$

6. The place values in base 5 are $...5^4, \ 5^3, \ 5^2, \ 5^1, \ 1$ or $...3125, \ 625, \ 125, \ 25, \ 5, \ 1$

$$\begin{array}{cccc}
4 & 2 & 0 & 2 \\
625\overline{)2763} & 125\overline{)263} & 25\overline{)13} & 5\overline{)13} \\
\underline{2500} & \underline{250} & \underline{0} & \underline{10} \\
263 & 13 & 13 & 3
\end{array}$$

$2763_{\text{ten}} = (4 \times 625) + (2 \times 125) + (0 \times 25) + (2 \times 5) + (3 \times 1) = (4 \times 5^4) + (2 \times 5^3) + (0 \times 5^2) + (2 \times 5^1) + (3 \times 1) = 42023_{\text{five}}$

Exercise Set 4.2

1. $(4 \times 5^1) + (3 \times 1) = 20 + 3 = 23$

3. $(5 \times 8^1) + (2 \times 1) = 40 + 2 = 42$

5. $(1 \times 4^2) + (3 \times 4^1) + (2 \times 1) = 16 + 12 + 2 = 30$

7. $(1 \times 2^3) + (0 \times 2^2) + (1 \times 2^1) + (1 \times 1) = 8 + 0 + 2 + 1 = 11$

9. $(2 \times 6^3) + (0 \times 6^2) + (3 \times 6^1) + (5 \times 1) = 432 + 0 + 18 + 5 = 455$

11. $(7 \times 8^4) + (0 \times 8^3) + (3 \times 8^2) + (5 \times 8^1) + (5 \times 1) = 28,672 + 0 + 192 + 40 + 5 = 28,909$

13. $(2 \times 16^3) + (0 \times 16^2) + (9 \times 16^1) + (6 \times 1) = 8192 + 0 + 144 + 6 = 8342$

15. $(1 \times 2^5) + (1 \times 2^4) + (0 \times 2^3) + (1 \times 2^2) + (0 \times 2^1) + (1 \times 1) = 32 + 16 + 0 + 4 + 0 + 1 = 53$

17. $(10 \times 16^3) + (12 \times 16^2) + (14 \times 16^1) + (5 \times 1) = 40,960 + 3072 + 224 + 5 = 44,261$

19. 12_{five}

21. 14_{seven}

23. 10_{two}

25. 31_{four}

27. 101_{six}

29.
$$25\overline{)87} \quad 5\overline{)12}$$
$$\underline{75} \quad\quad \underline{10}$$
$$12 \quad\quad\;\; 2$$
$$87 = 322_{\text{five}}$$

31.
$$64\overline{)108} \quad 16\overline{)44} \quad 4\overline{)12}$$
$$\underline{64} \quad\quad\; \underline{32} \quad\;\; \underline{12}$$
$$44 \quad\quad\;\; 12 \quad\quad 0$$
$$108 = 1230_{\text{four}}$$

33.
$$16\overline{)19} \quad 8\overline{)3} \quad 4\overline{)3} \quad 2\overline{)3}$$
$$\underline{16} \quad\;\; \underline{0} \quad\; \underline{0} \quad\; \underline{2}$$
$$3 \quad\quad 3 \quad\; 3 \quad\; 1$$
$$19 = 10011_{\text{two}}$$

35.
$$81\overline{)138} \quad 27\overline{)57} \quad 9\overline{)3} \quad 3\overline{)3}$$
$$\underline{81} \quad\quad\; \underline{54} \quad\;\; \underline{0} \quad\; \underline{3}$$
$$57 \quad\quad\;\; 3 \quad\quad 3 \quad\; 0$$
$$138 = 12010_{\text{three}}$$

37.
$$216\overline{)386} \quad 36\overline{)170} \quad 6\overline{)26}$$
$$\underline{216} \quad\quad\;\; \underline{144} \quad\quad \underline{24}$$
$$170 \quad\quad\;\; 26 \quad\quad\; 2$$
$$386 = 1442_{\text{six}}$$

39.
$$343\overline{)1599} \quad 49\overline{)227} \quad 7\overline{)31}$$
$$\underline{1372} \quad\quad\;\; \underline{196} \quad\quad \underline{28}$$
$$227 \quad\quad\;\; 31 \quad\quad\; 3$$
$$1599 = 4443_{\text{seven}}$$

41. 21 weeks = 147 days
    153 days = 21 weeks, 6 days

43. Example:
    $\$8.79 \div 0.25 = 35.16$
    35 quarters = $\$8.75$
    $\$8.79 = 35$ quarters + 4 pennies

45. Answers will vary.

47. Preceding: Following:
$$888_{\text{nine}} \quad\quad\quad 888_{\text{nine}}$$
$$\underline{-\;\; 1_{\text{nine}}} \quad\quad\; \underline{+\;\; 1_{\text{nine}}}$$
$$887_{\text{nine}} \quad\quad\; 1000_{\text{nine}}$$

49. 11111011_{two}

$(1 \times 2^7) + (1 \times 2^6) + (1 \times 2^5) + (1 \times 2^4) + (1 \times 2^3) + (0 \times 2^2) + (1 \times 2^1) + (1 \times 1) = 128 + 64 + 32 + 16 + 8 + 0 + 2 + 1 = 251$

$3A6_{twelve}$

$(3 \times 12^2) + (10 \times 12^1) + (6 \times 1) = 432 + 120 + 6 = 558$

673_{eight}

$(6 \times 8^2) + (7 \times 8^1) + (3 \times 1) = 384 + 56 + 3 = 443$

11111011_{two}, 673_{eight}, $3A6_{twelve}$

51. 37

53. 1000110111_{two}

Check Points 4.3

1.
$$\overset{1}{3}2_{five}$$
$$+44_{five}$$
$$\overline{131_{five}}$$

$2 + 4 = 6 = (1 \times 5^1) + (1 \times 1) = 11_{five}$

$1 + 3 + 4 = 8 = (1 \times 5^1) + (3 \times 1) = 13_{five}$

2.
$$\overset{11}{111}_{two}$$
$$+111_{two}$$
$$\overline{1110_{two}}$$

$1 + 1 = 2 = (1 \times 2^1) + (0 \times 1) = 10_{two}$

$1 + 1 + 1 = 3 = (1 \times 2^1) + (1 \times 1) = 11_{two}$

3.
$$\overset{3\,6}{4\,1}_{five}$$
$$-23_{five}$$
$$\overline{13_{five}}$$

4.
$$\overset{4\,8\,3\,11}{5\,1\,4\,4}_{seven}$$
$$-3236_{seven}$$
$$\overline{1605_{seven}}$$

5.
$$\overset{2}{4}5_{seven}$$
$$\times 3_{seven}$$
$$\overline{201_{seven}}$$

$3 \times 5 = 15 = (2 \times 7^1) + (1 \times 1) = 21_{seven}$

$(3 \times 4) + 2 = 14 = (2 \times 7^1) + (0 \times 1) = 20_{seven}$

6.
$$2_{four} \overline{)112_{four}} \quad \overset{23}{}$$
$$\underline{10}$$
$$12$$
$$\underline{12}$$
$$0$$
$$23_{four}$$

Exercise Set 4.3

Note: Numbers with no base specified are base 10.

1.
$$\overset{1}{2}3_{four}$$
$$+13_{four}$$
$$\overline{102_{four}}$$

$3 + 3 = 6 = (1 \times 4^1) + (2 \times 1) = 12_{four}$

$1 + 2 + 1 = 4 = (1 \times 4^1) + (0 \times 1) = 10_{four}$

3.
$$\overset{1}{1}1_{two}$$
$$+11_{two}$$
$$\overline{110_{two}}$$

$1 + 1 + 1 = 3 = (1 \times 2^1) + (1 \times 1) = 11_{two}$

5.

$\begin{array}{r} \overset{1\ 1}{342_{\text{five}}} \\ +413_{\text{five}} \\ \hline 1310_{\text{five}} \end{array}$

$2+3=5=(1\times 5^1)+(0\times 1)=10_{\text{five}}$

$1+4+1=6=(1\times 5^1)+(1\times 1)=11_{\text{five}}$

$1+3+4=8=(1\times 5^1)+(3\times 1)=13_{\text{five}}$

7.

$\begin{array}{r} \overset{1\ 1}{645_{\text{seven}}} \\ +324_{\text{seven}} \\ \hline 1302_{\text{seven}} \end{array}$

$5+4=9=(1\times 7^1)+(2\times 1)=12_{\text{seven}}$

$1+4+2=7=(1\times 7^1)+(0\times 1)=10_{\text{seven}}$

$1+6+3=10=(1\times 7^1)+(3\times 1)=13_{\text{seven}}$

9.

$\begin{array}{r} \overset{1\ 1\ 1}{6784_{\text{nine}}} \\ +7865_{\text{nine}} \\ \hline 15760_{\text{nine}} \end{array}$

$4+5=9=(1\times 9^1)+(0\times 1)=10_{\text{nine}}$

$1+8+6=15=(1\times 9^1)+(6\times 1)=16_{\text{nine}}$

$1+7+8=16=(1\times 9^1)+(7\times 1)=17_{\text{nine}}$

$1+6+7=14=(1\times 9^1)+(5\times 1)=15_{\text{nine}}$

11.

$\begin{array}{r} \overset{1\ 1}{14632_{\text{seven}}} \\ +5604_{\text{seven}} \\ \hline 23536_{\text{seven}} \end{array}$

$6+6=12=(1\times 7^1)+(5\times 1)=15_{\text{seven}}$

$1+4+5=10=(1\times 7^1)+(3\times 1)=13_{\text{seven}}$

13.

$\begin{array}{r} \overset{2\ 6}{3\cancel{2}_{\text{four}}} \\ -13_{\text{four}} \\ \hline 13_{\text{four}} \end{array}$

15.

$\begin{array}{r} \overset{1\ 8}{2\cancel{3}_{\text{five}}} \\ -14_{\text{five}} \\ \hline 4_{\text{five}} \end{array}$

17.

$\begin{array}{r} \overset{6\ 13}{47\cancel{5}_{\text{eight}}} \\ -267_{\text{eight}} \\ \hline 206_{\text{eight}} \end{array}$

19.

$\begin{array}{r} \overset{4\ 12\quad 10}{\cancel{5}\ \cancel{6}\ \cancel{3}_{\text{seven}}} \\ -164_{\text{seven}} \\ \hline 366_{\text{seven}} \end{array}$

21.

$\begin{array}{r} \overset{0\ 1\ 2}{1\cancel{0}\cancel{0}1_{\text{two}}} \\ -111_{\text{two}} \\ \hline 10_{\text{two}} \end{array}$

23.

$\begin{array}{r} \overset{1\ 2\ 3}{12\cancel{0}\cancel{0}_{\text{three}}} \\ -1012_{\text{three}} \\ \hline 111_{\text{three}} \end{array}$

25.

$\begin{array}{r} \overset{3}{25_{\text{six}}} \\ \times 4_{\text{six}} \\ \hline 152_{\text{six}} \end{array}$

$(2_{\text{six}}\times 4_{\text{six}})+3_{\text{six}}=8_{\text{ten}}+3_{\text{six}}$

$=12_{\text{six}}+3_{\text{six}}$

$=15_{\text{six}}$

27.

$\begin{array}{r} 11_{\text{two}} \\ \times \ 1_{\text{two}} \\ \hline 11_{\text{two}} \end{array}$

29.

$\begin{array}{r} \overset{3\ 2}{543_{\text{seven}}} \\ \times \ 5_{\text{seven}} \\ \hline 4011_{\text{seven}} \end{array}$

$3\times 5=15=(2\times 7^1)+(1\times 1)=21_{\text{seven}}$

$(4\times 5)+2=22=(3\times 7^1)+(1\times 1)=31_{\text{seven}}$

$(5\times 5)+3=28=(4\times 7^1)+(0\times 1)=40_{\text{seven}}$

31.

$$\overset{1\ 1}{6\,2\,3}_{eight}$$
$$\times\quad 4_{eight}$$
$$\overline{3114_{eight}}$$

$$(3_{eight} \times 4_{eight}) = 12_{ten}$$
$$= (1 \times 8^1) + (4 \times 1)$$
$$= 14_{eight}$$
$$(2_{eight} \times 4_{eight}) + 1_{eight} = 8_{ten} + 1_{ten}$$
$$= 9_{ten}$$
$$= (1 \times 8^1) + (1 \times 1)$$
$$= 11_{eight}$$
$$(6_{eight} \times 4_{eight}) + 1_{eight} = 24_{ten} + 1_{ten}$$
$$= 25_{ten}$$
$$= (3 \times 8^1) + (1 \times 1)$$
$$= 31_{eight}$$

33.

$$21_{four}$$
$$\times 12_{four}$$
$$\overline{102}$$
$$\underline{210}$$
$$312_{four}$$

$$21_{four} \times 2_{four} = 9_{ten} \times 2_{ten}$$
$$= 18$$
$$= (1 \times 4^2) + (0 \times 4) + (2 \times 1)$$
$$= 102_{four}$$
$$21_{four} \times 1_{four} = 21_{four}$$

35.

$$\begin{array}{r} 20 \\ 2_{four}\overline{)100_{four}} \\ \underline{10} \\ 00 \end{array}$$

$$20_{four}$$

37.

$$\begin{array}{r} 41 \\ 3_{five}\overline{)224_{five}} \\ \underline{22} \\ 04 \\ \underline{3} \\ 1 \end{array}$$

$$41_{five} \text{ remainder of } 1$$

39-41. Answers will vary.

43. 1367_{eight}

45. $12F_{sixteen}$

Check Points 4.4

1. $100,000 + 100,000 + 100,000 + 100 + 100 + 10 + 10 + 1 + 1 = 300,222$

2. $2563 = 1000 + 1000 + 100 + 100 + 100 + 100 + 100 + 10 + 10 + 10 + 10 + 10 + 10 + 1 + 1 + 1$

☲☲&#x0001;&#x0001;&#x0001;&#x0001;&#x0001;&#x0001;∩∩∩∩∩∩|||

3. $MCCCLXI = 1000 + 100 + 100 + 100 + 50 + 10 + 1 = 1361$

4. $MCDXLVII = \overset{M}{1000} + \overset{CD}{(500-100)} + \overset{XL}{(50-10)} + \overset{V}{5} + \overset{I}{1} + \overset{I}{1} = 1000 + 400 + 40 + 5 + 1 + 1 = 1447$

5. $399 = 100 + 100 + 100 + 90 + 9 = \overset{C}{100} + \overset{C}{100} + \overset{C}{100} + \overset{XC}{(100-10)} + \overset{IX}{(10-1)} = CCCXCIX$

6. $2693 = 2000 + 600 + 90 + 3$

二
千
六
百
九
十
三

7. $\omega\pi\varepsilon = 800 + 80 + 5 = 885$

Exercise Set 4.4

1. 322

3. 300,423

5. 132

7. $423 = (4\times100) + (2\times10) + (3\times1)$

𓍢𓍢𓍢𓍢∩∩|||

9. $1846 = (1\times1000) + (8\times100) + (4\times10) + (6\times1)$

𓍢𓍢𓍢𓍢𓍢𓍢𓍢𓍢∩∩∩∩||||||

11. $23,547 = (2\times10,000) + (3\times1000) +$
$\qquad\qquad (5\times100) + (4\times10) + (7\times1)$

13. $XI = 11$

15. $XVI = 16$

17. $XL = 40$

19. $LIX = 59$

21. $CXLVI = 146$

23. $MDCXXI = 1621$

25. $MMDCLXXVII = 2677$

27. $43 = XLIII$

29. $129 = CXXIX$

31. $1896 = MDCCCXCVI$

33. $6892 = MMMMMMDCCCXCII$

35. $80 + 8 = 88$

$\left.\begin{array}{r}8\\10\end{array}\right\}80$
$8\}8$

37. $500 + 20 + 7 = 527$

$\left.\begin{array}{r}5\\100\end{array}\right\}500$
$\left.\begin{array}{r}2\\10\end{array}\right\}20$
$7\}7$

39. $2000+700+70+6=2776$

$\left.\begin{array}{r}2\\1000\end{array}\right\}2000$
$\left.\begin{array}{r}7\\100\end{array}\right\}700$
$\left.\begin{array}{r}7\\10\end{array}\right\}70$
$6\}6$

41.

四
十
三

43.

五
百
八
十
三

45.

四
千
八
百
七
十

47. $\iota\beta=12$

49. $\sigma\lambda\delta = 234$

51. $43 = \mu\gamma$

53. $483 = \upsilon\pi\gamma$

55. $MDCCLXXVI = 1776$
Declaration of Independence

57-61. Answers will vary.

63. Preceding: 𓍢𓍢∩∩∩∩∩∩∩∩|||||||
Following: 𓍢𓍢𓍢

Chapter 4 Review Exercises

1. $11^2 = 11 \times 11 = 121$

2. $7^3 = 7 \times 7 \times 7 = 343$

3. $472 = (4 \times 10^2) + (7 \times 10^1) + (2 \times 1) = (4 \times 100) + (7 \times 10) + (2 \times 1)$

4. $8076 = (8 \times 10^3) + (0 \times 10^2) + (7 \times 10^1) + (6 \times 1) = (8 \times 1000) + (0 \times 100) + (7 \times 10) + (6 \times 1)$

5. $70,329 = (7 \times 10^4) + (0 \times 10^3) + (3 \times 10^2) + (2 \times 10^1) + (9 \times 1) = (7 \times 10,000) + (0 \times 1000) + (3 \times 100) + (2 \times 10) + (9 \times 1)$

6. $706,953$

7. $740,000,306$

8. $<\vee \quad <\vee\vee\vee = (10+1) \times 60^1 + (10+1+1+1) \times 1 = (11 \times 60^1) + (13 \times 1) = 660 + 13 = 673$

9. $\vee\vee \quad << \quad <<< = (1+1) \times 60^2 + (10+10) \times 60^1 + (10+10+10) \times 1$

   $= (2 \times 60^2) + (20 \times 60) + (30 \times 1) = (2 \times 3600) + 1200 + 30 = 7200 + 1230 = 8430$

10. $6 \times 360 = 2160$
    $8 \times 20 \ = \ \ 160$
    $11 \times 1 \ = \ \underline{\ \ \ 11}$
    $ 2331$

11. $9 \times 7200 = 64,800$
    $2 \times 360 = \ \ \ \ 720$
    $0 \times 20 = \ \ \ \ \ \ \ \ 0$
    $16 \times 1 = \ \ \underline{\ \ \ \ \ 16}$
    $ 65,536$

12. Answers will vary.

13. $34_{\text{five}} = (3 \times 5^1) + (4 \times 1) = 15 + 4 = 19$

14. $110_{\text{two}} = (1 \times 2^2) + (1 \times 2^1) + (0 \times 1) = 4 + 2 + 0 = 6$

15. $643_{\text{seven}} = (6 \times 7^2) + (4 \times 7^1) + (3 \times 1) = 294 + 28 + 3 = 325$

16. $1081_{\text{nine}} = (1 \times 9^3) + (0 \times 9^2) + (8 \times 9^1) + (1 \times 1) = 729 + 0 + 72 + 1 = 805$

17. $FD3_{\text{sixteen}} = (15 \times 16^2) + (13 \times 16^1) + (3 \times 1) = 3840 + 208 + 3 = 4051$

18. $202202_{\text{three}} = (2 \times 3^5) + (0 \times 3^4) + (2 \times 3^3) + (2 \times 3^2) + (0 \times 3^1) + (2 \times 1) = 486 + 0 + 54 + 18 + 0 + 2 = 560$

19. $89 = (3 \times 5^2) + (2 \times 5^1) + (4 \times 1) = 324_{\text{five}}$

20. $21 = (1 \times 2^4) + (0 \times 2^3) + (1 \times 2^2) + (0 \times 2^1) + (1 \times 1) = 10101_{\text{two}}$

21. $473 = (1 \times 3^5) + (2 \times 3^4) + (2 \times 3^3) + (1 \times 3^2) + (1 \times 3^1) + (2 \times 1) = 243 + 162 + 54 + 9 + 3 + 2 = 122112_{three}$

22. $7093 = (2 \times 7^4) + (6 \times 7^3) + (4 \times 7^2) + (5 \times 7^1) + (2 \times 1) = 4802 + 2058 + 196 + 35 + 2 = 26452_{seven}$

23. $9348 = (1 \times 6^5) + (1 \times 6^4) + (1 \times 6^3) + (1 \times 6^2) + (4 \times 6^1) + (0 \times 1) = 7776 + 1296 + 216 + 36 + 24 = 111140_{six}$

24. $554 = (3 \times 12^2) + (A \times 12^1) + (2 \times 1) = 3A2_{twelve}$

25.
$$\begin{array}{r} \overset{1}{}4\,6_{seven} \\ +\,5\,3_{seven} \\ \hline 1\,3\,2_{seven} \end{array}$$

26.
$$\begin{array}{r} \overset{1\ \ 1}{5\,7\,4_{eight}} \\ +\,6\,0\,5_{eight} \\ \hline 1\,4\,0\,1_{eight} \end{array}$$

27.
$$\begin{array}{r} \overset{1\,1\,1\,1}{1\,1\,0\,1\,1_{two}} \\ 1\,0\,1\,0\,1_{two} \\ \hline 1\,1\,0\,0\,0\,0_{two} \end{array}$$

28.
$$\begin{array}{r} \overset{1}{}4\,3\,C_{sixteen} \\ +\,6\,9\,4_{sixteen} \\ \hline A\,D\,0_{sixteen} \end{array}$$

29.
$$\begin{array}{r} \overset{2\ \ 10}{3\,4_{six}} \\ 2\,5_{six} \\ \hline 5_{six} \end{array}$$

30.
$$\begin{array}{r} \overset{5\ \ 8\ \ 11}{6\,2\,4_{seven}} \\ -\,2\,4\,6_{seven} \\ \hline 3\,4\,5_{seven} \end{array}$$

31.
$$\begin{array}{r} \overset{0\ \ 1\ \ 2}{1\,0\,0\,1_{two}} \\ -\ \ 1\,1\,0_{two} \\ \hline 1\,1_{two} \end{array}$$

32.
$$\begin{array}{r} \overset{3\ 6\ 1\ \ 6}{4\,1\,2\,1_{five}} \\ -1\,3\,1\,2_{five} \\ \hline 2\,3\,0\,4_{five} \end{array}$$

33.
$$\begin{array}{r} \overset{1}{}3\,2_{four} \\ \times\ \ 3_{four} \\ \hline 2\,2\,2_{four} \end{array}$$

34.
$$\begin{array}{r} \overset{2}{4}\,3_{seven} \\ \times\ \ 6_{seven} \\ \hline 3\,5\,4_{seven} \end{array}$$

35.
$$\begin{array}{r} \overset{2\,2}{1\,2\,3_{five}} \\ 4_{five} \\ \hline 1\,1\,0\,2_{five} \end{array}$$

36.
$$\begin{array}{r} 133 \\ 2_{four} \overline{)332_{four}} \\ \underline{2} \\ 13 \\ \underline{12} \\ 12 \\ \underline{12} \\ 0 \end{array}$$

133_{four}

37.
$$\begin{array}{r} 12 \\ 4_{five} \overline{)103_{five}} \\ \underline{4} \\ 13 \\ \underline{13} \\ 0 \end{array}$$

12_{five}

38. 1246

39. 12,432

40. 2486
$= (2 \times 1000) + (4 \times 100) + (8 \times 10) + (6 \times 1)$
⛣⛣⛣𝟙𝟙𝟙𝟙𝟙∩∩∩∩∩∩∩∩|||||

41. 34,573
$= (3 \times 10,000) + (4 \times 1000) +$
$ (5 \times 100) + (7 \times 10) + (3 \times 1)$
𝅘𝅘𝅘⛣⛣⛣⛣𝟙𝟙𝟙𝟙𝟙∩∩∩∩∩∩∩|||

42. DDCCCBAAAA = 2314

43. 5492 = DDDDDCCCCBBBBBBBBBAA

44. Answers will vary.

45. CLXIII = 163

46. MXXXIV = 1034

47. MCMXC = 1990

48. 49 = XLIX

49. 2965 = MMCMLXV

50. Answers will vary.

51. $500 + 50 + 4 = 554$

$\left.\begin{array}{c}5\\100\end{array}\right\}500$

$\left.\begin{array}{c}5\\10\end{array}\right\}50$

$4\}4$

52. $8000 + 200 + 50 + 3 = 8253$

$\left.\begin{array}{c}8\\1000\end{array}\right\}8000$

$\left.\begin{array}{c}2\\100\end{array}\right\}200$

$\left.\begin{array}{c}5\\10\end{array}\right\}50$

$3\}3$

53.
二
百
七
十
四

54.
三
千
五
百
八
十
七

55. 365

56. 4520

57. G
Y
I
X
C

58. F
Z
H
Y
E
X
D

59. Answers will vary.

60. $\chi\nu\gamma = 653$

61. $\chi o\eta = 678$

62. $453 = \upsilon\nu\gamma$

63. $902 = \pi\beta$

64. UNG = 357

65. mhZRD = 37,894

66. rXJH = 80,618

67. 597 = WRG

68. $25,483 = lfVQC$

69. Answers will vary.

Chapter 4 Test

1. $9 \times 9 \times 9 = 729$

2. $567 = (5 \times 10^2) + (6 \times 10^1) + (7 \times 1)$
$= (5 \times 100) + (6 \times 10) + (7 \times 1)$

3. $63,028 = (6 \times 10^4) + (3 \times 10^3) +$
$\qquad (0 \times 10^2) + (2 \times 10^1) + (8 \times 1)$
$= (6 \times 10,000) + (3 \times 1000) +$
$\qquad (0 \times 100) + (2 \times 10) + (8 \times 1)$

4. $7000 + 400 + 90 + 3 = 7493$

5. $400,000 + 200 + 6 = 400,206$

6-7. Answers will vary.

8. $<< <\vee\vee\ <\vee = (10+10)\times60^2 + (10+1+1)\times60^1 + (10+1)\times1 = (20\times60^2) + (12\times60) + (11\times1) = 72{,}000 + 720 + 11 = 72{,}731$

9. $\begin{aligned} 4\times360 &= 1440 \\ 6\times20 &= 120 \\ 0\times1 &= \underline{0} \\ &\ 1560 \end{aligned}$

10. $423_{\text{five}} = (4\times5^2) + (2\times5^1) + (3\times1) = 4\times25 + 10 + 3 = 100 + 10 + 3 = 113$

11. $267_{\text{nine}} = (2\times9^2) + (6\times9^1) + (7\times1) = 2\times81 + 54 + 7 = 162 + 54 + 7 = 223$

12. $110101_{\text{two}} = (1\times2^5) + (1\times2^4) + (0\times2^3) + (1\times2^2) + (0\times2^1) + (1\times1) = 32 + 16 + 0 + 4 + 0 + 1 = 53$

13. $77 = (2\times3^3) + (2\times3^2) + (1\times3^1) + (2\times1) = 2212_{\text{three}}$

14. $56 = (1\times2^5) + (1\times2^4) + (1\times2^3) + (0\times2^2) + (0\times2^1) + (0\times1) = 111000_{\text{two}}$

15. $1844 = (2\times5^4) + (4\times5^3) + (3\times5^2) + (3\times5^1) + (4\times1) = 1250 + 500 + 75 + 15 + 4 = 24334_{\text{five}}$

16. $\begin{array}{r} {\scriptstyle 1\ 1} \\ 234_{\text{five}} \\ +423_{\text{five}} \\ \hline 1212_{\text{five}} \end{array}$

17. $\begin{array}{r} {\scriptstyle 5\ 9} \\ 5\,6\,2_{\text{seven}} \\ -145_{\text{seven}} \\ \hline 414_{\text{seven}} \end{array}$

18. $\begin{array}{r} {\scriptstyle 2} \\ 54_{\text{six}} \\ \times\ 3_{\text{six}} \\ \hline 250_{\text{six}} \end{array}$

19. $\begin{array}{r} 221 \\ 3_{\text{five}}\overline{)\,1213_{\text{five}}} \\ \underline{11} \\ 11 \\ \underline{11} \\ 03 \\ \underline{3} \\ 0 \end{array}$

221_{five}

20. 20,303

21. $32{,}634 = (3\times10{,}000) + (2\times1000) + (6\times100) + (3\times10) + (4\times1)$ 𝄍𝄍𝄍 ⚷⚷ 𝟿𝟿𝟿𝟿𝟿𝟿 ∩∩∩||||

22. $\text{MCMXCIV} = \overset{M}{1000} + \overset{CM}{(1000-100)} + \overset{XC}{(100-10)} + \overset{IV}{(5-1)} = 1000 + 900 + 90 + 4 = 1994$

23. $459 = \overset{CD}{(500-100)} + \overset{L}{50} + \overset{IX}{(10-1)} = \text{CDLIX}$

24. Answers will vary.

Chapter 5
Number Theory and the Real Number System

Check Points 5.1

1. The statement given in part (b) is true.

   a. False, 8 does not divide 48,324 because 8 does not divide 324.

   b. True, 6 divides 48,324 because both 2 and 3 divide 48,324. 2 divides 48,324 because the last digit is 4. 3 divides 48,324 because the sum of the digits, 21, is divisible by 3.

   c. False, 4 *does* divide 48,324 because the last two digits form 24 which is divisible by 4.

2.

   $$120 = 2^3 \cdot 3 \cdot 5$$

3. $225 = 3^2 \cdot 5^2$

   $825 = 3 \cdot 5^2 \cdot 11$

   Greatest Common Divisor: $3 \cdot 5^2 = 75$

4. $192 = 2^6 \cdot 3$

   $288 = 2^5 \cdot 3^2$

   Greatest Common Divisor: $2^5 \cdot 3 = 96$
   The largest number of people that can be placed in each singing group is 96.

5. $18 = 2 \cdot 3^2$

   $30 = 2 \cdot 3 \cdot 5$

   Least common multiple is: $90 = 2 \cdot 3^2 \cdot 5$

6. $40 = 2^3 \cdot 5$

   $60 = 2^2 \cdot 3 \cdot 5$

   Least common multiple is: $120 = 2^3 \cdot 3 \cdot 5$
   It will be 120 minutes, or 2 hours, until both movies begin again at the same time. The time will be 5:00 PM.

Exercise Set 5.1

1. 6944

---

a. Yes. The last digit is four.

b. No. The sum of the digits is 23, which is not divisible by 3.

c. Yes. The last two digits form 44, which is divisible by 4.

d. No. The number does not end in 0 or 5.

e. No. The number is not divisible by both 2 and 3.

f. Yes. The last three digits form 944, which is divisible by 8.

g. No. The sum of the digits is 23, which is not divisible by 9.

h. No. The number does not end in 0.

i. No. The number is not divisible by both 3 and 4.

3. 21,408

   a. Yes. The last digit is eight.

   b. Yes. The sum of the digits is 15, which is divisible by 3.

   c. Yes. The last two digits form 08, which is divisible by 4.

   d. No. The number does not end in 0 or 5.

   e. Yes. The number is divisible by both 2 and 3.

   f. Yes. The last three digits form 408, which is divisible by 8.

   g. No. The sum of the digits is 15, which is not divisible by 9.

   h. No. The number does not end in 0.

   i. Yes. The number is divisible by both 3 and 4.

5. 26,428

   a. Yes. The last digit is 8.

   b. No. The sum of the digits is 22, which is not divisible by 3.

c. Yes. The last 2 digits form 28, which is divisible by 4.

d. No. The last digit is eight.

e. No. The number is not divisible by both two and three.

f. No. The last three digits form 428, which is not divisible by 8.

g. No. The sum of the digits is 22, which is not divisible by 9.

h. No. The number does not end in 0.

i. No. The number is not divisible by 3 and 4.

7. 374,832

  a. Yes. The last digit is 2.

  b. Yes. The sum of the digits is 27, which is divisible by 3.

  c. Yes. The last two digits form 32, which is divisible by 4.

  d. No. The last digit is two.

  e. Yes. The number is divisible by 2 and 3.

  f. Yes. The last 3 digits form 832, which is divisible by 8.

  g. Yes. The sum of the digits is 27, which is divisible by 9.

  h. No. The last digit is 2.

  i. Yes. The number is divisible by both 3 and 4.

9. 6,126,120

  a. Yes. The last digit is 0.

  b. Yes. The sum of the digits is 18, which is divisible by 3.

  c. Yes. The last two digits form 20, which is divisible by 4.

  d. Yes. The last digit is 0.

  e. Yes. The number is divisible by both 2 and 3.

  f. Yes. The last 3 digits form 120, which is divisible by 8.

g. Yes. The sum of the digits is 18, which is divisible by 9.

h. Yes. The last digit is 0.

i. Yes. The number is divisible by both 3 and 4.

11. True. $5958 \div 3 = 1986$
The sum of the digits is 27, which is divisible by 3.

13. True. $10,612 \div 4 = 2653$
The last two digits form 12, which is divisible by 4.

15. False

17. True. $104,538 \div 6 = 17,423$
The number is divisible by both 2 and 3.

19. True. $20,104 \div 8 = 2513$
The last three digits form 104, which is divisible by 8.

21. False

23. True. $517,872 \div 12 = 43,156$
The number is divisible by both 3 and 4.

25.

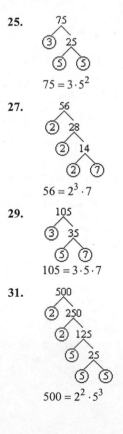

$75 = 3 \cdot 5^2$

27.

$56 = 2^3 \cdot 7$

29.

$105 = 3 \cdot 5 \cdot 7$

31.

$500 = 2^2 \cdot 5^3$

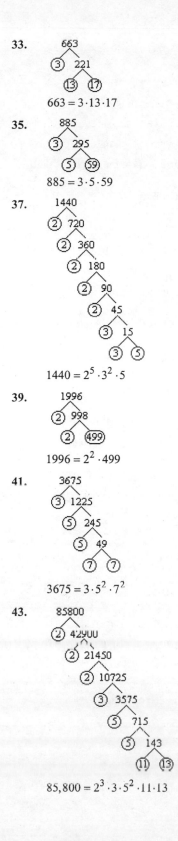

33. 663

③ 221

⑬ ⑰

$663 = 3 \cdot 13 \cdot 17$

35. 885

③ 295

⑤ ㊿

$885 = 3 \cdot 5 \cdot 59$

37. 1440

② 720

② 360

② 180

② 90

② 45

③ 15

③ ⑤

$1440 = 2^5 \cdot 3^2 \cdot 5$

39. 1996

② 998

② ㊾

$1996 = 2^2 \cdot 499$

41. 3675

③ 1225

⑤ 245

⑤ 49

⑦ ⑦

$3675 = 3 \cdot 5^2 \cdot 7^2$

43. 85800

② 42900

② 21450

② 10725

③ 3575

⑤ 715

⑤ 143

⑪ ⑬

$85,800 = 2^3 \cdot 3 \cdot 5^2 \cdot 11 \cdot 13$

45. $42 = 2 \cdot 3 \cdot 7$

$56 = 2^3 \cdot 7$

Greatest Common Divisor: $2 \cdot 7 = 14$

47. $16 = 2^4$

$42 = 2 \cdot 3 \cdot 7$

Greatest Common Divisor: 2

49. $60 = 2^2 \cdot 3 \cdot 5$

$108 = 2^2 \cdot 3^3$

Greatest Common Divisor: $2^2 \cdot 3 = 12$

51. $72 = 2^3 \cdot 3^2$

$120 = 2^3 \cdot 3 \cdot 5$

Greatest Common Divisor: $2^3 \cdot 3 = 24$

53. $324 = 2 \cdot 3^2 \cdot 19$

$380 = 2^2 \cdot 5 \cdot 19$

Greatest Common Divisor: $2 \cdot 19 = 38$

55. $240 = 2^4 \cdot 3 \cdot 5$

$285 = 3 \cdot 5 \cdot 19$

Greatest Common Divisor: $3 \cdot 5 = 15$

57. $42 = 2 \cdot 3 \cdot 7$

$56 = 2^3 \cdot 7$

Least Common Multiple: $2^3 \cdot 3 \cdot 7 = 168$

59. $16 = 2^4$

$42 = 2 \cdot 3 \cdot 7$

Least Common Multiple: $2^4 \cdot 3 \cdot 7 = 336$

61. $60 = 2^2 \cdot 3 \cdot 5$

$108 = 2^2 \cdot 3^3$

Least Common Multiple: $2^2 \cdot 3^3 \cdot 5 = 540$

63. $72 = 2^3 \cdot 3^2$

$120 = 2^3 \cdot 3 \cdot 5$

Least Common Multiple: $2^3 \cdot 3^2 \cdot 5 = 360$

65. $342 = 2 \cdot 3^2 \cdot 19$

$380 = 2^2 \cdot 5 \cdot 19$

Least Common Multiple: $2^2 \cdot 3^2 \cdot 5 \cdot 19 = 3420$

67. $240 = 2^4 \cdot 3 \cdot 5$

$285 = 3 \cdot 5 \cdot 19$

Least Common Multiple

$= 2^4 \cdot 3 \cdot 5 \cdot 19$

$= 4560$

69. $300 = 2^2 \cdot 3 \cdot 5^2$

$144 = 2^4 \cdot 3^2$

Greatest Common Divisor: $2^2 \cdot 3 = 12$

There would be 25 groups with 12 bottles of water each. There would be 12 groups with 12 cans of food each.

71. $310 = 2 \cdot 5 \cdot 31$

$460 = 2^2 \cdot 5 \cdot 23$

Greatest Common Divisor: $2 \cdot 5 = 10$

There would be 31 groups of 10 five-dollar bills. There would be 46 groups of 10 ten-dollar bills.

73. $6 = 2 \cdot 3$

$10 = 2 \cdot 5$

Least Common Multiple is: $2 \cdot 3 \cdot 5 = 30$

It will be 30 more nights until both have the evening off, or July 1.

75. $15 = 3 \cdot 5$

$18 = 2 \cdot 3^2$

Least Common Multiple is: $2 \cdot 3^2 \cdot 5 = 90$

It will be 90 minutes, or at 1:30 PM.

77-87. Answers will vary.

89. a. GCD = $2^{14} \cdot 3^{25} \cdot 5^{30}$

b. LCM = $2^{17} \cdot 3^{37} \cdot 5^{31}$

91. $85 + 15 = 100 = 2^2 \cdot 5^2$

$100 + 15 = 115 = 5 \cdot 23$

LCM = $2^2 \cdot 5 \cdot 23 = 2300$

The films will begin at the same time

$2300 \text{ min} \left(= 38\frac{1}{3}\text{hr} \right)$ after noon (today), or at 2:20 A.M. on the third day.

93. No, since $1 + 2 + 5 + 4 + 1 + 7 + 5 + 0 = 25$ is not divisible by 3, then 12,541,750 is not divisible by 3.

Check Points 5.2

1.

2. a. $6 > -7$ because 6 is to the right of –2 on the number line.

b. $-8 < -1$ because –8 is to the left of –1 on the number line.

c. $-25 < -2$ because –25 is to the left of –2 on the number line.

d. $-14 < 0$ because –14 is to the left of 0 on the number line.

3. a. $|-8| = 8$ because –8 is 8 units from 0.

b. $|6| = 6$ because 6 is 6 units from 0.

c. $|0| = 0$ because 0 is 0 units from itself.

4. a. $30 - (-7) = 30 + 7 = 37$

b. $-14 - (-10) = -14 + 10 = -4$

c. $-14 - 10 = -24$

5. The difference in elevation is
$8848 - (-10,915) = 8848 + 10,915 = 19,763$ m.

6. a. $8^2 = 8 \cdot 8 = 64$

b. $(-8)^2 = (-8)(-8) = 64$

c. $7^3 = 7 \cdot 7 \cdot 7 = 49 \cdot 7 = 343$

d. $(-7)^3 = (-7)(-7)(-7) = 49(-7) = -343$

e.

$(-3)^4 = (-3)(-3)(-3)(-3) = 9(-3)(-3) = (-27)(-3) = 81$

7.

$$7^2 - 48 \div 4^2 \cdot 5 + 2$$
$$= 49 - 48 \div 16 \cdot 5 + 2$$
$$= 49 - 3 \cdot 5 + 2$$
$$= 49 - 15 + 2$$
$$= 34 + 2$$
$$= 36$$

8.

$$(-8)^2 - (10 - 13)^2 (-2)$$
$$= (-8)^2 - (-3)^2 (-2)$$
$$= 64 - (9)(-2)$$
$$= 64 - (-18)$$
$$= 64 + (+18)$$
$$= 82$$

Exercise Set 5.2

1.

3.

5. $-2 < 7$ because -2 is to the left of 7 on the number line.

7. $-13 < -2$ because -13 is to the left of -2 on the number line.

9. $8 > -50$ because 8 is to the right of -50 on the number line.

11. $-100 < 0$ because -100 is to the left of 0 on the number line.

13. $|-14| = 14$ because -14 is 14 units from 0.

15. $|14| = 14$ because 14 is 14 units from 0.

17. $|-300,000| = 300,000$ because $-300,000$ is 300,000 units from 0.

19. $-7 + (-5) = -12$

21. $12 + (-8) = 4$

23. $6 + (-9) = -3$

25. $-9 + (+4) = -5$

27. $-9 + (-9) = -18$

29. $9 + (-9) = 0$

31. $13 - 8 = 5$

33. $8 - 15 = 8 + (-15) = -7$

35. $4 - (-10) = 4 + 10 = 14$

37. $-6 - (-17) = -6 + 17 = 11$

39. $-12 - (-3) = -12 + 3 = -9$

41. $-11 - 17 = -11 + (-17) = -28$

43. $6(-9) = -54$

45. $(-7)(-3) = 21$

47. $(-2)(6) = -12$

49. $(-13)(-1) = 13$

51. $0(-5) = 0$

53. $5^2 = 5 \cdot 5 = 25$

55. $(-5)^2 = (-5) \cdot (-5) = 25$

57. $4^3 = 4 \cdot 4 \cdot 4 = 64$

59. $(-5)^3 = (-5)(-5)(-5) = 25(-5) = -125$

61. $(-5)^4 = (-5)(-5)(-5)(-5) = 625$

63. $-3^4 = -[3 \cdot 3 \cdot 3 \cdot 3] = -81$

65. $(-3)^4 = (-3)(-3)(-3)(-3) - 81$

67. $\dfrac{-12}{4} = -3$

69. $\dfrac{21}{-3} = -7$

71. $\dfrac{-90}{-3} = 30$

73. $\dfrac{0}{-7} = 0$

75. $\dfrac{-7}{0}$ is undefined

77. $(-480) \div 24 = \dfrac{-480}{24} = -20$

79. $(465) \div (-15) = \dfrac{465}{-15} = -31$

81. $7 + 6 \cdot 3 = 7 + 18 = 25$

83. $(-5) - 6(-3) = -5 + 18 = 13$

85. $6 - 4(-3) - 5 = 6 - (-12) - 5$
$\qquad = 6 + 12 - 5$
$\qquad = 18 - 5$
$\qquad = 13$

87. $3 - 5(-4 - 2) = 3 - 5(-6)$
$\qquad = 3 - (-30)$
$\qquad = 3 + 30$
$\qquad = 33$

89. $(2 - 6)(-3 - 5) = (-4)(-8) = 32$

91. $3(-2)^2 - 4(-3)^2 = 3(4) - 4(9)$
$\qquad = 12 - 36$
$\qquad = -24$

93. $(2 - 6)^2 - (3 - 7)^2 = (-4)^2 - (-4)^2$
$\qquad = 16 - 16$
$\qquad = 0$

95. $6(3 - 5)^3 - 2(1 - 3)^3 = 6(-2)^3 - 2(-2)^3$
$\qquad = 6(-8) - 2(-8)$
$\qquad = -48 + 16$
$\qquad = -32$

97. $8^2 - 16 \div 2^2 \cdot 4 - 3 = 64 - 16 \div 4 \cdot 4 - 3$
$\qquad = 64 - 4 \cdot 4 - 3$
$\qquad = 64 - 16 - 3$
$\qquad = 45$

99. $-3°F$

101. The high temperature was $-56 + 100 = 44$ degrees.

103. The difference in elevation $= 19,321 - (-436)$
$19,321 + 436 = 19,757$ feet.

105. $1,453,100 - 1,560,500 = -107,400$ which represents a deficit.

107. $1,827,500 - 1,703,000 = 124,500$ which represents a surplus.

109. $520 - (-112) = 520 + 112 = 632$ thousand jobs.

111. $\$4.29 - \$3.66 = \$0.63$

113. $\$4.29 - \$2.55 = \$1.74$

115-123. Answers will vary.

125. $8 - 2 \cdot (3 - 4) = 10$

127. -7

129. 150

Check Points 5.3

1. $72 = 2^3 \cdot 3^2$
$90 = 2 \cdot 5 \cdot 3^2$
Greatest Common Divisor is $2 \cdot 3^2$ or 18.
$\dfrac{72}{90} = \dfrac{72 \div 18}{90 \div 18} = \dfrac{4}{5}$

2. a. $\dfrac{3}{8} = 0.375$

$$\begin{array}{r} 0.375 \\ 8\overline{)3.000} \\ \underline{24} \\ 60 \\ \underline{56} \\ 40 \\ \underline{40} \\ 0 \end{array}$$

b. $\dfrac{5}{11} = 0.\overline{45}$

$$\begin{array}{r} 0.4545\ldots \\ 11\overline{)5.0000} \\ \underline{44} \\ 60 \\ \underline{55} \\ 50 \\ \underline{44} \\ 60 \\ \underline{55} \\ 5 \end{array}$$

3. a. $0.9 = \dfrac{9}{10}$

b. $0.86 = \dfrac{86}{100} = \dfrac{86 \div 2}{100 \div 2} = \dfrac{43}{50}$

c. $0.053 = \dfrac{53}{1000}$

4. $n = 0.\overline{2}$

$n = 0.22222\ldots$

$10n = 2.22222\ldots$

$10n = 2.2222\ldots$

$-n = 0.2222\ldots$

$\overline{}$

$9n = 2.0$

$n = \dfrac{2}{9}$

5. $n = 0.\overline{79}$

$n = 0.7979\ldots$

$100n = 79.7979\ldots$

$100n = 79.7979\ldots$

$- \quad n = 0.7979\ldots$

$\overline{}$

$99n = 79$

$n = \dfrac{79}{99}$

6. $\dfrac{1}{5} + \dfrac{3}{4} = \dfrac{1}{5} \cdot \dfrac{4}{4} + \dfrac{3}{4} \cdot \dfrac{5}{5} = \dfrac{4}{20} + \dfrac{15}{20} = \dfrac{19}{20}$

7. $\dfrac{3}{10} - \dfrac{7}{12} = \dfrac{3}{10} \cdot \dfrac{6}{6} - \dfrac{7}{12} \cdot \dfrac{5}{5} = \dfrac{18}{60} - \dfrac{35}{60} = -\dfrac{17}{60}$

8. First, find the sum:

$\dfrac{1}{3} + \dfrac{1}{2} = \dfrac{1}{3} \cdot \dfrac{2}{2} + \dfrac{1}{2} \cdot \dfrac{3}{3} = \dfrac{2}{6} + \dfrac{3}{6} = \dfrac{5}{6}$

Next, divide by 2: $\dfrac{5}{6} \div \dfrac{2}{1} = \dfrac{5}{6} \cdot \dfrac{1}{2} = \dfrac{5}{12}$

9. First, find the sum of the other categories:

$\dfrac{1}{3} + \dfrac{7}{24} + \dfrac{1}{12} + \dfrac{1}{8} = \dfrac{8}{24} + \dfrac{7}{24} + \dfrac{2}{24} + \dfrac{3}{24} = \dfrac{20}{24} = \dfrac{5}{6}$

Next, subtract this result from 1: $1 - \dfrac{5}{6} = \dfrac{6}{6} - \dfrac{5}{6} = \dfrac{1}{6}$

Therefore $\dfrac{1}{6}$ of the student's weekday is spent in class.

Exercise Set 5.3

1. $10 = 2 \cdot 5$

$15 = 3 \cdot 5$

Greatest Common Divisor is 5.

$\dfrac{10}{15} = \dfrac{10 \div 5}{15 \div 5} = \dfrac{2}{3}$

3. $15 = 3 \cdot 5$

$18 = 2 \cdot 3^2$

Greatest Common Divisor is 3.

$\dfrac{15}{18} = \dfrac{15 \div 3}{18 \div 3} = \dfrac{5}{6}$

5. $24 = 2^3 \cdot 3$

$42 = 2 \cdot 3 \cdot 7$

Greatest Common Divisor is $2 \cdot 3$ or 6.

$\dfrac{24}{42} = \dfrac{24 \div 6}{42 \div 6} = \dfrac{4}{7}$

7. $60 = 2^2 \cdot 3 \cdot 5$

$108 = 2^2 \cdot 3^3$

Greatest Common Divisor is $2^2 \cdot 3$ or 12.

$\dfrac{60}{108} = \dfrac{60 \div 12}{108 \div 12} = \dfrac{5}{9}$

9. $342 = 2 \cdot 3^2 \cdot 19$

$380 = 2^2 \cdot 5 \cdot 19$

Greatest Common Divisor is $2 \cdot 19$ or 38.

$\dfrac{342}{380} = \dfrac{342 \div 38}{380 \div 38} = \dfrac{9}{10}$

11. $308 = 2^2 \cdot 7 \cdot 11$

$418 = 2 \cdot 11 \cdot 19$

Greatest Common Divisor is $2 \cdot 11$ or 22.

$\dfrac{308}{418} = \dfrac{308 \div 22}{418 \div 22} = \dfrac{14}{19}$

13. $\dfrac{3}{4} = 0.75$

$$\begin{array}{r} 0.75 \\ 4\overline{)3.00} \\ 28 \\ \overline{20} \\ 20 \\ \overline{0} \end{array}$$

15. $\dfrac{7}{20} = 0.35$

$$\begin{array}{r} 0.35 \\ 20\overline{)7.00} \\ 60 \\ \overline{100} \\ 100 \\ \overline{0} \end{array}$$

17. $\dfrac{7}{8} = 0.875$

$$\begin{array}{r} 0.875 \\ 8\overline{)7.000} \\ 64 \\ \overline{60} \\ 56 \\ \overline{40} \\ 40 \\ \overline{0} \end{array}$$

19. $\dfrac{9}{11} = 0.\overline{81}$

$$\begin{array}{r} 0.8181... \\ 11\overline{)9.0000} \\ 88 \\ \overline{20} \\ 11 \\ \overline{90} \\ 88 \\ \overline{20} \\ 11 \\ \overline{9} \end{array}$$

21. $\dfrac{22}{7} = 3.\overline{142857}$

$$\begin{array}{r} 3.142857... \\ 7\overline{)22.000000} \\ 21 \\ \overline{10} \\ 7 \\ \overline{30} \\ 28 \\ \overline{20} \\ 14 \\ \overline{60} \\ 56 \\ \overline{40} \\ 35 \\ \overline{50} \\ 49 \\ \overline{10} \end{array}$$

23. $\dfrac{2}{7} = 0.\overline{285714}$

$$\begin{array}{r} 0.2857142... \\ 7\overline{)2.000000} \\ 14 \\ \overline{60} \\ 56 \\ \overline{40} \\ 35 \\ \overline{50} \\ 49 \\ \overline{10} \\ 7 \\ \overline{30} \\ 28 \\ \overline{20} \\ 14 \\ \overline{6} \end{array}$$

25. $0.3 = \dfrac{3}{10}$

27. $0.4 = \dfrac{4}{10} = \dfrac{4 \div 2}{10 \div 2} = \dfrac{2}{5}$

29. $0.39 = \dfrac{39}{100}$

31. $0.82 = \dfrac{82}{100} = \dfrac{82 \div 2}{100 \div 2} = \dfrac{41}{50}$

33. $0.725 = \dfrac{725}{1000}$

$725 = 5^2 \cdot 29$

$1000 = 2^3 \cdot 5^3$

Greatest Common Divisor is 5^2 or 25.

$$\frac{725}{1000} = \frac{725 \div 25}{1000 \div 25} = \frac{29}{40}$$

35. $0.5399 = \dfrac{5399}{10,000}$

37. $n = 0.777\ldots$
$10n = 7.777\ldots$

$10n = 7.777\ldots$
$\underline{-n = 0.777\ldots}$
$9n = 7$
$n = \dfrac{7}{9}$

39. $n = 0.999\ldots$
$10n = 9.999\ldots$

$10n = 9.999\ldots$
$\underline{-n = 0.999\ldots}$
$9n = 9$
$n = 1$

41. $n = 0.3636\ldots$
$100n = 36.3636\ldots$

$100n = 36.3636\ldots$
$\underline{-n = 0.3636\ldots}$
$99n = 36$
$n = \dfrac{36}{99}$ or $\dfrac{4}{11}$

43. $n = 0.257257\ldots$
$1000n = 257.257257\ldots$

$1000n = 257.257257\ldots$
$\underline{-n = .257257\ldots}$
$999n = 257$
$n = \dfrac{257}{999}$

45. $\dfrac{3}{8} \cdot \dfrac{7}{11} = \dfrac{3 \cdot 7}{8 \cdot 11} = \dfrac{21}{88}$

47. $\left(-\dfrac{1}{10}\right)\left(\dfrac{7}{12}\right) = \dfrac{(-1)(7)}{10 \cdot 12} - \dfrac{-7}{120} = -\dfrac{7}{120}$

49. $\left(-\dfrac{2}{3}\right)\left(-\dfrac{9}{4}\right) = \dfrac{(-2)(-9)}{3 \cdot 4} = \dfrac{18}{12} = \dfrac{3}{2}$

51. $\dfrac{5}{4} \div \dfrac{3}{8} = \dfrac{5}{4} \cdot \dfrac{8}{3} = \dfrac{5 \cdot 8}{4 \cdot 3} = \dfrac{40}{12} = \dfrac{10}{3}$

53. $-\dfrac{7}{8} \div \dfrac{15}{16} = -\dfrac{7}{8} \cdot \dfrac{16}{15}$
$= \dfrac{(-7)(16)}{8 \cdot 15}$
$= \dfrac{-112}{120}$
$= -\dfrac{14}{15}$

55. $\dfrac{2}{11} + \dfrac{3}{11} = \dfrac{2+3}{11} = \dfrac{5}{11}$

57. $\dfrac{5}{6} - \dfrac{1}{6} = \dfrac{5-1}{6} = \dfrac{4}{6} = \dfrac{2}{3}$

59. $\dfrac{7}{12} - \left(-\dfrac{1}{12}\right) = \dfrac{7}{12} + \dfrac{1}{12} = \dfrac{7+1}{12} = \dfrac{8}{12} = \dfrac{2}{3}$

61. $\dfrac{1}{2} + \dfrac{1}{5} = \left(\dfrac{1}{2}\right)\left(\dfrac{5}{5}\right) + \left(\dfrac{1}{5}\right)\left(\dfrac{2}{2}\right)$
$= \dfrac{5}{10} + \dfrac{2}{10}$
$= \dfrac{5+2}{10}$
$= \dfrac{7}{10}$

63. $\dfrac{3}{4} + \dfrac{3}{20} = \left(\dfrac{3}{4}\right)\left(\dfrac{5}{5}\right) + \dfrac{3}{20}$
$= \dfrac{15}{20} + \dfrac{3}{20}$
$= \dfrac{15+3}{20}$
$= \dfrac{18}{20}$
$= \dfrac{9}{10}$

65. $\dfrac{5}{24} + \dfrac{7}{30} = \left(\dfrac{5}{24}\right)\left(\dfrac{5}{5}\right) + \left(\dfrac{7}{30}\right)\left(\dfrac{4}{4}\right)$

$= \dfrac{25}{120} + \dfrac{28}{120}$

$= \dfrac{25 + 28}{120}$

$= \dfrac{53}{120}$

67. $\dfrac{13}{18} - \dfrac{2}{9} = \dfrac{13}{18} - \dfrac{2}{9}\left(\dfrac{2}{2}\right)$

$= \dfrac{13}{18} - \dfrac{4}{18}$

$= \dfrac{13 - 4}{18}$

$= \dfrac{9}{18}$

$= \dfrac{1}{2}$

69. $\dfrac{4}{3} - \dfrac{3}{4} = \dfrac{4}{3}\left(\dfrac{4}{4}\right) - \dfrac{3}{4}\left(\dfrac{3}{3}\right)$

$= \dfrac{16}{12} - \dfrac{9}{12}$

$= \dfrac{16 - 9}{12}$

$= \dfrac{7}{12}$

71. $\dfrac{1}{15} - \dfrac{27}{50}$

$15 = 3 \cdot 5$

$50 = 2 \cdot 5^2$

Least Common Multiple is $2 \cdot 3 \cdot 5^2 = 6 \cdot 25 = 150$

$\dfrac{1}{15}\left(\dfrac{10}{10}\right) - \dfrac{27}{50}\left(\dfrac{3}{3}\right) = \dfrac{10}{150} - \dfrac{81}{150}$

$= \dfrac{10 - 81}{150}$

$= \dfrac{-71}{150}$

73. $\left(\dfrac{1}{2} - \dfrac{1}{3}\right) \div \dfrac{5}{8} = \left[\left(\dfrac{1}{2}\right)\left(\dfrac{3}{3}\right) - \dfrac{1}{3}\left(\dfrac{2}{2}\right)\right] \div \dfrac{5}{8}$

$= \left(\dfrac{3}{6} - \dfrac{2}{6}\right) \div \dfrac{5}{8}$

$= \dfrac{1}{6} \div \dfrac{5}{8}$

$= \dfrac{1}{6} \cdot \dfrac{8}{5}$

$= \dfrac{1 \cdot 8}{6 \cdot 5}$

$= \dfrac{8}{30}$

$= \dfrac{4}{15}$

75. $\dfrac{1}{4} + \dfrac{1}{3} = \left(\dfrac{1}{4}\right)\left(\dfrac{3}{3}\right) + \left(\dfrac{1}{3}\right)\left(\dfrac{4}{4}\right)$

$= \dfrac{3}{12} + \dfrac{4}{12}$

$= \dfrac{3 + 4}{12}$

$= \dfrac{7}{12}$

$\dfrac{7}{12} \div 2 = \dfrac{7}{12} \cdot \dfrac{1}{2} = \dfrac{7}{24}$

77. $\dfrac{1}{2} + \dfrac{2}{3} = \left(\dfrac{1}{2}\right)\left(\dfrac{3}{3}\right) + \left(\dfrac{2}{3}\right)\left(\dfrac{2}{2}\right)$

$= \dfrac{3}{6} + \dfrac{4}{6}$

$= \dfrac{3 + 4}{6}$

$= \dfrac{7}{6}$

$\dfrac{7}{6} \div 2 = \dfrac{7}{6} \cdot \dfrac{1}{2} = \dfrac{7}{12}$

79. $-\dfrac{2}{3} + \left(-\dfrac{5}{6}\right) = \left(-\dfrac{2}{3}\right)\left(\dfrac{2}{2}\right) - \dfrac{5}{6}$

$= \dfrac{-4}{6} - \dfrac{5}{6}$

$= \dfrac{-4 - 5}{6}$

$= -\dfrac{9}{6}$

$-\dfrac{9}{6} \div 2 = -\dfrac{9}{6} \cdot \dfrac{1}{2} = -\dfrac{9}{12} = -\dfrac{3}{4}$

81. $\dfrac{13}{4}+\dfrac{13}{9}=\dfrac{13\cdot9}{4\cdot9}+\dfrac{13\cdot4}{9\cdot4}$

$\qquad\qquad=\dfrac{117}{36}+\dfrac{52}{36}$

$\qquad\qquad=\dfrac{117+52}{36}$

$\qquad\qquad=\dfrac{169}{36}$

$\quad\dfrac{13}{4}\times\dfrac{13}{9}=\dfrac{13\cdot13}{4\cdot9}$

$\qquad\qquad=\dfrac{169}{36}$

Both are equal to $\dfrac{169}{36}$

83. $512=2^9$

$800=2^5\cdot5^2$

Greatest Common Divisor: $2^5=32$

$\dfrac{512}{800}=\dfrac{512\div32}{800\div32}=\dfrac{16}{25}$

85. $300-186=114=2\cdot3\cdot19$

$300=2^2\cdot3\cdot5^2$

Greatest Common Divisor: $2\cdot3=6$

$\dfrac{114}{300}=\dfrac{114\div6}{300\div6}=\dfrac{19}{50}$

87. $\dfrac{19}{20}$ planned to go to college, 300 African American teenagers.

$\dfrac{19}{20}\cdot300=\dfrac{19\cdot300}{20}=\dfrac{5700}{20}=285$

89. Find $\dfrac{1}{2}$ of $\dfrac{3}{4}$ or divide $\dfrac{3}{4}$ by 2:

$\dfrac{3}{4}\div2=\dfrac{3}{4}\cdot\dfrac{1}{2}=\dfrac{3}{8}$ cup

91. $1-\dfrac{5}{12}-\dfrac{1}{4}=\dfrac{12}{12}-\dfrac{5}{12}-\dfrac{3}{12}=\dfrac{4}{12}=\dfrac{1}{3}$ ownership.

93. 40 hours at $12 rate

6 hours at $\left(\dfrac{3}{2}\right)$\$12 rate or $\left(\dfrac{3}{2}\right)$\$12$=\dfrac{\$36}{2}=\$18$

$40\cdot\$12+6\cdot\$18=\$480+\$108=\$588$

95-103. Answers will vary.

105. Find a rational number halfway between the two numbers $\dfrac{1}{4}$ and $\dfrac{1}{2}$ (since 3 is halfway between 2 and 4).

$\dfrac{1}{4}+\dfrac{1}{2}=\dfrac{1}{4}+\dfrac{1}{2}\cdot\dfrac{2}{2}=\dfrac{1}{4}+\dfrac{2}{4}=\dfrac{3}{4}$

$\dfrac{3}{4}\div2=\dfrac{3}{4}\cdot\dfrac{1}{2}=\dfrac{3\cdot1}{4\cdot2}=\dfrac{3}{8}$

The amount of salt needed for 3 servings is $\dfrac{3}{8}$ teaspoon.

107. a. $\dfrac{197}{800}=0.24625$

b. $\dfrac{4539}{3125}=1.45248$

c. $\dfrac{7}{6250}=0.00112$

Check Points 5.4

1. a. $\sqrt{12}=\sqrt{4\cdot3}=\sqrt{4}\cdot\sqrt{3}=2\sqrt{3}$

b. $\sqrt{60}=\sqrt{4\cdot15}=\sqrt{4}\cdot\sqrt{15}=2\sqrt{15}$

c. $\sqrt{55}$ cannot be simplified.

2. a. $\sqrt{3}\cdot\sqrt{10}=\sqrt{3\cdot10}=\sqrt{30}$

b. $\sqrt{10}\cdot\sqrt{10}=\sqrt{10\cdot10}=\sqrt{100}=10$

c. $\sqrt{6}\cdot\sqrt{2}=\sqrt{6\cdot2}=\sqrt{12}=\sqrt{4}\cdot\sqrt{3}=2\sqrt{3}$

3. a. $\dfrac{\sqrt{80}}{\sqrt{5}}=\sqrt{\dfrac{80}{5}}=\sqrt{16}=4$

b. $\dfrac{\sqrt{48}}{\sqrt{6}}=\sqrt{\dfrac{48}{6}}=\sqrt{8}=\sqrt{4}\cdot\sqrt{2}=2\sqrt{2}$

4. a. $8\sqrt{3}+10\sqrt{3}=(8+10)\sqrt{3}=18\sqrt{3}$

b. $4\sqrt{13}-9\sqrt{13}=(4-9)\sqrt{13}=-5\sqrt{13}$

c. $7\sqrt{10}+2\sqrt{10}-\sqrt{10}=(7+2-1)\sqrt{10}=8\sqrt{10}$

5. a. $\sqrt{3}+\sqrt{12}=\sqrt{3}+\sqrt{4}\cdot\sqrt{3}=\sqrt{3}+2\sqrt{3}=3\sqrt{3}$

b. $4\sqrt{8}-7\sqrt{18}$
$=4\sqrt{4\cdot2}-7\sqrt{9\cdot2}$
$=4\cdot2\sqrt{2}-7\cdot3\sqrt{2}$
$=8\sqrt{2}-21\sqrt{2}$
$=(8-21)\sqrt{2}$
$=-13\sqrt{2}$

6. a. $\dfrac{25}{\sqrt{10}}=\dfrac{25}{\sqrt{10}}\cdot\dfrac{\sqrt{10}}{\sqrt{10}}=\dfrac{25\sqrt{10}}{\sqrt{100}}=\dfrac{25\sqrt{10}}{10}=\dfrac{5\sqrt{10}}{2}$

b. $\sqrt{\dfrac{2}{7}}=\dfrac{\sqrt{2}}{\sqrt{7}}=\dfrac{\sqrt{2}}{\sqrt{7}}\cdot\dfrac{\sqrt{7}}{\sqrt{7}}=\dfrac{\sqrt{14}}{\sqrt{49}}=\dfrac{\sqrt{14}}{7}$

c. $\dfrac{5}{\sqrt{18}}=\dfrac{5}{\sqrt{18}}\cdot\dfrac{\sqrt{2}}{\sqrt{2}}=\dfrac{5\sqrt{2}}{\sqrt{36}}=\dfrac{5\sqrt{2}}{6}$

Exercise Set 5.4

1. $\sqrt{9}=3$ because $3^2=9$.

3. $\sqrt{25}=5$ because $5^2=25$.

5. $\sqrt{64}=8$ because $8^2=64$.

7. $\sqrt{121}=11$ because $11^2=121$.

9. $\sqrt{169}=13$ because $13^2=169$.

11. a. $\sqrt{173}\approx13.2$

b. $\sqrt{173}\approx13.15$

c. $\sqrt{173}\approx13.153$

13. a. $\sqrt{17,761}\approx133.3$

b. $\sqrt{17,761}\approx133.27$

c. $\sqrt{17,761}\approx133.270$

15. a. $\sqrt{\pi}\approx1.8$

b. $\sqrt{\pi}\approx1.77$

c. $\sqrt{\pi}\approx1.772$

17. $\sqrt{20}=\sqrt{4\cdot5}=\sqrt{4}\cdot\sqrt{5}=2\sqrt{5}$

19. $\sqrt{80}=\sqrt{16\cdot5}=\sqrt{16}\cdot\sqrt{5}=4\sqrt{5}$

21. $\sqrt{250}=\sqrt{25\cdot10}=\sqrt{25}\cdot\sqrt{10}=5\sqrt{10}$

23. $7\sqrt{28}=7\sqrt{4\cdot7}$
$=7\sqrt{4}\cdot\sqrt{7}$
$=7\cdot2\cdot\sqrt{7}$
$=14\sqrt{7}$

25. $\sqrt{7}\cdot\sqrt{6}=\sqrt{7\cdot6}=\sqrt{42}$

27. $\sqrt{6}\cdot\sqrt{6}=\sqrt{6\cdot6}=\sqrt{36}=6$

29. $\sqrt{3}\cdot\sqrt{6}=\sqrt{3\cdot6}$
$=\sqrt{18}$
$=\sqrt{9\cdot2}$
$=\sqrt{9}\cdot\sqrt{2}$
$=3\sqrt{2}$

31. $\sqrt{2}\cdot\sqrt{26}=\sqrt{2\cdot26}$
$=\sqrt{52}$
$=\sqrt{4\cdot13}$
$=\sqrt{4}\cdot\sqrt{13}$
$=2\sqrt{13}$

33. $\dfrac{\sqrt{54}}{\sqrt{6}}=\sqrt{\dfrac{54}{6}}=\sqrt{9}=3$

35. $\dfrac{\sqrt{90}}{\sqrt{2}}=\sqrt{\dfrac{90}{2}}$
$=\sqrt{45}$
$=\sqrt{9\cdot5}$
$=\sqrt{9}\cdot\sqrt{5}$
$=3\sqrt{5}$

37. $\dfrac{-\sqrt{96}}{\sqrt{2}}=-\sqrt{\dfrac{96}{2}}$
$=-\sqrt{48}$
$=-\sqrt{16\cdot3}$
$=-\sqrt{16}\cdot\sqrt{3}$
$=-4\sqrt{3}$

39. $7\sqrt{3} + 6\sqrt{3} = (7+6)\sqrt{3} = 13\sqrt{3}$

41. $4\sqrt{13} - 6\sqrt{13} = (4-6)\sqrt{13} = -2\sqrt{13}$

43. $\sqrt{5} + \sqrt{5} = 1\sqrt{5} + 1\sqrt{5} = (1+1)\sqrt{5} = 2\sqrt{5}$

45. $4\sqrt{2} - 5\sqrt{2} + 8\sqrt{2} = (4-5+8)\sqrt{2} = 7\sqrt{2}$

47. $\sqrt{5} + \sqrt{20} = 1\sqrt{5} + \sqrt{4} \cdot \sqrt{5}$
$= 1\sqrt{5} + 2\sqrt{5}$
$= (1+2)\sqrt{5}$
$= 3\sqrt{5}$

49. $\sqrt{50} - \sqrt{18} = \sqrt{25} \cdot \sqrt{2} - \sqrt{9} \cdot \sqrt{2}$
$= 5\sqrt{2} - 3\sqrt{2}$
$= (5-3)\sqrt{2}$
$= 2\sqrt{2}$

51. $3\sqrt{18} + 5\sqrt{50} = 3\sqrt{9} \cdot \sqrt{2} + 5\sqrt{25} \cdot \sqrt{2}$
$= 3 \cdot 3 \cdot \sqrt{2} + 5 \cdot 5\sqrt{2}$
$= 9\sqrt{2} + 25\sqrt{2}$
$= (9+25)\sqrt{2}$
$= 34\sqrt{2}$

53. $\frac{1}{4}\sqrt{12} - \frac{1}{2}\sqrt{48} = \frac{1}{4}\sqrt{4} \cdot \sqrt{3} - \frac{1}{2}\sqrt{16} \cdot \sqrt{3}$
$= \frac{1}{4} \cdot 2 \cdot \sqrt{3} - \frac{1}{2} \cdot 4 \cdot \sqrt{3}$
$= \frac{1}{2}\sqrt{3} - \frac{4}{2}\sqrt{3}$
$= \left(\frac{1}{2} - \frac{4}{2}\right)\sqrt{3}$
$= -\frac{3}{2}\sqrt{3}$

55. $3\sqrt{75} + 2\sqrt{12} - 2\sqrt{48}$
$= 3 \cdot \sqrt{25} \cdot \sqrt{3} + 2 \cdot \sqrt{4} \cdot \sqrt{3} - 2 \cdot \sqrt{16} \cdot \sqrt{3}$
$= 3 \cdot 5 \cdot \sqrt{3} + 2 \cdot 2 \cdot \sqrt{3} - 2 \cdot 4 \cdot \sqrt{3}$
$= 15\sqrt{3} + 4\sqrt{3} - 8\sqrt{3}$
$= (15+4-8)\sqrt{3}$
$= 11\sqrt{3}$

57. $\frac{5}{\sqrt{3}} = \frac{5}{\sqrt{3}} \cdot \frac{\sqrt{3}}{\sqrt{3}} = \frac{5\sqrt{3}}{\sqrt{9}} = \frac{5\sqrt{3}}{3}$

59. $\frac{21}{\sqrt{7}} = \frac{21}{\sqrt{7}} \cdot \frac{\sqrt{7}}{\sqrt{7}} = \frac{21\sqrt{7}}{\sqrt{49}} = \frac{21\sqrt{7}}{7} = 3\sqrt{7}$

61. $\frac{12}{\sqrt{30}} = \frac{12\sqrt{30}}{\sqrt{30}\sqrt{30}}$
$= \frac{12\sqrt{30}}{\sqrt{900}}$
$= \frac{12\sqrt{30}}{30}$
$= \frac{2\sqrt{30}}{5}$

63. $\frac{15}{\sqrt{12}} = \frac{15}{\sqrt{4 \cdot 3}}$
$= \frac{15}{\sqrt{4}\sqrt{3}}$
$= \frac{15}{2\sqrt{3}}$
$= \frac{15\sqrt{3}}{2\sqrt{3}\sqrt{3}}$
$= \frac{15\sqrt{3}}{2\sqrt{9}}$
$= \frac{15\sqrt{3}}{2 \cdot 3}$
$= \frac{15\sqrt{3}}{6}$
$= \frac{5\sqrt{3}}{2}$

65. $\sqrt{\frac{2}{5}} = \frac{\sqrt{2}}{\sqrt{5}} = \frac{\sqrt{2}}{\sqrt{5}} \cdot \frac{\sqrt{5}}{\sqrt{5}} = \frac{\sqrt{10}}{\sqrt{25}} = \frac{\sqrt{10}}{5}$

67. $\frac{7\sqrt{2 \cdot 2 \cdot 3}}{6} = \frac{7 \cdot \sqrt{4} \cdot \sqrt{3}}{6}$
$= \frac{7 \cdot 2 \cdot \sqrt{3}}{6}$
$= \frac{14 \cdot \sqrt{3}}{6}$
$= \frac{7}{3}\sqrt{3}$

69. $s = 4\sqrt{r}$

$s = 4\sqrt{12}$

$s = 4 \cdot \sqrt{4 \cdot 3}$

$s = 4 \cdot 2\sqrt{3}$

$s = 8\sqrt{3}$

$s \approx 13.9$

The greatest speed is about 13.9 mph.

71. $t = \sqrt{\dfrac{x}{16}}$

$t = \sqrt{\dfrac{640}{16}}$

$t = \sqrt{40}$

$t = \sqrt{4} \cdot \sqrt{10}$

$t = 2\sqrt{10}$

$t \approx 6.3$

It will take about 6.3 seconds.

73. $h = 2.9\sqrt{x} + 20.1$

$h = 2.9\sqrt{60} + 20.1 \approx 42.6$ inches

The formula models the actual median height well.

75. $P = 6.85\sqrt{t} + 19$

$P = 6.85\sqrt{4} + 19 = 32.7\%$ online in 2001.

The formula models the actual data well.

77. $P = 6.85\sqrt{t} + 19$

$P = 6.85\sqrt{13} + 19 \approx 44\%$ online in 2010.

79-83. Answers will vary.

85. c is true

$\sqrt{\sqrt{16}} = \sqrt{4} = 2$

87. $-\pi \approx -3.14$

$-\pi > -3.5$

89. The square root is multiplied by $\sqrt{2}$.

91. $\sqrt{2} + \sqrt{\dfrac{1}{2}} = \sqrt{2} + \dfrac{\sqrt{1}}{\sqrt{2}} = \sqrt{2} + \dfrac{\sqrt{1}}{\sqrt{2}} \cdot \dfrac{\sqrt{2}}{\sqrt{2}}$

$= \sqrt{2} + \dfrac{\sqrt{2}}{2} = \dfrac{2\sqrt{2}}{2} + \dfrac{\sqrt{2}}{2}$

$= \dfrac{2\sqrt{2} + \sqrt{2}}{2}$

$= \dfrac{(2+1)\sqrt{2}}{2} = \dfrac{3\sqrt{2}}{2}$

Check Points 5.5

1. $\left\{ -9, \ -1.3, \ 0, \ 0.\overline{3}, \ \dfrac{\pi}{2}, \ \sqrt{9}, \ \sqrt{10} \right\}$

    a. Natural numbers: $\sqrt{9}$ because $\sqrt{9} = 3$

    b. Whole numbers: $0, \ \sqrt{9}$

    c. Integers: $-9, \ 0, \ \sqrt{9}$

    d. Rational numbers: $-9, \ -1.3, \ 0, \ 0.\overline{3}, \ \sqrt{9}$

    e. Irrational numbers: $\dfrac{\pi}{2}, \ \sqrt{10}$

    f. Real numbers: All numbers in this set.

2. **a.** Associative property of multiplication

    b. Commutative property of addition

    c. Distributive property of multiplication over addition

    d. Commutative property of multiplication

3. **a.** Yes, the natural numbers are closed with respect to multiplication.

    b. No, the integers are not closed with respect to division. Example: $3 \div 5 = 0.6$ which is not an integer.

Exercise Set 5.5

1. $\left\{ -9, \ -\dfrac{4}{5}, \ 0, \ 0.25, \ \sqrt{3}, \ 9.2, \ \sqrt{100} \right\}$

    a. Natural numbers: $\sqrt{100}$ because $\sqrt{100} = 10$

    b. Whole numbers: $0, \ \sqrt{100}$

c. Integers: $-9, 0, \sqrt{100}$

d. Rational numbers: $-9, -\dfrac{4}{5}, 0, 0.25, 9.2, \sqrt{100}$

e. Irrational numbers: $\sqrt{3}$

f. Real numbers: All numbers in this set.

3. $\left\{ -11, -\dfrac{5}{6}, 0, 0.75, \sqrt{5}, \pi, \sqrt{64} \right\}$

   a. Natural numbers: $\sqrt{64}$ because $\sqrt{64} = 8$

   b. Whole numbers: 0 and $\sqrt{64}$

   c. Integers: $-11, 0, \sqrt{64}$

   d. Rational numbers: $-11, -\dfrac{5}{6}, 0, 0.75, \sqrt{64}$

   e. Irrational numbers: $\sqrt{5}, \pi$

   f. Real numbers: All numbers in this set.

5. 0 is the only whole number that is not a natural number.

7. Answers will vary. Possible answer: 0.5

9. Answers will vary. Possible answer: 7

11. Answers will vary. Possible answer: $\sqrt{3}$

13. $3 + (4 + 5) = 3 + (5 + 4)$

15. $9 \cdot (6 + 2) = 9 \cdot (2 + 6)$

17. $(4 \cdot 5) \cdot 3 = 4 \cdot (5 \cdot 3)$

19. $7 \cdot (4 + 5) = 7 \cdot 4 + 7 \cdot 5$

21. $5(6 + \sqrt{2}) = 5 \cdot 6 + 5 \cdot \sqrt{2} = 30 + 5\sqrt{2}$

23. $\sqrt{7}(3 + \sqrt{2}) = \sqrt{7} \cdot 3 + \sqrt{7} \cdot \sqrt{2} = 3\sqrt{7} + \sqrt{14}$

25. $\sqrt{3}(5 + \sqrt{3}) = \sqrt{3} \cdot 5 + \sqrt{3} \cdot \sqrt{3} = 5\sqrt{3} + \sqrt{9}$
$$= 5\sqrt{3} + 3$$

27. $\sqrt{6}(\sqrt{2} + \sqrt{6}) = \sqrt{6} \cdot \sqrt{2} + \sqrt{6} \cdot \sqrt{6}$
$$= \sqrt{12} + \sqrt{36}$$
$$= 2\sqrt{3} + 6$$

29. $6 + (-4) = (-4) + 6$
Commutative property of addition.

31. $6 + (2 + 7) = (6 + 2) + 7$
Associative property of addition.

33. $(2 + 3) + (4 + 5) = (4 + 5) + (2 + 3)$
Commutative property of addition.

35. $2(-8 + 6) = -16 + 12$
Distributive property of multiplication over addition.

37. $(2\sqrt{3}) \cdot \sqrt{5} = 2(\sqrt{3} \cdot \sqrt{5})$
Associative property of multiplication

39. Answers will vary.
Example: $1 - 2 = -1$

41. Answers will vary.
Example: $\dfrac{-2}{8} = -\dfrac{1}{4}$

43. Answers will vary.
Example: $\sqrt{5}\sqrt{5} = \sqrt{25} = 5$

45. No; the result is not the same if first you took a shower, then you got undressed.

47. Answers will vary.
Possible answer: First put on left glove, then put on right glove.

49-55. Answers will vary.

57. c is true

Check Points 5.6

1. a. $19^0 = 1$

   b. $(3\pi)^0 = 1$

   c. $(-14)^0 = 1$

2. a. $9^{-2} = \dfrac{1}{9^2} = \dfrac{1}{81}$

b. $6^{-3} = \dfrac{1}{6^3} = \dfrac{1}{216}$

c. $12^{-1} = \dfrac{1}{12}$

3. a. $7.4 \times 10^9 = 7,400,000,000$

  b. $3.017 \times 10^{-6} = 0.000003017$

4. a. $7,410,000,000 = 7.41 \times 10^9$

  b. $0.000000092 = 9.2 \times 10^{-8}$

5.

$410 \times 10^7 = (4.1 \times 10^2) \times 10^7 = 4.1 \times 10^{2+7} = 4.1 \times 10^9$

6. $(1.3 \times 10^7) \times (4 \times 10^{-2}) = (1.3 \times 4) \times (10^7 \times 10^{-2})$
$$= 5.2 \times 10^{7+(-2)}$$
$$= 5.2 \times 10^5$$
$$= 520,000$$

7. $\dfrac{6.9 \times 10^{-8}}{3 \times 10^{-2}} = \left(\dfrac{6.9}{3}\right) \times \left(\dfrac{10^{-8}}{10^{-2}}\right)$
$$= 2.3 \times 10^{-8-(-2)}$$
$$= 2.3 \times 10^{-6}$$
$$= 0.0000023$$

8. a. $0.0036 \times 5,200,000$
$$= 3.6 \times 10^{-3} \times 5.2 \times 10^6$$
$$= (3.6 \times 5.2) \times (10^{-3} \times 10^6)$$
$$= 18.72 \times 10^3$$
$$= 1.872 \times 10 \times 10^3$$
$$= 1.872 \times 10^4$$

  b. Based on part (a):
  $0.0036 \times 5,200,000$
$$= 1.872 \times 10^4$$
$$= 18720$$

9. $\dfrac{3.6 \times 10^9}{2.8 \times 10^8} = \left(\dfrac{3.6}{2.8}\right) \times \left(\dfrac{10^9}{10^8}\right)$
$$\approx 1.3 \times 10$$
$$\approx 13$$
$$\approx \$13 \text{ per American}$$

Exercise Set 5.6

1. $2^2 \cdot 2^3 = 2^{2+3} = 2^5 = 32$

3. $4 \cdot 4^2 = 4^1 \cdot 4^2 = 4^{1+2} = 4^3 = 64$

5. $(2^2)^3 = 2^{2 \cdot 3} = 2^6 = 64$

7. $(1^4)^5 = 1^{4 \cdot 5} = 1^{20} = 1$

9. $\dfrac{4^7}{4^5} = 4^{7-5} = 4^2 = 16$

11. $\dfrac{2^8}{2^4} = 2^{8-4} = 2^4 = 16$

13. $3^0 = 1$

15. $(-3)^0 = 1$

17. $2^{-2} = \dfrac{1}{2^2} = \dfrac{1}{4}$

19. $4^{-3} = \dfrac{1}{4^3} = \dfrac{1}{64}$

21. $2^{-5} = \dfrac{1}{2^5} = \dfrac{1}{32}$

23. $4^{-1} = \dfrac{1}{4^1} = \dfrac{1}{4}$

25. $3^4 \cdot 3^{-2} = 3^{4+(-2)} = 3^2 = 9$

27. $3^{-3} \cdot 3 = 3^{-3} \cdot 3^1 = 3^{-3+1} = 3^{-2} = \dfrac{1}{3^2} = \dfrac{1}{9}$

29. $\dfrac{2^3}{2^7} = 2^{3-7} = 2^{-4} = \dfrac{1}{2^4} = \dfrac{1}{16}$

31. $2.7 \times 10^2 = 270$

33. $9.12 \times 10^5 = 912,000$

35. $8 \times 10^7 = 8.0 \times 10^7 = 80,000,000$

37. $1 \times 10^5 = 1.0 \times 10^5 = 100,000$

39. $7.9 \times 10^{-1} = 0.79$

41. $2.15 \times 10^{-2} = 0.0215$

43. $7.86 \times 10^{-4} = 0.000786$

45. $3.18 \times 10^{-6} = 0.00000318$

47. $370 = 3.7 \times 10^2$

49. $3600 = 3.6 \times 10^3$

51. $32,000 = 3.2 \times 10^4$

53. $220,000,000 = 2.2 \times 10^8$

55. $0.027 = 2.7 \times 10^{-2}$

57. $0.0037 = 3.7 \times 10^{-3}$

59. $0.00000293 = 2.93 \times 10^{-6}$

61. $(2 \times 10^3)(3 \times 10^2) = (2 \times 3) \times (10^{3+2})$
$$= 6 \times 10^5$$
$$= 600,000$$

63. $(2 \times 10^9)(3 \times 10^{-5}) = (2 \times 3) \times (10^{9-5})$
$$= 6 \times 10^4$$
$$= 60,000$$

65. $(4.1 \times 10^7)(3 \times 10^{-4}) = (4.1 \times 3) \times (10^{2-4})$
$$= 12.3 \times 10^{-2}$$
$$= 1.23 \times 10 \times 10^{-2}$$
$$= 1.23 \times 10^{-1}$$
$$= 0.123$$

67. $\dfrac{12 \times 10^6}{4 \times 10^2} = \left(\dfrac{12}{4}\right) \times \left(\dfrac{10^6}{10^2}\right)$
$$= 3 \times 10^{6-2}$$
$$= 3 \times 10^4$$
$$= 30,000$$

69. $\dfrac{15 \times 10^4}{5 \times 10^{-2}} = \left(\dfrac{15}{5}\right) \times \left(\dfrac{10^4}{10^{-2}}\right)$
$$= 3 \times 10^{4-(-2)}$$
$$= 3 \times 10^6$$
$$= 3,000,000$$

71. $\dfrac{6 \times 10^3}{2 \times 10^5} = \left(\dfrac{6}{2}\right) \times \left(\dfrac{10^3}{10^5}\right)$
$$= 3 \times 10^{3-5}$$
$$= 3 \times 10^{-2}$$
$$= 0.03$$

73. $\dfrac{6.3 \times 10^{-6}}{3 \times 10^{-3}} = \left(\dfrac{6.3}{3}\right) \times \left(\dfrac{10^{-6}}{10^{-3}}\right)$
$$= 2.1 \times 10^{-6-(-3)}$$
$$= 2.1 \times 10^{-3}$$
$$= 0.0021$$

75. $(82,000,000)(3,000,000,000)$
$$= (8.2 \times 10^7)(3.0 \times 10^9)$$
$$= (8.2 \times 3.0) \times (10^{7+9})$$
$$= 24.6 \times 10^{16}$$
$$= 2.46 \times 10 \times 10^{16}$$
$$= 2.46 \times 10^{17}$$

77. $(0.0005)(6,000,000)$
$$= (5.0 \times 10^{-4})(6.0 \times 10^6)$$
$$= (5.0 \times 6.0)(10^{-4+6})$$
$$= 30 \times 10^2$$
$$= 3 \times 10 \times 10^2$$
$$= 3 \times 10^3$$

79. $\dfrac{9,500,000}{500} = \dfrac{9.5 \times 10^6}{5 \times 10^2}$

$\qquad = \left(\dfrac{9.5}{5}\right) \times (10^{6-2})$

$\qquad = 1.9 \times 10^4$

81. $\dfrac{0.00008}{200} = \dfrac{8 \times 10^{-5}}{2 \times 10^2}$

$\qquad = \left(\dfrac{8}{2}\right) \times (10^{-5-2})$

$\qquad = 4 \times 10^{-7}$

83. $\dfrac{480,000,000,000}{0.00012} = \dfrac{4.8 \times 10^{11}}{1.2 \times 10^{-4}}$

$\qquad = \left(\dfrac{4.8}{1.2}\right) \times (10^{11-(-4)})$

$\qquad = 4 \times 10^{15}$

85. $53.3 \times 10^6 = 5.33 \times 10 \times 10^6 = 5.33 \times 10^7$

87. $(77.0 - 35.3) \times 10^6$

$\qquad = 41.7 \times 10^6$

$\qquad = 4.17 \times 10 \times 10^6$

$\qquad = 4.17 \times 10^7$

89. 25% of 1.6×10^{12}

$\quad 0.25 \times (1.6 \times 10^{12})$

$\qquad = (0.25 \times 1.6) \times 10^{12}$

$\qquad = 0.4 \times 10^{12}$

$\qquad = 4.0 \times 10^{-1} \times 10^{12}$

$\qquad = 4.0 \times 10^{11}$

$\qquad = \$4 \times 10^{11}$

91. $\dfrac{1.9 \times 10^{12}}{2.8 \times 10^8} = \left(\dfrac{1.9}{2.8}\right) \times \left(\dfrac{10^{12}}{10^8}\right)$

$\qquad \approx 0.68 \times 10^4$

$\qquad \approx 6.8 \times 10^{-1} \times 10^4$

$\qquad \approx 6.8 \times 10^3$

$\qquad \approx \$6800$ per American

93. $4000 \times 2.8 \times 10^8$

$\qquad = 4 \times 10^3 \times 2.8 \times 10^8$

$\qquad = (4 \times 2.8) \times (10^3 \times 10^8)$

$\qquad = 11.2 \times 10^{11}$

$\qquad = 1.12 \times 10 \times 10^{11}$

$\qquad = \$1.12 \times 10^{12}$

95. $20,000 \times 5.3 \times 10^{-23}$

$\qquad = 2 \times 10^4 \times 5.3 \times 10^{-23}$

$\qquad = (2 \times 5.3) \times (10^4 \times 10^{-23})$

$\qquad = 10.6 \times 10^{-19}$

$\qquad = 1.06 \times 10 \times 10^{-19}$

$\qquad = 1.06 \times 10^{-18}$ grams

97-105. Answers will vary.

107. b is true

109. Answers will vary. Possible answer:

$\quad 2.0 \times 10^0 = 2.0 \times 1 = 2$

There is no advantage here since $10^0 = 1$.

Check Points 5.7

1. $100, 100 + 20 = 120, 120 + 20 = 140, 140 + 20 = 160, 160 + 20 = 180, 180 + 20 = 200$

100, 120, 140, 160, 180, and 200

2. $8, 8 - 3 = 5, 5 - 3 = 2, 2 - 3 = -1, -1 - 3 = -4, -4 - 3 = -7$

8, 5, 2, −1, −4, and −7

3. $a_n = a_1 + (n-1)d$

$\quad a_9 = 6 + (9-1)(-5)$

$\qquad = 6 + 8(-5)$

$\qquad = 6 - 40$

$\qquad = -34$

4. $a_n = a_1 + (n-1)d$ with $a_1 = 12,808$ and $d = 2350$.

$\quad$ **a.** $\quad a_n = a_1 + (n-1)d$ with $a_1 = 12,808$ and $d = 2350$.

$\qquad a_n = 12,808 + (n-1)(2350)$

$\qquad\quad = 12,808 + 2350n - 2350$

$\qquad\quad = 2350n + 10,458$

b. $a_n = 10458 + 2350n$ with
$n = 2010 - 1983 = 27$
$a_{27} = 10,458 + 2350(27)$
$\qquad = 73,908$
U.S. travelers will spend about $73,908 million by 2010.

$12, \ 6, \ 3, \ \dfrac{3}{2}, \ \dfrac{3}{4}, \ \dfrac{3}{8}$

6. $a_n = a_1 r^{n-1}$ with $a_1 = 5$, $r = -3$, and $n = 7$
$a_7 = 5(-3)^{7-1} = 5(-3)^6 = 5(729) = 3645$

5.

$12, \ 12 \cdot \dfrac{1}{2} = 6, \ 6 \cdot \dfrac{1}{2} = 3, \ 3 \cdot \dfrac{1}{2} = \dfrac{3}{2}, \ \dfrac{3}{2} \cdot \dfrac{1}{2} = \dfrac{3}{4}, \ \dfrac{3}{4} \cdot \dfrac{1}{2} = \dfrac{3}{8}$

7. $a_n = a_1 r^{n-1}$ with $a_1 = 3$, $r = \dfrac{6}{3} = 2$
$a_n = 3(2)^{n-1} = 3(2)^{8-1} = 3(2)^7 = 3(128) = 384$

Exercise Set 5.7

1. $8, \ 8 + 2 = 10, \ 10 + 2 = 12, \ 12 + 2 = 14, \ 14 + 2 = 16, \ 16 + 2 = 18$
8, 10, 12, 14, 16, and 18

3. $200, \ 200 + 20 = 220, \ 220 + 20 = 240, \ 240 + 20 = 260, \ 260 + 20 = 280, \ 280 + 20 = 300$
200, 220, 240, 260, 280, and 300

5. $-7, -7 + 4 = -3, -3 + 4 = 1, 1 + 4 = 5, 5 + 4 = 9, 9 + 4 = 13$
−7, −3, 1, 5, 9, and 13

7. $-400, \ -400 + 300 = -100, \ -100 + 300 = 200, \ 200 + 300 = 500, \ 500 + 300 = 800, \ 800 + 300 = 1100$
−400, −100, 200, 500, 800, and 1100

9. $7, \ 7 - 3 = 4, \ 4 - 3 = 1, \ 1 - 3 = -2, \ -2 - 3 = -5, \ -5 - 3 = -8$
7, 4, 1, −2, −5, and −8

11. $200, \ 200 - 60 - 140, \ 140 - 60 = 80, \ 80 - 60 = 20, \ 20 - 60 = -40, \ -40 - 60 = -100$
200, 140, 80, 20, −40, and −100

13. $\dfrac{5}{2}, \quad \dfrac{5}{2} + \dfrac{1}{2} = \dfrac{6}{2} = 3, \quad \dfrac{6}{2} + \dfrac{1}{2} = \dfrac{7}{2}, \quad \dfrac{7}{2} + \dfrac{1}{2} = \dfrac{8}{2} = 4, \quad \dfrac{8}{2} + \dfrac{1}{2} = \dfrac{9}{2}, \quad \dfrac{9}{2} + \dfrac{1}{2} = \dfrac{10}{2} = 5$

$\dfrac{5}{2}, 3, \dfrac{7}{2}, 4, \dfrac{9}{2}$, and 5

15. $\dfrac{3}{2}, \quad \dfrac{6}{4} + \dfrac{1}{4} = \dfrac{7}{4}, \quad \dfrac{7}{4} + \dfrac{1}{4} = \dfrac{8}{4} = 2, \quad \dfrac{8}{4} + \dfrac{1}{4} = \dfrac{9}{4}, \quad \dfrac{9}{4} + \dfrac{1}{4} = \dfrac{10}{4} = \dfrac{5}{2}, \quad \dfrac{10}{4} + \dfrac{1}{4} = \dfrac{11}{4}$

$\dfrac{3}{2}, \dfrac{7}{4}, 2, \dfrac{9}{4}, \dfrac{5}{2}$, and $\dfrac{11}{4}$

17. $4.25, \ 4.25 + 0.3 = 4.55, \ 4.55 + 0.3 = 4.85, \ 4.85 + 0.3 = 5.15, \ 5.15 + 0.3 = 5.45, \ 5.45 + 0.3 = 5.75$
4.25, 4.55, 4.85, 5.15, 5.45, and 5.75

19. $4.5, \ 4.5 - 0.75 = 3.75, \ 3.75 - 0.75 = 3, \ 3 - 0.75 = 2.25, \ 2.25 - 0.75 = 1.5, \ 1.5 - 0.75 = 0.75$
4.5, 3.75, 3, 2.25, 1.5, and 0.75

21. $a_1 = 13, \ d = 4$
$a_6 = 13 + (6-1)(4)$
$\qquad = 13 + 5(4)$
$\qquad = 13 + 20$
$\qquad = 33$

23. $a_1 = 7, d = 5$

$a_{50} = 7 + (50 - 1)(5)$

$= 7 + 49(5)$

$= 7 + 245$

$= 252$

25. $a_1 = -5, d = 9$

$a_9 = -5 + (9 - 1)(9)$

$= -5 + 8(9)$

$= -5 + 72$

$= 67$

27. $a_1 = -40, d = 5$

$a_{200} = -40 + (200 - 1)(5)$

$= -40 + 199(5)$

$= -40 + 995$

$= 955$

29. $a_1 = -8, d = 10$

$a_{10} = -8 + (10 - 1)(10)$

$= -8 + 9(10)$

$= -8 + 90$

$= 82$

31. $a_1 = 35, d = -3$

$a_{60} = 35 + (60 - 1)(-3)$

$= 35 + 59(-3)$

$= 35 + (-177)$

$= -142$

33. $a_1 = 12, d = -5$

$a_{12} = 12 + (12 - 1)(-5)$

$= 12 + 11(-5)$

$= 12 + (-55)$

$= -43$

35. $a_1 = -70, d = -2$

$a_{90} = -70 + (90 - 1)(-2)$

$= -70 + 89(-2)$

$= -70 + (-178)$

$= -248$

37. $a_1 = 6, d = \dfrac{1}{2}$

$a_{12} = 6 + (12 - 1)\left(\dfrac{1}{2}\right)$

$= 6 + 11\left(\dfrac{1}{2}\right)$

$= \dfrac{12}{2} + \dfrac{11}{2}$

$= \dfrac{23}{2}$

39. $a_1 = 14, d = -0.25$

$a_{50} = 14 + (50 - 1)(-0.25)$

$= 14 + 49(-0.25)$

$= 14 + (-12.25)$

$= 1.75$

41. $a_1 = 4, r = 2$

$4, 4 \cdot 2 = 8, 8 \cdot 2 = 16, 16 \cdot 2 = 32, 32 \cdot 2 = 64,$

$64 \cdot 2 = 128$

4, 8, 16, 32, 64, 128

43. $a_1 = 1000, r = 1$

$1000, 1000 \cdot 1 = 1000, 1000 \cdot 1 = 1000, \ldots$

1000, 1000, 1000, 1000, 1000, 1000

45. $a_1 = 3, r = -2$

$3, 3(-2) = -6, -6(-2) = 12, 12(-2) = -24, -24(-2) =$
$48, 48(-2) = -96$

3, −6, 12, −24, 48, −96

47. $a_1 = 10, r = -4$

$10, 10(-4) = -40, -40(-4) = 160,$
$160(-4) = -640, -640(-4) = 2560,$
$2560(-4) = -10{,}240$

10, −40, 160, −640, 2560, −10,240

49. $a_1 = 2000, r = -1$

$2000, 2000(-1) = -2000,$
$-2000(-1) = 2000, \ldots$

2000, −2000, 2000, −2000, 2000, −2000

51. $a_1 = -2, r = -3$

$-2, -2(-3) = 6, 6(-3) = -18, -18(-3) = 54, 54(-3) =$
$-162, -162(-3) = 486$

−2, 6, −18, 54, −162, 486

53. $a_1 = -6, r = -5$

$-6, -6(-5) = 30, 30(-5) = -150,$

$-150(-5) = 750, 750(-5) = -3750,$

$-3750(-5) = 18,750$

$-6, 30, -150, 750, -3750, 18750$

55. $a_1 = \dfrac{1}{4}, r = 2$

$\dfrac{1}{4}, \dfrac{1}{4} \cdot 2 = \dfrac{1}{2}, \dfrac{1}{2} \cdot 2 = 1, 1 \cdot 2 = 2, 2 \cdot 2 = 4,$

$4 \cdot 2 = 8$

$\dfrac{1}{4}, \dfrac{1}{2}, 1, 2, 4, 8$

57. $a_1 = \dfrac{1}{4}, r = \dfrac{1}{2}$

$\dfrac{1}{4}, \dfrac{1}{4} \cdot \dfrac{1}{2} = \dfrac{1}{8}, \dfrac{1}{8} \cdot \dfrac{1}{2} = \dfrac{1}{16}, \dfrac{1}{16} \cdot \dfrac{1}{2} = \dfrac{1}{32},$

$\dfrac{1}{32} \cdot \dfrac{1}{2} = \dfrac{1}{64}, \dfrac{1}{64} \cdot \dfrac{1}{2} = \dfrac{1}{128}$

$\dfrac{1}{4}, \dfrac{1}{8}, \dfrac{1}{16}, \dfrac{1}{32}, \dfrac{1}{64}, \dfrac{1}{128}$

59. $a_1 = -\dfrac{1}{16}, r = -4$

$-\dfrac{1}{16}, -\dfrac{1}{16} \cdot (-4) = \dfrac{1}{4}, \dfrac{1}{4} \cdot (-4) = -1,$

$-1(-4) = 4, 4(-4) = -16, -16(-4) = 64$

$-\dfrac{1}{16}, \dfrac{1}{4}, -1, 4, -16, 64$

61. $a_1 = 2, r = 0.1$

$2, 2(0.1) = 0.2, 0.2(0.1) = 0.02,$

$0.02(0.1) = 0.002, 0.002(0.1) = 0.0002, 0.0002(0.1)$

$= 0.00002.$

$2, 0.2, 0.02, 0.002, 0.0002, 0.00002$

63. $a_1 = 4, r = 2$

$a_7 = 4(2)^{7-1}$

$= 4(2)^6$

$= 4(64)$

$= 256$

65. $a_1 = 2, r = 3$

$a_{20} = 2(3)^{20-1}$

$= 2(3)^{19}$

$= 2,324,522,934$

$\approx 2.32 \times 10^9$

67. $a_1 = 50, r = 1$

$a_{100} = 50(1)^{100-1}$

$= 50(1)^{99}$

$= 50$

69. $a_1 = 5, r = -2$

$a_7 = 5(-2)^{7\ 1}$

$= 5(-2)^6$

$= 320$

71. $a_1 = 2, r = -1$

$a_{30} = 2(-1)^{30-1}$

$= 2(-1)^{29}$

$= -2$

73. $a_1 = -2, r = -3$

$a_6 = -2(-3)^{6-1}$

$= -2(-3)^5$

$= 486$

75. $a_1 = 6, r = \dfrac{1}{2}$

$a_8 = 6\left(\dfrac{1}{2}\right)^{8-1}$

$= 6\left(\dfrac{1}{2}\right)^7$

$= \dfrac{6}{128}$

$= \dfrac{3}{64}$

77. $a_1 = 18,\ r = -\dfrac{1}{3}$

$$a_6 = 18\left(-\dfrac{1}{3}\right)^{6-1}$$

$$= 18\left(-\dfrac{1}{3}\right)^{5}$$

$$= -\dfrac{18}{243}$$

$$= -\dfrac{2}{27}$$

79. $a_1 = 1000,\ r = -\dfrac{1}{2}$

$$a_{40} = 1000\left(-\dfrac{1}{2}\right)^{40-1}$$

$$= 1000\left(-\dfrac{1}{2}\right)^{39}$$

$$\approx -1.82 \times 10^{-9}$$

81. $a_1 = 1,000,000,\ r = 0.1$

$$a_8 = 1,000,000(0.1)^{8-1}$$

$$= 1,000,000(0.1)^{7}$$

$$= 0.1$$

83. The common difference of the arithmetic sequence is 4.
$2 + 4 = 6,\ 6 + 4 = 10,\ 10 + 4 = 14,$
$14 + 4 = 18,\ 18 + 4 = 22$
$2,\ 6,\ 10,\ 14,\ 18,\ 22,\ \ldots$

85. The common ratio of the geometric sequence is 3.
$5 \cdot 3 = 15,\ 15 \cdot 3 = 45,\ 45 \cdot 3 = 135,\ \ 5,\ 15,\ 45,\ 135,$
$135 \cdot 3 = 405,\ 405 \cdot 3 = 1215$
$405,\ 1215,\ \ldots$

87. The common difference of the arithmetic sequence is 5.
$-7 + 5 = -2,\ -2 + 5 = 3,\ 3 + 5 = 8,$
$8 + 5 = 13,\ 13 + 5 = 18.$
$-7,\ -2,\ 3,\ 8,\ 13,\ 18,\ \ldots$

89. The common ratio of the geometric sequence is $\dfrac{1}{2}$.

$3 \cdot \dfrac{1}{2} = \dfrac{3}{2},\ \dfrac{3}{2} \cdot \dfrac{1}{2} = \dfrac{3}{4},\ \dfrac{3}{4} \cdot \dfrac{1}{2} = \dfrac{3}{8},\ \dfrac{3}{8} \cdot \dfrac{1}{2} = \dfrac{3}{16},$

$\dfrac{3}{16} \cdot \dfrac{1}{2} = \dfrac{3}{32}$

$3,\ \dfrac{3}{2},\ \dfrac{3}{4},\ \dfrac{3}{8},\ \dfrac{3}{16},\ \dfrac{3}{32},\ \ldots$

91. The common difference of the arithmetic sequence is $\dfrac{1}{2}$.

$\dfrac{1}{2} + \dfrac{1}{2} = 1,\ 1 + \dfrac{1}{2} = \dfrac{3}{2},\ \dfrac{3}{2} + \dfrac{1}{2} = 2,\ 2 + \dfrac{1}{2} = \dfrac{5}{2},$

$\dfrac{5}{2} + \dfrac{1}{2} = 3$

$\dfrac{1}{2},\ 1,\ \dfrac{3}{2},\ 2,\ \dfrac{5}{2},\ 3,\ \ldots$

93. The common ratio of the geometric sequence is -1.
$7(-1) = -7,\ -7(-1) = 7,\ 7(-1) = -7,$
$-7(-1) = 7,\ 7(-1) = -7$
$7,\ -7,\ 7,\ -7,\ 7,\ -7,\ \ldots$

95. The common difference of the arithmetic sequence is -14.
$7 - 14 = -7,\ -7 - 14 = -21,\ -21 - 14 = -35,\ -35 - 14 = -49,\ -49 - 14 = -63$
$7,\ -7,\ -21,\ -35,\ -49,\ -63,\ \ldots$

97. The common ratio of the geometric sequence is $\sqrt{5}$.
$\sqrt{5} \cdot \sqrt{5} = 5,\ 5 \cdot \sqrt{5} = 5\sqrt{5},\ 5\sqrt{5} \cdot \sqrt{5} = 25,$
$25 \cdot \sqrt{5} = 25\sqrt{5},\ 25\sqrt{5} \cdot \sqrt{5} = 125$
$\sqrt{5},\ 5,\ 5\sqrt{5},\ 25,\ 25\sqrt{5},\ 125,\ \ldots$

99. $a_1 = 126,424,\ d = 1265,$ and $n = 2005 - 1989 = 16$

 a. $a_n = 126,424 + (n-1)(1265)$
 $= 126,424 + 1265n - 1265$
 $= 1265n + 125,159$

 b. $a_n = 1265n + 125,159$
 $= 1265(16) + 125,159$
 $= 145,399$
 There will be 145,399,000 employees in 2005.

101. Company A: $a_{10} = 24000 + (10-1)1600 = 38,400$
 Company B: $b_{10} = 28000 + (10-1)1000 = 37,000$
 Company A will pay $1400 more in year 10.

103. $a_1 = 1, r = 2$

$a_{15} = 1(2)^{15-1}$

$\qquad = 2^{14}$

$\qquad = 16,384$

On the 15th day you will put aside \$16,384.

105. $a_7 = \$3,000,000(1.04)^{7-1}$

$\qquad \approx \$3,795,957$ salary in year 7.

107. a. $\dfrac{21.36}{20.60} \approx 1.04; \quad \dfrac{22.19}{21.36} \approx 1.04;$

$\dfrac{23.02}{22.19} \approx 1.04; \qquad r \approx 1.04$

b. $a_n = 20.60(1.04)^{n-1}$

c. $a_{11} \approx 20.60(1.04)^{11-1}$

$\qquad \approx 30.49$ million in 2005.

109-115. Answers will vary.

117. d is true.

119. $a_1 = 7721, d = -905$

$a_{327} = 7721 + (327 - 1)(-905)$

$\qquad = 7721 - 295,030$

$\qquad = -287,309$

121. $a_1 = 7721, r = 5$

$a_{32} = 7721(5)^{32-1}$

$\qquad = 7721(5)^{31}$

$\qquad \approx 3.595 \times 10^{25}$

Chapter 5 Review Exercises

1. 238,632

2: Yes; The last digit is 2.

3: Yes; The sum of the digits is 24, which is divisible by 3.

4: Yes; The last two digits form 32, which is divisible by 4.

5: No; The last number does not end in 0 or 5.

6: Yes; The number is divisible by both 2 and 3.

8: Yes; The last three digits form 632, which is divisible by 8.

9: No; The sum of the digits is 24, which is not divisible by 9.

10: No; the last digit is not 0.

12: Yes; The number is divisible by both 3 and 4.

The number is divisible by 2, 3, 4, 6, 8, 12.

2. 421,153,470

2: Yes; The last digit is 0.

3: Yes; The sum of the digits is 27, which is divisible by 3.

4: No; The last two digits form 70, which is not divisible by 4.

5: Yes; The number ends in 0.

6: Yes; The number is divisible by both 2 and 3.

8: No; The last three digits form 470, which is not divisible by 8.

9: Yes; The sum of the digits is 27, which is divisible by 9.

10: Yes; The number ends in 0.

12: No; The number is not divisible by both 3 and 4.

The number is divisible by 2, 3, 5, 6, 9, 10.

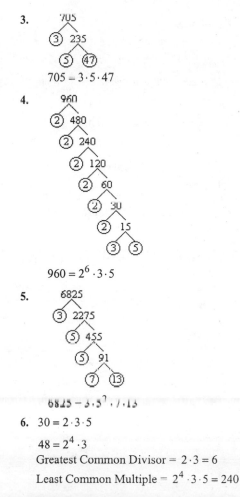

3.

$705 = 3 \cdot 5 \cdot 47$

4.

$960 = 2^6 \cdot 3 \cdot 5$

5.

$6825 = 3 \cdot 5^2 \cdot 7 \cdot 13$

6. $30 = 2 \cdot 3 \cdot 5$

$48 = 2^4 \cdot 3$

Greatest Common Divisor $= 2 \cdot 3 = 6$

Least Common Multiple $= 2^4 \cdot 3 \cdot 5 = 240$

7. $36 = 2^2 \cdot 3^2$

$150 = 2 \cdot 3 \cdot 5^2$

Greatest Common Divisor $= 2 \cdot 3 = 6$

Least Common Multiple $= 2^2 \cdot 3^2 \cdot 5^2 = 900$

8. $216 = 2^3 \cdot 3^3$

$254 = 2 \cdot 127$

Greatest Common Divisor = 2

Least Common Multiple = $2^3 \cdot 3^3 \cdot 127$

$\qquad\qquad\qquad\qquad = 27,432$

9. $24 = 2^3 \cdot 3$

$60 = 2^2 \cdot 3 \cdot 5$

Greatest Common Divisor = $2^2 \cdot 3 = 12$

There can be 12 people placed on each team.

10. $42 = 2 \cdot 3 \cdot 7$

$56 = 2^3 \cdot 7$

Least Common Multiple = $2^3 \cdot 3 \cdot 7 = 168$

$168 \div 60 = 2.8$ or 2 hours and 48 minutes. They will begin again at 11:48 A.M.

11. $-93 < 17$ because -93 is to the left of 17 on the number line.

12. $-2 > -200$ because -2 is to the right of -200 on the number line.

13. $|-860| = 860$ because -860 is 860 units from 0 on the number line.

14. $|53| = 53$ because 53 is 53 units from 0 on the number line.

15. $|0| = 0$ because 0 is 0 units from 0 on the number line.

16. $8 + (-11) = -3$

17. $-6 + (-5) = -11$

18. $-7 - 8 = -7 + (-8) = -15$

19. $-7 - (-8) = -7 + 8 = 1$

20. $(-9)(-11) = 99$

21. $5(-3) = -15$

22. $\dfrac{-36}{-4} = 9$

23. $\dfrac{20}{-5} = -4$

24. $-40 \div 5 \cdot 2 = -8 \cdot 2 = -16$

25. $-6 + (-2) \cdot 5 = -6 + (-10) = -16$

26. $6 - 4(-3 + 2) = 6 - 4(-1) = 6 + 4 = 10$

27. $28 \div (2 - 4^2) = 28 \div (2 - 16)$

$\qquad\qquad\qquad\quad = 28 \div (-14)$

$\qquad\qquad\qquad\quad = -2$

28. $36 - 24 \div 4 \cdot 3 - 1 = 36 - 6 \cdot 3 - 1$

$\qquad\qquad\qquad\qquad = 36 - 18 - 1$

$\qquad\qquad\qquad\qquad = 18 - 1$

$\qquad\qquad\qquad\qquad = 17$

29. $2 - (-19) = 2 + 19 = 21$ degree difference.

30. $-12 - (-19) = -12 + 19 = 7$ degrees warmer.

31. $40 = 2^3 \cdot 5$

$75 = 3 \cdot 5^2$

Greatest Common Divisor is 5.

$\dfrac{40}{75} = \dfrac{40 \div 5}{75 \div 5} = \dfrac{8}{15}$

32. $36 = 2^2 \cdot 3^2$

$150 = 2 \cdot 3 \cdot 5^2$

Greatest Common Divisor is $2 \cdot 3$ or 6.

$\dfrac{36}{150} = \dfrac{36 \div 6}{150 \div 6} = \dfrac{6}{25}$

33. $165 = 3 \cdot 5 \cdot 11$

$180 = 2^2 \cdot 3^2 \cdot 5$

Greatest Common Divisor is $3 \cdot 5$ or 15.

$\dfrac{165}{180} = \dfrac{165 \div 15}{180 \div 15} = \dfrac{11}{12}$

34. $\dfrac{4}{5} = 0.8$

$$
\begin{array}{r}
0.8 \\
5\overline{)4.0} \\
\underline{40} \\
0
\end{array}
$$

35. $\frac{3}{7} = 0.\overline{428571}$

$$
\begin{array}{r}
0.4285714 \\
7\overline{)3.0000000} \\
28 \\
\overline{20} \\
14 \\
\overline{60} \\
56 \\
\overline{40} \\
35 \\
\overline{50} \\
49 \\
\overline{10} \\
7 \\
\overline{30} \\
28 \\
\overline{2}
\end{array}
$$

36. $\frac{5}{8} = 0.625$

$$
\begin{array}{r}
0.625 \\
8\overline{)5.000} \\
48 \\
\overline{20} \\
16 \\
\overline{40} \\
40 \\
\overline{0}
\end{array}
$$

37. $\frac{9}{16} = 0.5625$

$$
\begin{array}{r}
0.5625 \\
16\overline{)9.0000} \\
80 \\
\overline{100} \\
96 \\
\overline{40} \\
32 \\
\overline{80} \\
80 \\
\overline{0}
\end{array}
$$

38. $0.6 = \frac{6}{10} = \frac{6 \div 2}{10 \div 2} = \frac{3}{5}$

39. $0.68 = \frac{68}{100}$

$68 = 2^2 \cdot 17$

$100 = 2^2 \cdot 5^2$

Greatest Common Divisor is 2^2 or 4.

$\frac{68 \div 4}{100 \div 4} = \frac{17}{25}$

40. $0.588 = \frac{588}{1000}$

$588 = 2^2 \cdot 3 \cdot 7^2$

$1000 = 2^3 \cdot 5^3$

Greatest Common Divisor is 2^2 or 4.

$\frac{588 \div 4}{1000 \div 4} = \frac{147}{250}$

41. $0.0084 = \frac{84}{10,000}$

$84 = 2^2 \cdot 3 \cdot 7$

$10,000 = 2^4 \cdot 5^4$

Greatest Common Divisor is 2^2 or 4.

$\frac{84 \div 4}{10,000 \div 4} = \frac{21}{2500}$

42. $n = 0.555\ldots$
$10n = 5.555\ldots$

$$
\begin{array}{r}
10n = 5.555\ldots \\
- \quad n = 0.555\ldots \\
\hline
9n = 5
\end{array}
$$

$n = \frac{5}{9}$

43. $n = 0.3434\ldots$
$100n = 34.3434\ldots$

$$
\begin{array}{r}
100n = 34.3434\ldots \\
- \quad n = 0.3434\ldots \\
\hline
99n = 34
\end{array}
$$

$n = \frac{34}{99}$

44.
$$n = 0.113113 \ldots$$
$$1000n = 113.113113 \ldots$$

$$1000n = 113.113113\ldots$$
$$- \quad\; n = \quad 0.113113\ldots$$
$$\overline{\quad 999n = 113 \quad}$$

$$n = \frac{113}{999}$$

45. $\dfrac{3}{5} \cdot \dfrac{7}{10} = \dfrac{3 \cdot 7}{5 \cdot 10} = \dfrac{21}{50}$

46. $\dfrac{4}{5} \div \dfrac{3}{10} = \dfrac{4}{5} \cdot \dfrac{10}{3} = \dfrac{4 \cdot 10}{5 \cdot 3} = \dfrac{40}{15} = \dfrac{8}{3}$

47. $\dfrac{2}{9} + \dfrac{4}{9} = \dfrac{2+4}{9} = \dfrac{6}{9} = \dfrac{2}{3}$

48. $\dfrac{7}{9} + \dfrac{5}{12} = \dfrac{7}{9} \cdot \dfrac{4}{4} + \dfrac{5}{12} \cdot \dfrac{3}{3}$

$$= \dfrac{28}{36} + \dfrac{15}{36}$$

$$= \dfrac{28+15}{36}$$

$$= \dfrac{43}{36}$$

49. $\dfrac{3}{4} - \dfrac{2}{15} = \dfrac{3}{4} \cdot \dfrac{15}{15} - \dfrac{2}{15} \cdot \dfrac{4}{4}$

$$= \dfrac{45}{60} - \dfrac{8}{60}$$

$$= \dfrac{45-8}{60}$$

$$= \dfrac{37}{60}$$

50. $\dfrac{1}{3} + \dfrac{1}{2} \cdot \dfrac{4}{5} = \dfrac{1}{3} + \dfrac{1 \cdot 4}{2 \cdot 5}$

$$= \dfrac{1}{3} + \dfrac{4}{10}$$

$$= \dfrac{1}{3} + \dfrac{2}{5}$$

$$= \dfrac{1}{3} \cdot \dfrac{5}{5} + \dfrac{2}{5} \cdot \dfrac{3}{3}$$

$$= \dfrac{5}{15} + \dfrac{6}{15}$$

$$= \dfrac{11}{15}$$

51. $\dfrac{3}{8}\left(\dfrac{1}{2} + \dfrac{1}{3}\right) = \dfrac{3}{8}\left(\dfrac{1}{2} \cdot \dfrac{3}{3} + \dfrac{1}{3} \cdot \dfrac{2}{2}\right)$

$$= \dfrac{3}{8}\left(\dfrac{3}{6} + \dfrac{2}{6}\right)$$

$$= \dfrac{3}{8}\left(\dfrac{5}{6}\right)$$

$$= \dfrac{15}{48}$$

$$= \dfrac{5}{16}$$

52. $\dfrac{1}{7} + \dfrac{1}{8} = \dfrac{1}{7} \cdot \dfrac{8}{8} + \dfrac{1}{8} \cdot \dfrac{7}{7}$

$$= \dfrac{8}{56} + \dfrac{7}{56}$$

$$= \dfrac{15}{56}$$

$$\dfrac{15}{56} \div 2 = \dfrac{15}{56} \cdot \dfrac{1}{2} = \dfrac{15}{112}$$

53. $\dfrac{3}{4} + \dfrac{3}{5} = \dfrac{3}{4} \cdot \dfrac{5}{5} + \dfrac{3}{5} \cdot \dfrac{4}{4}$

$$= \dfrac{15}{20} + \dfrac{12}{20}$$

$$= \dfrac{27}{20}$$

$$\dfrac{27}{20} \div 2 = \dfrac{27}{20} \cdot \dfrac{1}{2} = \dfrac{27}{40}$$

54. $1 - \left(\dfrac{1}{4} + \dfrac{1}{3}\right) = 1 - \left(\dfrac{1}{4} \cdot \dfrac{3}{3} + \dfrac{1}{3} \cdot \dfrac{4}{4}\right)$

$$= 1 - \left(\dfrac{3}{12} + \dfrac{4}{12}\right)$$

$$= \dfrac{12}{12} - \dfrac{7}{12}$$

$$= \dfrac{5}{12}$$

At the end of the second day, $\dfrac{5}{12}$ of the tank is filled with gas.

55. $\sqrt{28} = \sqrt{4 \cdot 7} = \sqrt{4} \cdot \sqrt{7} = 2\sqrt{7}$

56. $\sqrt{72} = \sqrt{36 \cdot 2} = \sqrt{36} \cdot \sqrt{2} = 6\sqrt{2}$

57. $\sqrt{150} = \sqrt{25 \cdot 6} = \sqrt{25} \cdot \sqrt{6} = 5\sqrt{6}$

58. $\sqrt{300} = \sqrt{100 \cdot 3} = \sqrt{100} \cdot \sqrt{3} = 10\sqrt{3}$

59. $\sqrt{6} \cdot \sqrt{8} = \sqrt{6 \cdot 8} = \sqrt{48} = \sqrt{16} \cdot \sqrt{3} = 4\sqrt{3}$

60. $\sqrt{10} \cdot \sqrt{5} = \sqrt{10 \cdot 5}$
$= \sqrt{50}$
$= \sqrt{25} \cdot \sqrt{2}$
$= 5\sqrt{2}$

61. $\dfrac{\sqrt{24}}{\sqrt{2}} = \sqrt{\dfrac{24}{2}} = \sqrt{12} = \sqrt{4} \cdot \sqrt{3} = 2\sqrt{3}$

62. $\dfrac{\sqrt{27}}{\sqrt{3}} = \sqrt{\dfrac{27}{3}} = \sqrt{9} = 3$

63. $\sqrt{5} + 4\sqrt{5} = 1\sqrt{5} + 4\sqrt{5} = (1+4)\sqrt{5} = 5\sqrt{5}$

64. $7\sqrt{11} - 13\sqrt{11} = (7-13)\sqrt{11} = -6\sqrt{11}$

65. $\sqrt{50} + \sqrt{8} = \sqrt{25} \cdot \sqrt{2} + \sqrt{4} \cdot \sqrt{2}$
$= 5\sqrt{2} + 2\sqrt{2}$
$= (5+2)\sqrt{2}$
$= 7\sqrt{2}$

66. $\sqrt{3} - 6\sqrt{27} = \sqrt{3} - 6\sqrt{9} \cdot \sqrt{3}$
$= \sqrt{3} - 6 \cdot 3\sqrt{3}$
$= 1\sqrt{3} - 18\sqrt{3}$
$= (1-18)\sqrt{3}$
$= -17\sqrt{3}$

67. $2\sqrt{18} + 3\sqrt{8} = 2\sqrt{9} \cdot \sqrt{2} + 3\sqrt{4} \cdot \sqrt{2}$
$= 2 \cdot 3 \cdot \sqrt{2} + 3 \cdot 2 \cdot \sqrt{2}$
$= 6\sqrt{2} + 6\sqrt{2}$
$= (6+6)\sqrt{2}$
$= 12\sqrt{2}$

68. $\dfrac{30}{\sqrt{5}} = \dfrac{30}{\sqrt{5}} \cdot \dfrac{\sqrt{5}}{\sqrt{5}} = \dfrac{30\sqrt{5}}{\sqrt{25}} = \dfrac{30\sqrt{5}}{5} = 6\sqrt{5}$

69. $\sqrt{\dfrac{2}{3}} = \dfrac{\sqrt{2}}{\sqrt{3}} = \dfrac{\sqrt{2}}{\sqrt{3}} \cdot \dfrac{\sqrt{3}}{\sqrt{3}} = \dfrac{\sqrt{6}}{\sqrt{9}} = \dfrac{\sqrt{6}}{3}$

70. $2\sqrt{5(40)} = 2\sqrt{200} = 2\sqrt{100 \cdot 2}$
$= 2 \cdot 10\sqrt{2} = 20\sqrt{2} \approx 28$ mph

71. $\left\{ -17, -\dfrac{9}{13}, 0, 0.75, \sqrt{2}, \pi, \sqrt{81} \right\}$

   a. Natural numbers:
      $\sqrt{81}$ because $\sqrt{81} = 9$

   b. Whole numbers: $0, \sqrt{81}$

   c. Integers: $-17, 0, \sqrt{81}$

   d. Rational numbers:
      $-17, -\dfrac{9}{13}, 0, 0.75, \sqrt{81}$

   e. Irrational numbers: $\sqrt{2}, \pi$

   f. Real numbers: All numbers in this set.

72. Answers will vary. Example: -3

73. Answers will vary. Example: $\dfrac{1}{2}$

74. Answers will vary: Example: $\sqrt{2}$

75. $3 + 17 = 17 + 3$
Commutative property of addition

76. $(6 \cdot 3) \cdot 9 = 6 \cdot (3 \cdot 9)$
Associative property of multiplication

77. $\sqrt{3}\left(\sqrt{5} + \sqrt{3}\right) = \sqrt{15} + 3$
Distributive property of multiplication over addition.

78. $(6 \cdot 9) \cdot 2 = 2 \cdot (6 \cdot 9)$
Commutative property of multiplication

79. $\sqrt{3}\left(\sqrt{5} + \sqrt{3}\right) = \left(\sqrt{5} + \sqrt{3}\right)\sqrt{3}$
Commutative property of multiplication

80. $(3 \cdot 7) + (4 \cdot 7) = (4 \cdot 7) + (3 \cdot 7)$
Commutative property of addition

81. Answers will vary. Example: $\dfrac{1}{2}$

82. Answers will vary. Example: $4 - 5 = -1$

83. $6 \cdot 6^2 = 6^1 \cdot 6^2 = 6^{1+2} = 6^3 = 216$

84. $2^3 \cdot 2^3 = 2^{3+3} = 2^6 = 64$

85. $(2^2)^2 = 2^{2 \cdot 2} = 2^4 = 16$

86. $(3^3)^2 = 3^{3 \cdot 2} = 3^6 = 729$

87. $\dfrac{5^6}{5^4} = 5^{6-4} = 5^2 = 25$

88. $7^0 = 1$

89. $(-7)^0 = 1$

90. $6^{-3} = \dfrac{1}{6^3} = \dfrac{1}{216}$

91. $2^{-4} = \dfrac{1}{2^4} = \dfrac{1}{16}$

92. $\dfrac{7^4}{7^6} = 7^{4-6} = 7^{-2} = \dfrac{1}{7^2} = \dfrac{1}{49}$

93. $3^5 \cdot 3^{-2} = 3^{5-2} = 3^3 = 27$

94. $4.6 \times 10^2 = 460$

95. $3.74 \times 10^4 = 37,400$

96. $2.55 \times 10^{-3} = 0.00255$

97. $7.45 \times 10^{-5} = 0.0000745$

98. $7520 = 7.52 \times 10^3$

99. $3,590,000 = 3.59 \times 10^6$

100. $0.00725 = 7.25 \times 10^{-3}$

101. $0.000000409 = 4.09 \times 10^{-7}$

102. $(3 \times 10^7)(1.3 \times 10^{-5}) = (3 \times 1.3) \times 10^{7-5}$
$$= 3.9 \times 10^2$$
$$= 390$$

103. $(5 \times 10^3)(2.3 \times 10^2) = (5 \times 2.3) \times 10^{3+2}$
$$= 11.5 \times 10^5$$
$$= 1.15 \times 10 \times 10^5$$
$$= 1.15 \times 10^6$$
$$= 1,150,000$$

104. $\dfrac{6.9 \times 10^3}{3 \times 10^5} = \left(\dfrac{6.9}{3}\right) \times 10^{3-5}$
$$= 2.3 \times 10^{-2}$$
$$= 0.023$$

105. $\dfrac{2.4 \times 10^{-4}}{6 \times 10^{-6}} = \left(\dfrac{2.4}{6}\right) \times 10^{-4-(-6)}$
$$= 0.4 \times 10^{-4+6}$$
$$= 0.4 \times 10^2$$
$$= 40$$

106. $(60,000)(540,000) = (6.0 \times 10^4)(5.4 \times 10^5)$
$$= (6.0 \times 5.4) \times 10^{4+5}$$
$$= 32.4 \times 10^9$$
$$= 3.24 \times 10 \times 10^9$$
$$= 3.24 \times 10^{10}$$

107. $(91,000)(0.0004) = (9.1 \times 10^4)(4 \times 10^{-4})$
$$= (9.1 \times 4) \times 10^{4-4}$$
$$= 36.4 \times 10^0$$
$$= 3.64 \times 10^1$$

108. $\dfrac{8,400,000}{4000} = \dfrac{8.4 \times 10^6}{4 \times 10^3}$
$$= \left(\dfrac{8.4}{4}\right) \times 10^{6-3}$$
$$= 2.1 \times 10^3$$

109. $\dfrac{0.000003}{0.00000006} = \dfrac{3 \times 10^{-6}}{6 \times 10^{-8}}$
$$= \left(\dfrac{3}{6}\right) \times 10^{-6-(-8)}$$
$$= 0.5 \times 10^2$$
$$= 5 \times 10^{-1} \times 10^2$$
$$= 5 \times 10^1$$

110. $\dfrac{10^9}{10^6} = 10^{9-6} = 10^3 = 1000$ years

111. $(2.8 \times 10^8) \times 150 = (2.8 \times 10^8)(1.5 \times 10^2)$
$\qquad = (2.8 \times 1.5) \times 10^{8+2}$
$\qquad = \$4.2 \times 10^{10}$

112. $2 \times 6.1 \times 10^9 = 12.2 \times 10^9$
$\qquad = 1.22 \times 10 \times 10^9$
$\qquad = 1.22 \times 10^{10}$

113. $a_1 - 7, d - 4$
$7, \quad 7 + 4 = 11, \quad 11 + 4 = 15, \quad 15 + 4 = 19,$
$19 + 4 = 23, \quad 23 + 4 = 27$
$7, 11, 15, 19, 23, 27$

114. $a_1 = -4, d = -5$
$-4, \quad -4 - 5 = -9, \quad -9 - 5 = -14,$
$-14 - 5 = -19, \quad -19 - 5 = -24, \quad -24 - 5 = -29$
$-4, -9, -14, -19, -24, -29$

115. $a_1 = \dfrac{3}{2}, d = -\dfrac{1}{2}$
$\dfrac{3}{2}, \quad \dfrac{3}{2} - \dfrac{1}{2} = \dfrac{2}{2} = 1, \quad \dfrac{2}{2} - \dfrac{1}{2} = \dfrac{1}{2},$
$\dfrac{1}{2} - \dfrac{1}{2} = 0, \quad 0 - \dfrac{1}{2} = -\dfrac{1}{2}, \quad -\dfrac{1}{2} - \dfrac{1}{2} = -1$
$\dfrac{3}{2}, 1, \dfrac{1}{2}, 0, -\dfrac{1}{2}, -1$

116. $a_1 = 5, d = 3$
$a_6 = 5 + (6 - 1)(3)$
$\qquad = 5 + 5(3)$
$\qquad = 5 + 15$
$\qquad = 20$

117. $a_1 = 8, d = 2$
$a_{12} = -8 + (12 - 1)(-2)$
$\qquad = -8 + 11(-2)$
$\qquad = -8 + (-22)$
$\qquad = -30$

118. $a_1 = 14, d = -4$
$a_{14} = 14 + (14 - 1)(-4)$
$\qquad = 14 + 13(-4)$
$\qquad = 14 + (-52)$
$\qquad = -38$

119. $a_1 = 3, r = 2$
$3, 3 \cdot 2 = 6, 6 \cdot 2 = 12, 12 \cdot 2 = 24,$
$24 \cdot 2 = 48, 48 \cdot 2 = 96$
$3, 6, 12, 24, 48, 96$

120. $a_1 = \dfrac{1}{2}, r = \dfrac{1}{2}$
$\dfrac{1}{2}, \dfrac{1}{2} \cdot \dfrac{1}{2} = \dfrac{1}{4}, \dfrac{1}{4} \cdot \dfrac{1}{2} = \dfrac{1}{8}, \dfrac{1}{8} \cdot \dfrac{1}{2} = \dfrac{1}{16},$
$\dfrac{1}{16} \cdot \dfrac{1}{2} = \dfrac{1}{32}, \dfrac{1}{32} \cdot \dfrac{1}{2} = \dfrac{1}{64}$
$\dfrac{1}{2}, \dfrac{1}{4}, \dfrac{1}{8}, \dfrac{1}{16}, \dfrac{1}{32}, \dfrac{1}{64}$

121. $a_1 = 16, r = -\dfrac{1}{2}$
$16, 16\left(-\dfrac{1}{2}\right) = -8, -8\left(-\dfrac{1}{2}\right) = 4, \quad 4\left(-\dfrac{1}{2}\right) = -2,$
$-2\left(-\dfrac{1}{2}\right) = 1, 1\left(-\dfrac{1}{2}\right) = -\dfrac{1}{2}$
$16, -8, 4, -2, 1, -\dfrac{1}{2}$

122. $a_1 = 2, r = 3$
$a_4 = 2(3)^{4-1}$
$\qquad = 2(3)^3$
$\qquad = 2(27)$
$\qquad = 54$

123. $a_1 = 16, r = \dfrac{1}{2}$
$a_6 = 16\left(\dfrac{1}{2}\right)^{6-1}$
$\qquad = 16\left(\dfrac{1}{2}\right)^5$
$\qquad = \dfrac{16}{32}$
$\qquad = \dfrac{1}{2}$

124. $a_1 = -3, r = 2$
$a_5 = -3(2)^{5-1}$
$\qquad = -3(2)^4$
$\qquad = -3(16)$
$\qquad = -48$

125. The common difference in the arithmetic sequence is 5.
$4 + 5 = 9, 9 + 5 = 14, 14 + 5 = 19,$
$19 + 5 = 24, 24 + 5 = 29$
$4, 9, 14, 19, 24, 29, \ldots$

126. The common ratio in the geometric sequence is 3.
$2 \cdot 3 = 6, 6 \cdot 3 = 18, 18 \cdot 3 = 54, 54 \cdot 3 = 162, 162 \cdot 3 = 486$
$2, 6, 18, 54, 162, 486, \ldots$

127. The common ratio in the geometric sequence is $\frac{1}{4}$.

$1 \cdot \frac{1}{4} = \frac{1}{4}, \frac{1}{4} \cdot \frac{1}{4} = \frac{1}{16}, \frac{1}{16} \cdot \frac{1}{4} = \frac{1}{64},$

$\frac{1}{64} \cdot \frac{1}{4} = \frac{1}{256}, \frac{1}{256} \cdot \frac{1}{4} = \frac{1}{1024}$

$1, \frac{1}{4}, \frac{1}{16}, \frac{1}{64}, \frac{1}{256}, \frac{1}{1024}, \ldots$

128. The common difference in the arithmetic sequence is –7.
$0 - 7 = -7, -7 - 7 = -14, -14 - 7 = -21, -21 - 7 = -28,$
$-28 - 7 = -35$
$0, -7, -14, -21, -28, -35, \ldots$

129. $a_1 = 27,966, \ d = 553,$ and $n = 2010 - 1984 = 26$

a. $a_n = 27,966 + (n-1)(553)$
$= 27,966 + 553n - 553$
$= 553n + 27,413$

b. $a_n = 553n + 27,413$
$= 553(26) + 27,413$
$= 41,791$
The approximate salary in 2010 will be $41,791.

130. $a_1 = 30,000, \ r = 1.08$
$a_n = a_1 r^{n-1}$
$a_{20} = 30,000(1.08)^{20-1}$
$= 30,000(1.08)^{19}$
$= 129,471$
The salary for the 20th year will be $129,471.

Chapter 5 Test

1. 391,248

2: Yes; the last digit is 8.

3: Yes; the sum of the digits is 27, which is divisible by 3.

4: Yes; the last two digits form 48, which is divisible by 4.

5: No; the number does not end in 0 or 5.

6: Yes; the number is divisible by both 2 and 3.

8: Yes; the last three digits form 248, which is divisible by 8.

9: Yes; the sum of the digits is 27, which is divisible by 9.

10: No; the number does not end in 0.

12: Yes; the number is divisible by both 3 and 4.

391, 248 is divisible by 2, 3, 4, 6, 8, 9, 12.

2.

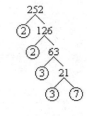

$252 = 2^2 \cdot 3^2 \cdot 7$

3. $48 = 2^4 \cdot 3$
$72 = 2^3 \cdot 3^2$
Greatest Common Divisor $= 2^3 \cdot 3 = 24$
Least Common Multiple $= 2^4 \cdot 3^2 = 144$

4. $-6 - (5 - 12) = -6 - (-7) = -6 + 7 = 1$

5. $(-3)(-4) \div (7 - 10) = (-3)(-4) \div (-3)$
$= 12 \div (-3)$
$= -4$

6. $(6 - 8)^2 (5 - 7)^3 = (-2)^2 (-2)^3$
$= 4(-8)$
$= -32$

7. $\frac{7}{12} = 0.58\overline{3}$

$$\begin{array}{r} 0.5833... \\ 12\overline{)7.0000} \\ 60 \\ \hline 100 \\ 96 \\ \hline 40 \\ 36 \\ \hline 40 \\ 36 \\ \hline 4 \end{array}$$

8. $n = 0.6464...$
$100n = 64.6464...$

$100n = 64.6464...$
$-\quad n = \;\;0.6464...$
$\overline{99n = 64}$

$n = \frac{64}{99}$

9. $\left(-\frac{3}{7}\right) \div \left(-\frac{15}{7}\right) = \left(-\frac{3}{7}\right) \cdot \left(-\frac{7}{15}\right)$

$= \frac{(-3)(-7)}{7 \cdot 15}$

$= \frac{21}{105}$

$= \frac{1}{5}$

10. $\frac{19}{24} - \frac{7}{40} = \frac{19}{24} \cdot \frac{5}{5} - \frac{7}{40} \cdot \frac{3}{3}$

$= \frac{95}{120} - \frac{21}{120}$

$= \frac{95 - 21}{120}$

$= \frac{74}{120}$

$= \frac{37}{60}$

11. $\frac{1}{2} - 8\left(\frac{1}{4} + 1\right) = \frac{1}{2} - 8\left(\frac{5}{4}\right)$

$= \frac{1}{2} - 10$

$= \frac{1}{2} - \frac{20}{2}$

$= -\frac{19}{2}$

12. $\frac{1}{2} + \frac{2}{3} = \frac{1}{2} \cdot \frac{3}{3} + \frac{2}{3} \cdot \frac{2}{2}$

$= \frac{3}{6} + \frac{4}{6}$

$= \frac{7}{6}$

$\frac{7}{6} \div 2 = \frac{7}{6} \cdot \frac{1}{2} = \frac{7}{12}$

13. $\sqrt{10} \cdot \sqrt{5} = \sqrt{10 \cdot 5}$

$= \sqrt{50}$

$= \sqrt{25 \cdot 2}$

$= \sqrt{25} \cdot \sqrt{2}$

$= 5\sqrt{2}$

14. $\sqrt{50} + \sqrt{32} = \sqrt{25} \cdot \sqrt{2} + \sqrt{16} \cdot \sqrt{2}$

$= 5\sqrt{2} + 4\sqrt{2}$

$= (5 + 4)\sqrt{2}$

$= 9\sqrt{2}$

15. $\frac{6}{\sqrt{2}} = \frac{6}{\sqrt{2}} \cdot \frac{\sqrt{2}}{\sqrt{2}} = \frac{6\sqrt{2}}{\sqrt{4}} = \frac{6\sqrt{2}}{2} = 3\sqrt{2}$

16. The rational numbers are $-7, -\frac{4}{5}, 0, 0.25, \sqrt{4}, \frac{22}{7}$.

17. Commutative property of addition

18. Distributive property of multiplication over addition

19. $3^3 \cdot 3^2 = 3^{3+2} = 3^5 = 243$

20. $\frac{4^6}{4^3} = 4^{6-3} = 4^3 = 64$

21. $8^{-2} = \frac{1}{8^2} = \frac{1}{64}$

22. $(3 \times 10^8)(2.5 \times 10^{-5}) = (3 \times 2.5) \times 10^{8-5}$
$$= 7.5 \times 10^3$$
$$= 7500$$

23. $\dfrac{49,000}{0.007} = \dfrac{4.9 \times 10^4}{7 \times 10^{-3}}$
$$= \left(\dfrac{4.9}{7}\right) \times 10^{4-(-3)}$$
$$= 0.7 \times 10^7$$
$$= 7 \times 10^{-1} \times 10^7$$
$$= 7 \times 10^6$$

24. $\dfrac{3 \times 10^{10}}{7.5 \times 10^9} = \left(\dfrac{3}{7.5}\right) \times \left(\dfrac{10^{10}}{10^9}\right) = 0.4 \times 10 = 4$

25. $a_1 = 1, d = -5$
$1, 1 - 5 = -4, -4 - 5 = -9,$
$-9 - 5 = -14, -14 - 5 = -19,$
$-19 - 5 = -24$
$1, -4, -9, -14, -19, -24$

26. $a_1 = -2, d = 3$
$a_9 = -2 + (9 - 1)(3)$
$$= -2 + 8(3)$$
$$= -2 + 24$$
$$= 22$$

27. $a_1 = 16, r = \dfrac{1}{2}$
$16, \quad 16 \cdot \dfrac{1}{2} = 8, \quad 8 \cdot \dfrac{1}{2} = 4, \quad 4 \cdot \dfrac{1}{2} = 2,$
$2 \cdot \dfrac{1}{2} = 1, \quad 1 \cdot \dfrac{1}{2} = \dfrac{1}{2}$
$16, 8, 4, 2, 1, \dfrac{1}{2}$

28. $a_1 = 5, r = 2$
$a_7 = 5(2)^{7-1}$
$$= 5(2)^6$$
$$= 5(64)$$
$$= 320$$

Chapter 6
Algebra: Equations and Inequalities

Check Points 6.1

1. $6x + 9 = 6 \cdot 12 + 9 = 72 + 9 = 81$

2. $x^2 + 4x - 7 = (-5)^2 + 4(-5) - 7 = 25 - 20 - 7 = -2$

3. $-3x^2 + 4xy - y^2 = -3(5)^2 + 4(5)(6) - (6)^2$
$$= -3 \cdot 25 + 120 - 36$$
$$= -75 + 120 - 36$$
$$= 9$$

4. $81 - 0.6x = 81 - 0.6(50)$
$$= 81 - 30$$
$$= 51$$
About 51% of Americans smoked cigarettes in 1950.

5. Using $0.6(220 - a)$
$0.6(220 - 40) = 0.6(180) = 108$
Using $132 - 0.6a$
$132 - 0.6(40) = 132 - 24 = 108$

6. a. $4y + 15y = (4 + 15)y = 19y$

b. $7x - 20x = (7 - 20)x = -13x$

c. $y + 99y = (1 + 99)y = 100y$

d. $5x^2 - 15x^2 = (5 - 15)x^2 = -10x^2$

7. a. $8x + 7 + 10x + 3$
$$= 8x + 10x + 7 + 3$$
$$= 18x + 10$$

b. $9x + 6y + 15 - 3x - 20y - 16$
$$= 9x - 3x + 6y - 20y + 15 - 16$$
$$= 6x - 14y - 1$$

8. $7(2x - 3) - 11x = 7 \cdot 2x - 7 \cdot 3 - 11x$
$$= 14x - 21 - 11x$$
$$= 3x - 21$$

9. $7x - (15x + 2y - 1) - 7x - 15x - 2y + 1$
$$= -8x - 2y + 1$$

10. $P = 0.72x^2 + 9.4x + 783$
$P = 0.72(40)^2 + 9.4(40) + 783$
$P = 0.72(1600) + 9.4(40) + 783$
$P = 1152 + 376 + 783$
$P = 2311$
The formula models the data quite well.

Exercise Set 6.1

1. $5x + 7 = 5 \cdot 4 + 7 = 20 + 7 = 27$

3. $-7x - 5 = -7(-4) - 5 = 28 - 5 = 23$

5. $x^2 + 4 = 5^2 + 4 = 25 + 4 = 29$

7. $x^2 - 6 = (-2)^2 - 6 = 4 - 6 = -2$

9. $x^2 + 4x = (10)^2 + 4 \cdot 10 = 100 + 40 = 140$

11. $8x^2 + 17 = 8(5)^2 + 17$
$$= 8(25) + 17$$
$$= 200 + 17$$
$$= 217$$

13. $x^2 - 5x = (-11)^2 - 5(-11)$
$$= 121 + 55$$
$$= 176$$

15. $x^2 + 5x - 6 = 4^2 + 5 \cdot 4 - 6$
$$= 16 + 20 - 6$$
$$= 30$$

17. $2x^2 - 5x - 6 = 2(-3)^2 - 5(-3) - 6$
$$= 2(9) - 5(-3) - 6$$
$$= 18 + 15 - 6$$
$$= 27$$

19. $-5x^2 - 4x - 11 = -5(-1)^2 - 4(-1) - 11$
$$= -5(1) - 4(-1) - 11$$
$$= -5 + 4 - 11$$
$$= -12$$

21. $4xy = 4(-3)(-1) = 12$

23. $x^2 - 4xy + y^2 = (-1)^2 - 4(-1)(-3) + (-3)^2$
$$= 1 - 4(-1)(-3) + 9$$
$$= 1 - 12 + 9$$
$$= -2$$

25. $3x^2 + 2xy + 5y^2 = 3(2)^2 + 2(2)(3) + 5(3)^2$
$$= 3(4) + 2(2)(3) + 5(9)$$
$$= 12 + 12 + 45$$
$$= 69$$

27. $-x^2 - 4xy + 3y^2$
$= -(-1)^2 - 4(-1)(-2) + 3(-2)^2$
$= -(1) - 4(-1)(-2) + 3(4)$
$= -1 - 8 + 12$
$= 3$

29. $x^3 - 5x^2 + 2 = (-4)^3 - 5(-4)^2 + 2$
$= -64 - 5(16) + 2$
$= -64 - 80 + 2$
$= -142$

31. $3(x + 5) = 3 \cdot x + 3 \cdot 5 = 3x + 15$

33. $8(2x + 3) = 8 \cdot 2x + 8 \cdot 3 = 16x + 24$

35. $\frac{1}{3}(12 + 6r) = \frac{1}{3} \cdot 12 + \frac{1}{3} \cdot 6r = 4 + 2r$

37. $5(x + y) = 5 \cdot x + 5 \cdot y = 5x + 5y$

39. $3(x - 2) = 3 \cdot x - 3 \cdot 2 = 3x - 6$

41. $2(4x - 5) = 2 \cdot 4x - 2 \cdot 5 = 8x - 10$

43. $-4(2x - 3) = -4(2x) - 4(-3) = -8x + 12$

45. $-3(-2x + 4) = -3(-2x) - 3(4) = 6x - 12$

47. $7x + 10x = (7 + 10)x = 17x$

49. $11x - 3x = (11 - 3)x = 8x$

51. $x + 9x = 1x + 9x = (1 + 9)x = 10x$

53. $5y - 7y = (5 - 7)y = -2y$

55. $-5y + y = -5y + 1y = (-5 + 1)y = -4y$

57. $2x^2 + 5x^2 = (2 + 5)x^2 = 7x^2$

59. $5x^2 - 8x^2 = (5 - 8)x^2 = -3x^2$

61. $5x - 3 + 6x = 5x + 6x - 3 = 11x - 3$

63. $11y + 12 - 3y - 2 = 11y - 3y + 12 - 2$
$= 8y + 10$

65. $3x + 10y + 12 + 7x - 2y - 14$
$= 3x + 7x + 10y - 2y + 12 - 14$
$= 10x + 8y - 2$

67. $5(3x + 4) - 4 = 5(3x) + 5(4) - 4$
$= 15x + 20 - 4$
$= 15x + 16$

69. $5(3x - 2) + 12x = 5(3x) + 5(-2) + 12x$
$= 15x - 10 + 12x$
$= 27x - 10$

71. $7(3y - 5) + 2(4y + 3)$
$= 7(3y) + 7(-5) + 2(4y) + 2(3)$
$= 21y - 35 + 8y + 6$
$= 29y - 29$

73. $4x - (7x - 12y - 2)$
$= 4x - 7x + 12y + 2$
$= -3x + 12y + 2$

75. $5(3y - 2) - (7y + 2)$
$= 15y - 10 - 7y - 2$
$= 8y - 12$

77. $15x = 15(20) = 300$
One can tan for 300 minutes or 5 hours without burning with a number 15 lotion.

79. $W = 1.5x + 7$
$= 1.5(4) + 7$
$= 6 + 7$
$= 13$ pounds
The point on the graph is (4, 13).

81. $E = 0.22t + 71$
$= 0.22(50) + 71$
$= 11 + 71$
$= 82$ years
The formula models the data quite well..

83. $N = 0.2x^2 - 1.2x + 2$
$= 0.2(10)^2 - 1.2(10) + 2$
$= 20 - 12 + 2$
$= 10$ virus infections per 1000
The formula models the data quite well..

85-91. Answers will vary.

93. d is true.

95. $\dfrac{0.5x + 5000}{x}$

   a. $x = 100$

$$\frac{0.5(100) + 5000}{100} = \$50.50$$

$x = 1000$

$$\frac{0.5(1000) + 5000}{1000} = \$5.50$$

$x = 10{,}000$

$$\frac{0.5(10{,}000) + 5000}{10{,}000} = \$1$$

   b. No; the business must produce at least 10,000 clocks each week to be competitive.

97. $N = -1.65x^2 + 51.8x + 111.44$

$= -1.65(9)^2 + 51.8(9) + 111.44$

$= -133.65 + 466.2 + 111.44$

$= 443.99$ thousand deaths

$= 443{,}990$ deaths

The formula models the data quite well.

Check Points 6.2

1. $x - 5 = 12$

$x - 5 + 5 = 12 + 5$

$x = 17$

The solution set is $\{17\}$.

2. $x + 11 = -4$

$x + 11 - 11 = -4 - 11$

$x = -15$

The solution set is $\{-15\}$.

3. $\dfrac{x}{3} = 12$

$3 \cdot \dfrac{x}{3} = 3 \cdot 12$

$x = 36$

The solution set is $\{36\}$.

4. $4x = 84$

$\dfrac{4x}{4} = \dfrac{84}{4}$

$x = 21$

The solution set is $\{21\}$.

5. $4x + 5 = 29$

$4x + 5 - 5 = 29 - 5$

$4x = 24$

$\dfrac{4x}{4} = \dfrac{24}{4}$

$x = 6$

Check:

$4x + 5 = 29$

$4(6) + 5 = 29$

$24 + 5 = 29$

$29 = 29$

The solution set is $\{6\}$.

6. $6(x - 3) + 7x = -57$

$6x - 18 + 7x = -57$

$13x - 18 = -57$

$13x - 18 + 18 = -57 + 18$

$13x = -39$

$\dfrac{13x}{13} = \dfrac{-39}{13}$

$x = -3$

The solution set is $\{-3\}$.

7. $8x - 3 = 2x + 9$

$8x - 3 - 2x = 2x + 9 - 2x$

$6x - 3 = 9$

$6x - 3 + 3 = 9 + 3$

$6x = 12$

$\dfrac{6x}{6} = \dfrac{12}{6}$

$x = 2$

The solution set is $\{2\}$.

8. $4(2x + 1) - 29 = 3(2x - 5)$

$8x + 4 - 29 = 6x - 15$

$8x - 25 = 6x - 15$

$8x - 25 - 6x = 6x - 15 - 6x$

$2x - 25 = -15$

$2x - 25 + 25 = -15 + 25$

$2x = 10$

$\dfrac{2x}{2} = \dfrac{10}{2}$

$x = 5$

The solution set is $\{5\}$.

9.
$$\frac{x}{4} = \frac{2x}{3} + \frac{5}{6}$$
$$12 \cdot \frac{x}{4} = 12 \cdot \left(\frac{2x}{3} + \frac{5}{6}\right)$$
$$12 \cdot \frac{x}{4} = 12 \cdot \frac{2x}{3} + 12 \cdot \frac{5}{6}$$
$$3 \cdot x = 4 \cdot 2x + 2 \cdot 5$$
$$3x = 8x + 10$$
$$3x - 8x = 8x + 10 - 8x$$
$$-5x = 10$$
$$\frac{-5x}{-5} = \frac{10}{-5}$$
$$x = -2$$
The solution set is $\{-2\}$.

10.
$$W = 0.3x + 46.6$$
$$55.9 = 0.3x + 46.6$$
$$55.9 - 46.6 = 0.3x + 46.6 - 46.6$$
$$9.3 = 0.3x$$
$$\frac{9.3}{0.3} = \frac{0.3x}{0.3}$$
$$31 = x$$
The formula indicates that we will average 55.9 hours of work per week 31 years after 1980, or in 2011.

11.
$$4x + 9y = 27$$
$$4x - 4x + 9y = -4x + 27$$
$$9y = -4x + 27$$
$$\frac{9y}{9} = \frac{-4x + 27}{9}$$
$$y = \frac{-4x + 27}{9}$$
$$y = \frac{-4x}{9} + \frac{27}{9}$$
$$y = -\frac{4}{9}x + 3$$

12.
$$T = D + pm$$
$$T - D = D - D + pm$$
$$T - D = pm$$
$$\frac{T - D}{p} = \frac{pm}{p}$$
$$\frac{T - D}{p} = m$$

13.
$$3x + 7 = 3(x + 1)$$
$$3x + 7 = 3x + 3$$
$$3x + 7 - 3x = 3x + 3 - 3x$$
$$7 = 3$$
There is no solution, $\varnothing$.

14.
$$3(x - 1) + 9 = 8x + 6 - 5x$$
$$3x - 3 + 9 = 8x + 6 - 5x$$
$$3x + 6 = 3x + 6$$
$$3x + 6 - 3x = 3x + 6 - 3x$$
$$6 = 6$$
True for all real numbers.

Exercise Set 6.2

1.
$$x - 7 = 3$$
$$x - 7 + 7 = 3 + 7$$
$$x = 10$$
The solution set is $\{10\}$.

3.
$$x + 5 = -12$$
$$x + 5 - 5 = -12 - 5$$
$$x = -17$$
The solution set is $\{-17\}$.

5.
$$\frac{x}{3} = 4$$
$$3\left(\frac{x}{3}\right) = 3(4)$$
$$x = 12$$
The solution set is $\{12\}$.

7.
$$5x = 45$$
$$\frac{5x}{5} = \frac{45}{5}$$
$$x = 9$$
The solution set is $\{9\}$.

9.
$$8x = -24$$
$$\frac{8x}{8} = \frac{-24}{8}$$
$$x = -3$$
The solution set is $\{-3\}$.

11.
$$-8x = 2$$
$$\frac{-8x}{-8} = \frac{2}{-8}$$
$$x = -\frac{1}{4}$$
The solution set is $\left\{-\frac{1}{4}\right\}$.

PROPERTY OF TUTORING LAB INTERNATIONAL COLLEGE

13.
$$5x + 3 = 18$$
$$5x + 3 - 3 = 18 - 3$$
$$5x = 15$$
$$\frac{5x}{5} = \frac{15}{5}$$
$$x = 3$$
The solution set is $\{3\}$.

15.
$$6x - 3 = 63$$
$$6x - 3 + 3 = 63 + 3$$
$$6x = 66$$
$$\frac{6x}{6} = \frac{66}{6}$$
$$x = 11$$
The solution set is $\{11\}$.

17.
$$4x - 14 = -82$$
$$4x - 14 + 14 = -82 + 14$$
$$4x = -68$$
$$\frac{4x}{4} = \frac{-68}{4}$$
$$x = -17$$
The solution set is $\{-17\}$.

19.
$$14 - 5x = -41$$
$$14 - 5x - 14 = -41 - 14$$
$$-5x = -55$$
$$\frac{-5x}{-5} = \frac{-55}{-5}$$
$$x = 11$$
The solution set is $\{11\}$.

21.
$$9(5x - 2) = 45$$
$$45x - 18 = 45$$
$$45x - 18 + 18 = 45 + 18$$
$$45x = 63$$
$$\frac{45x}{45} = \frac{63}{45}$$
$$x = \frac{7}{5}$$
The solution set is $\left\{\frac{7}{5}\right\}$.

23.
$$5x - (2x - 10) = 35$$
$$5x - 2x + 10 = 35$$
$$3x + 10 = 35$$
$$3x + 10 - 10 = 35 - 10$$
$$3x = 25$$
$$\frac{3x}{3} = \frac{25}{3}$$
$$x = \frac{25}{3}$$
The solution set is $\left\{\frac{25}{3}\right\}$.

25.
$$3x + 5 = 2x + 13$$
$$3x + 5 - 5 = 2x + 13 - 5$$
$$3x = 2x + 8$$
$$3x - 2x = 2x + 8 - 2x$$
$$x = 8$$
The solution set is $\{8\}$.

27.
$$8x - 2 = 7x - 5$$
$$8x - 2 + 2 = 7x - 5 + 2$$
$$8x = 7x - 3$$
$$8x - 7x = 7x - 3 - 7x$$
$$x = -3$$
The solution set is $\{-3\}$.

29.
$$7x + 4 = x + 16$$
$$7x + 4 - 4 = x + 16 - 4$$
$$7x = x + 12$$
$$7x - x = x + 12 - x$$
$$6x = 12$$
$$\frac{6x}{6} = \frac{12}{6}$$
$$x = 2$$
The solution set is $\{2\}$.

31.
$$8y - 3 = 11y + 9$$
$$8y - 3 + 3 = 11y + 9 + 3$$
$$8y = 11y + 12$$
$$8y - 11y = 11y + 12 - 11y$$
$$-3y = 12$$
$$\frac{-3y}{-3} = \frac{12}{-3}$$
$$y = -4$$
The solution set is $\{-4\}$.

33.
$$2(4 - 3x) = 2(2x + 5)$$
$$8 - 6x = 4x + 10$$
$$8 - 6x - 8 = 4x + 10 - 8$$
$$-6x = 4x + 2$$
$$-6x - 4x = 4x + 2 - 4x$$
$$-10x = 2$$
$$\frac{-10x}{-10} = \frac{2}{-10}$$
$$x = -\frac{1}{5}$$
The solution set is $\left\{-\frac{1}{5}\right\}$.

35.
$$8(y+2) = 2(3y+4)$$
$$8y+16 = 6y+8$$
$$8y+16-16 = 6y+8-16$$
$$8y = 6y-8$$
$$8y-6y = 6y-8-6y$$
$$2y = -8$$
$$\frac{2y}{2} = \frac{-8}{2}$$
$$y = -4$$
The solution set is $\{-4\}$.

37.
$$3(x+1) = 7(x-2)-3$$
$$3x+3 = 7x-14-3$$
$$3x+3 = 7x-17$$
$$3x+3-3 = 7x-17-3$$
$$3x = 7x-20$$
$$3x-7x = 7x-20-7x$$
$$-4x = -20$$
$$\frac{-4x}{-4} = \frac{-20}{-4}$$
$$x = 5$$
The solution set is $\{5\}$.

39.
$$5(2x-8)-2 = 5(x-3)+3$$
$$10x-40-2 = 5x-15+3$$
$$10x-42 = 5x-12$$
$$10x-42+42 = 5x-12+42$$
$$10x = 5x+30$$
$$10x-5x = 5x+30-5x$$
$$5x = 30$$
$$\frac{5x}{5} = \frac{30}{5}$$
$$x = 6$$
The solution set is $\{6\}$.

41.
$$5(x-2)-2(2x+1) = 2+5x$$
$$5x-10-4x-2 = 2+5x$$
$$x-12 = 2+5x$$
$$x-12+12 = 2+5x+12$$
$$x = 14+5x$$
$$x-5x = 14+5x-5x$$
$$-4x = 14$$
$$\frac{-4x}{-4} = \frac{14}{-4}$$
$$x = -\frac{7}{2}$$
The solution set is $\left\{-\frac{7}{2}\right\}$.

43
$$\frac{x}{3}+\frac{x}{2} = \frac{5}{6}$$
$$6\left(\frac{x}{3}+\frac{x}{2}\right) = 6\left(\frac{5}{6}\right)$$
$$6\left(\frac{x}{3}\right)+6\left(\frac{x}{2}\right) = 6\left(\frac{5}{6}\right)$$
$$2x+3x = 5$$
$$5x = 5$$
$$\frac{5x}{5} = \frac{5}{5}$$
$$x = 1$$
The solution set is $\{1\}$.

45.
$$\frac{x}{2} = 20-\frac{x}{3}$$
$$6\left(\frac{x}{2}\right) = 6\left(20-\frac{x}{3}\right)$$
$$6\left(\frac{x}{2}\right) = 6(20)-6\left(\frac{x}{3}\right)$$
$$3x = 120-2x$$
$$3x+2x = 120-2x+2x$$
$$5x = 120$$
$$\frac{5x}{5} = \frac{120}{5}$$
$$x = 24$$
The solution set is $\{24\}$.

47.
$$\frac{3y}{4}-3 = \frac{y}{2}+2$$
$$4\left(\frac{3y}{4}-3\right) = 4\left(\frac{y}{2}+2\right)$$
$$4\cdot\frac{3y}{4}-4\cdot3 = 4\cdot\frac{y}{2}+4\cdot2$$
$$3y-12 = 2y+8$$
$$3y-12+12 = 2y+8+12$$
$$3y = 2y+20$$
$$3y-2y = 2y+20-2y$$
$$y = 20$$
The solution set is $\{20\}$.

49.

$$\frac{3x}{5} - x = \frac{x}{10} - \frac{5}{2}$$

$$10\left(\frac{3x}{5} - x\right) = 10\left(\frac{x}{10} - \frac{5}{2}\right)$$

$$10\left(\frac{3x}{5}\right) - 10(x) = 10\left(\frac{x}{10}\right) - 10\left(\frac{5}{2}\right)$$

$$6x - 10x = x - 25$$

$$-4x = x - 25$$

$$-4x - x = x - 25 - x$$

$$-5x = -25$$

$$\frac{-5x}{-5} = \frac{-25}{-5}$$

$$x = 5$$

The solution set is $\{5\}$.

51.

$$3x + y = 6$$

$$3x + y - 3x = -3x + 6$$

$$y = -3x + 6$$

53.

$$x + 2y = 6$$

$$x + 2y - x = -x + 6$$

$$2y = -x + 6$$

$$\frac{2y}{2} = \frac{-x + 6}{2}$$

$$y = -\frac{x}{2} + \frac{6}{2}$$

$$y = -\frac{1}{2}x + 3$$

55.

$$2x + 6y = 12$$

$$2x + 6y - 2x = -2x + 12$$

$$6y = -2x + 12$$

$$\frac{6y}{6} = \frac{-2x + 12}{6}$$

$$y = \frac{-2x}{6} + \frac{12}{6}$$

$$y = -\frac{1}{3}x + 2$$

57.

$$-2x + 4y = 0$$

$$-2x + 4y + 2x = 0 + 2x$$

$$4y = 2x$$

$$\frac{4y}{4} = \frac{2x}{4}$$

$$y = \frac{1}{2}x$$

59.

$$2x - 3y = 5$$

$$2x - 3y - 2x = -2x + 5$$

$$-3y = -2x + 5$$

$$\frac{-3y}{-3} = \frac{-2x + 5}{-3}$$

$$y = \frac{-2x}{-3} + \frac{5}{-3}$$

$$y = \frac{2}{3}x - \frac{5}{3}$$

61. $A = LW$ for L

$$\frac{A}{W} = \frac{LW}{W}$$

$$\frac{A}{W} = L \text{ or } L = \frac{A}{W}$$

63. $A = \frac{1}{2}bh$ for b

$$2(A) = 2\left(\frac{1}{2}bh\right)$$

$$2A = bh$$

$$\frac{2A}{h} = \frac{bh}{h}$$

$$\frac{2A}{h} = b \text{ or } b = \frac{2A}{h}$$

65. $I = Prt$ for P

$$\frac{I}{rt} = \frac{Prt}{rt}$$

$$\frac{I}{rt} = P \text{ or } P = \frac{I}{rt}$$

67. $E = mc^2$ for m

$$\frac{E}{c^2} = \frac{mc^2}{c^2}$$

$$\frac{E}{c^2} = m \text{ or } m = \frac{E}{c^2}$$

69. $y = mx + b$ for m

$$y - h = mx + b - b$$

$$y - b = mx$$

$$\frac{y - b}{x} = \frac{mx}{x}$$

$$\frac{y - b}{x} = m \text{ or } m = \frac{y - b}{x}$$

71. $A = \frac{1}{2}(a+b)$ for a

$2 \cdot A = 2 \cdot \frac{1}{2}(a+b)$

$2A = a+b$

$2A - b = a + b - b$

$2A - b = a$ or $a = 2A - b$

73. $S = P + Prt$ for r

$S - P = P + Prt - P$

$S - P = Prt$

$\frac{S-P}{Pt} = \frac{Prt}{Pt}$

$\frac{S-P}{Pt} = r$ or $r = \frac{S-P}{Pt}$

75. $10x - 2(4+5x) = -8$

$10x - 8 - 10x = -8$

$-8 = -8$ True

True for all real numbers.

77. $10x - 2(4+5x) = 8$

$10x - 8 - 10x = 8$

$-8 = 8$ False

There is no solution, $\varnothing$.

79. $2(3x+4) - 4 = 9x + 4 - 3x$

$6x + 8 - 4 = 9x + 4 - 3x$

$6x + 4 = 6x + 4$

$6x + 4 - 6x = 6x + 4 - 6x$

$4 = 4$ True

True for all real numbers.

81. $F = 10(x-65) + 50$

$250 = 10(x-65) + 50$

$250 = 10x - 650 + 50$

$250 = 10x - 600$

$250 + 600 = 10x - 600 + 600$

$850 = 10x$

$\frac{850}{10} = \frac{10x}{10}$

$85 = x$

The speed was 85 miles per hour.

83. $\frac{c}{2} + 80 = 2F$

$\frac{c}{2} + 80 = 2 \cdot 70$

$\frac{c}{2} + 80 = 140$

$\frac{c}{2} + 80 - 80 = 140 - 80$

$\frac{c}{2} = 60$

$2 \cdot \frac{c}{2} = 2 \cdot 60$

$c = 120$ chirps per minute

85. $p = 15 + \frac{5d}{11}$

$201 = 15 + \frac{5d}{11}$

$201 - 15 = 15 + \frac{5d}{11} - 15$

$186 = \frac{5d}{11}$

$\frac{11}{5} \cdot 186 = \frac{11}{5} \cdot \frac{5d}{11}$

$409.2 = d$

$d = 409.2$ feet

87. $R = 143 - 0.65A$

$117 = 143 - 0.65A$

$117 - 143 = 143 - 0.65A - 143$

$-26 = -0.65A$

$\frac{-26}{-0.65} = \frac{-0.65A}{-0.65}$

$40 = A$

The woman is 40 years old.

The point on the graph is (40, 117).

89. $P = -0.22t + 9.6$

$0.5 = -0.22t + 9.6$

$0.5 - 9.6 = -0.22t + 9.6 - 9.6$

$-9.1 = -0.22t$

$\frac{-9.1}{-0.22} = \frac{-0.22t}{-0.22}$

$41.36 = t$

41 years after 1960 is 2001.

91-99. Answers will vary.

101. c is true

103. Answers will vary.

105.
$$2.24y - 9.28 = 5.74y + 5.42$$
$$2.24y - 9.28 + 9.28 = 5.74y + 5.42 + 9.28$$
$$2.24y = 5.74y + 14.7$$
$$2.24y - 5.74y = 5.74y + 14.7 - 5.74y$$
$$-3.5y = 14.7$$
$$\frac{-3.5y}{-3.5} = \frac{14.7}{-3.5}$$
$$y = -4.2$$

Check Points 6.3

1. Let x = the number.
$$6x - 4 = 68$$
$$6x - 4 + 4 = 68 + 4$$
$$6x = 72$$
$$\frac{6x}{6} = \frac{72}{6}$$
$$x = 12$$

2. Let x = miles traveled.
$$125 + 0.20x = 335$$
$$125 + 0.20x - 125 = 335 - 125$$
$$0.20x = 210$$
$$\frac{0.20x}{0.20} = \frac{210}{0.20}$$
$$x = 1050 \text{ miles}$$
You can travel 1050 miles in one week for $335.

3. Let x = the number of *Saturday Night Fever* albums sold (in millions).
Let $x + 5$ = the number of *Jagged Little Pill* albums sold (in millions).
$$x + x + 5 = 27$$
$$2x + 5 = 27$$
$$2x + 5 - 5 = 27 - 5$$
$$2x = 22$$
$$\frac{2x}{2} = \frac{22}{2}$$
$$x = 11$$
$$x + 5 = 16$$
11 million *Saturday Night Fever* albums and 16 million *Jagged Little Pill* albums have been sold.

4. Let x = the number years until total costs for solar heating and electric heating will be the same.
$$\overbrace{29,700 + 150x}^{\text{solar heating}} = \overbrace{5000 + 1100x}^{\text{electric heating}}$$
$$29,700 + 150x - 29,700 = 5000 + 1100x - 29,700$$
$$150x = 1100x - 24,700$$
$$150x - 1100x = 1100x - 24,700 - 1100x$$
$$-950x = -24,700$$
$$\frac{-950x}{-950} = \frac{-24,700}{-950}$$
$$x = 26 \text{ years}$$
Cost for solar heating:
$$= \$29,700 + \$150x$$
$$= \$29,700 + \$150(26)$$
$$= \$29,700 + \$3900$$
$$= \$33,600$$
Cost for electric heating:
$$= \$5000 + \$1100x$$
$$= \$5000 + \$1100(26)$$
$$= \$5000 + \$28,600$$
$$= \$33,600$$

Exercise Set 6.3

1. $x + 9$

3. $20 - x$

5. $8 - 5x$

7. $\dfrac{15}{x}$

9. $2x + 20$

11. $7x - 30$

13. $4(x + 12)$

15.
$$x + 40 = 450$$
$$x + 40 - 40 = 450 - 40$$
$$x = 410$$
The number is 410.

17.
$$x - 13 = 123$$
$$x - 13 + 13 = 123 + 13$$
$$x = 136$$
The number is 136.

19.
$$7x = 91$$
$$\frac{7x}{7} = \frac{91}{7}$$
$$x = 13$$
The number is 13.

21.
$$\frac{x}{18} = 6$$
$$18 \cdot \frac{x}{18} = 18 \cdot 6$$
$$x = 108$$
The number is 108.

23.
$$4 + 2x = 36$$
$$4 + 2x - 4 = 36 - 4$$
$$2x = 32$$
$$\frac{2x}{2} = \frac{32}{2}$$
$$x = 16$$
The number is 16.

25.
$$5x - 7 = 123$$
$$5x - 7 + 7 = 123 + 7$$
$$5x = 130$$
$$\frac{5x}{5} = \frac{130}{5}$$
$$x = 26$$
The number is 26.

27.
$$x + 5 = 2x$$
$$x + 5 - x = 2x - x$$
$$5 = x$$
The number is 5.

29.
$$2(4 + x) = 36$$
$$8 + 2x = 36$$
$$8 + 2x - 8 = 36 - 8$$
$$2x = 28$$
$$\frac{2x}{2} = \frac{28}{2}$$
$$x = 14$$
The number is 14.

31.
$$9x = 3x + 30$$
$$9x - 3x = 3x + 30 - 3x$$
$$6x = 30$$
$$\frac{6x}{6} = \frac{30}{6}$$
$$x = 5$$
The number is 5.

33.
$$\frac{3x}{5} + 4 = 34$$
$$\frac{3x}{5} + 4 - 4 = 34 - 4$$
$$\frac{3x}{5} = 30$$
$$\frac{5}{3} \cdot \frac{3x}{5} = \frac{5}{3} \cdot 30$$
$$x = 50$$
The number is 50.

35. Let x = number of miles.
$$200 + 0.15x = 320$$
$$200 + 0.15x - 200 = 320 - 200$$
$$0.15x = 120$$
$$\frac{0.15x}{0.15} = \frac{120}{0.15}$$
$$x = 800$$
You can travel 800 miles in one week for \$320.

37. Let x = number of years since 1990.
$$28 + 0.6x = 37$$
$$28 + 0.6x - 28 = 37 - 28$$
$$0.6x = 9$$
$$\frac{0.6x}{0.6} = \frac{9}{0.6}$$
$$x = 15$$
$$1990 + 15 = 2005$$
37% of babies will be born out of wedlock in 2005.

39. Let x = number of years since 1980.
$$41.78 - 0.19x = 37.22$$
$$41.78 - 0.19x - 41.78 = 37.22 - 41.78$$
$$-0.19x = -4.56$$
$$\frac{-0.19x}{-0.19} = \frac{-4.56}{-0.19}$$
$$x = 24$$
$$1980 + 24 = 2004$$
If this trend continues, the winning time will be 37.22 seconds in 2004.

41. Let x = cost (in millions) to make *Waterworld*, then
$x + 40$ = cost (in millions) to make *Titanic*.

$$x + x + 40 = 360$$
$$2x + 40 = 360$$
$$2x + 40 - 40 = 360 - 40$$
$$2x = 320$$
$$\frac{2x}{2} = \frac{320}{2}$$
$$x = 160$$
$$x + 40 = 200$$

It cost \$160 million to make *Waterworld* and \$200 million to make *Titanic*.

43. Let x = cost (in billions) of Hurricane Hugo, then
$x + 5.5$ = cost (in billions) of the Northridge earthquake and
$x + 13$ = cost (in billions) of Hurricane Andrew.

$$x + x + 5.5 + x + 13 = 39.5$$
$$3x + 18.5 = 39.5$$
$$3x + 18.5 - 18.5 = 39.5 - 18.5$$
$$3x = 21$$
$$\frac{3x}{3} = \frac{21}{3}$$
$$x = 7$$
$$x + 5.5 = 12.5$$
$$x + 13 = 20$$

Hurricane Hugo, the Northridge earthquake, and Hurricane Andrew cost \$7, \$12.5, and \$20 billion respectively.

45. Let x = the losing score, then
$x + 1$ = the winning score.

$$x + x + 1 = 39$$
$$2x + 1 = 39$$
$$2x + 1 - 1 = 39 - 1$$
$$2x = 38$$
$$\frac{2x}{2} = \frac{38}{2}$$
$$x = 19$$
$$x + 1 = 20$$

The score of the 1991 Super Bowl was 20 to 19.

47. Let x = number of hours of labor

$$63 + 35x = 448$$
$$63 + 35x - 63 = 448 - 63$$
$$35x = 385$$
$$\frac{35x}{35} = \frac{385}{35}$$
$$x = 11$$

The shop took 11 hours to repair the car.

49. Let x = the number of minutes at which the costs of the two plans are the same.

$$\overbrace{15 + 0.08x}^{\text{Plan A}} = \overbrace{3 + 0.12x}^{\text{Plan B}}$$
$$15 + 0.08x - 15 = 3 + 0.12x - 15$$
$$0.08x = 0.12x - 12$$
$$0.08x - 0.12x = 0.12x - 12 - 0.12x$$
$$-0.04x = -12$$
$$\frac{-0.04x}{-0.04} = \frac{-12}{-0.04}$$
$$x = 300$$

The two plans are the same at 300 minutes.

51. Let x = the number of bus trips in the month.
Cost with coupon book: $21 + 0.50x$
Cost without coupon book: $1.25x$

$$21 + 0.50x = 1.25x$$
$$21 + 0.50x - 21 = 1.25x - 21$$
$$0.50x = 1.25x - 21$$
$$0.50x - 1.25x = 1.25x - 21 - 1.25x$$
$$-0.75x = -21$$
$$\frac{-0.75x}{-0.75} = \frac{-21}{-0.75}$$
$$x = 28$$

The monthly costs are the same after 28 bus trips.

53. Let x = the height of the bookcase, then
$3x$ = the length of the bookcase.
Note that the bookcase requires 3 vertical, and 4 horizontal boards.

$$3(x) + 4(3x) = 60$$
$$3x + 12x = 60$$
$$15x = 60$$
$$\frac{15x}{15} = \frac{60}{15}$$
$$x = 4$$
$$3x = 12$$

The length is 12 feet and the height is 4 feet.

55-57. Answers will vary.

59. Let x = number of minutes after first minute.

$$0.55 + 0.40x = 6.95$$
$$0.55 + 0.40x - 0.55 = 6.95 - 0.55$$
$$0.40x = 6.40$$
$$\frac{0.40x}{0.40} = \frac{6.40}{0.40}$$
$$x = 16$$

The call was 16 minutes plus the first minute, or 17 minutes long.

61. Let x = current age of woman, then
$3x$ = current age of "uncle."

$$2(x + 20) = 3x + 20$$
$$2x + 40 = 3x + 20$$
$$2x + 40 - 2x = 3x + 20 - 2x$$
$$40 = x + 20$$
$$40 - 20 = x + 20 - 20$$
$$20 = x$$
$$60 = 3x$$

The woman is 20 years and the "uncle" is 60 years.

Check Points 6.4

1. $\dfrac{160}{180} = \dfrac{160 \div 20}{180 \div 20} = \dfrac{8}{9}$

2. a. $\dfrac{10}{x} = \dfrac{2}{3}$

$$10 \cdot 3 = 2x$$
$$30 = 2x$$
$$\frac{30}{2} = \frac{2x}{2}$$
$$15 = x$$

The solution set is $\{15\}$.

b. $\dfrac{11}{910 - x} = \dfrac{2}{x}$

$$11x = 2(910 - x)$$
$$11x = 1820 - 2x$$
$$11x + 2x = 1820 - 2x + 2x$$
$$13x = 1820$$
$$\frac{13x}{13} = \frac{1820}{13}$$
$$x = 140$$

The solution set is $\{140\}$.

3. Let x = tax on a $112,500 house.

$$\frac{\$600}{\$45,000} = \frac{\$x}{\$112,500}$$
$$\frac{600}{45,000} = \frac{x}{112,500}$$
$$(600)(112,500) = 45,000x$$
$$67,500,000 = 45,000x$$
$$\frac{67,500,000}{45,000} = \frac{45,000x}{45,000}$$
$$1500 = x$$

The tax on the $112,500 house is $1500.

4. Let x = the number of deer in the refuge.

$$\frac{120 \text{ tagged deer}}{x} = \frac{25 \text{ tagged deer in sample}}{150 \text{ deer in sample}}$$
$$\frac{120}{x} = \frac{25}{150}$$
$$25x = (120)(150)$$
$$25x = 18,000$$
$$\frac{25x}{25} = \frac{18,000}{25}$$
$$x = 720$$

There are approximately 720 deer in the refuge.

5. Let x = the pressure (in pounds per square inch) at 330 feet below the surface.

$$\frac{25 \text{ pounds per square inch}}{60 \text{ feet}} = \frac{x}{330 \text{ feet}}$$
$$\frac{25}{60} = \frac{x}{330}$$
$$(25)(330) = 60x$$
$$8250 = 60x$$
$$\frac{8250}{60} = \frac{60x}{60}$$
$$137.5 = x$$

The pressure at 330 feet is 137.5 pounds per square inch.

6. Let x = the distance (in feet) required to stop a car traveling at 100 mph.

$$\frac{200 \text{ feet}}{60^2 \text{ miles per hour}} = \frac{x}{100^2 \text{ miles per hour}}$$

$$\frac{200}{60^2} = \frac{x}{100^2}$$

$$\frac{200}{3600} = \frac{x}{10,000}$$

$$(200)(10,000) = 3600x$$

$$2,000,000 = 3600x$$

$$\frac{2,000,000}{3600} = \frac{3600x}{3600}$$

$$556 \approx x$$

Approximately 556 feet are needed to stop at 100 mph.

7. Let x = pounds per square inch when the volume is 22 cubic inches.

$$\frac{8 \text{ cubic inches}}{\underbrace{x}_{\substack{\text{corresponds to} \\ \text{22 cubic inches}}}} = \frac{\overbrace{22 \text{ cubic inches}}{12 \text{ pounds per square inch}}}{\underbrace{}_{\substack{\text{corresponds to} \\ \text{8 cubic inches}}}}$$

$$\frac{8}{x} = \frac{22}{12}$$

$$(8)(12) = 22x$$

$$96 = 22x$$

$$\frac{96}{22} = \frac{22x}{22}$$

$$4.36 \approx x$$

The pressure is about 4.36 pounds per square inch.

Exercise Set 6.4

1. $\dfrac{24}{48} = \dfrac{24 \div 24}{48 \div 24} = \dfrac{1}{2}$

3. $\dfrac{48}{20} = \dfrac{48 \div 4}{20 \div 4} = \dfrac{12}{5}$

5. $\dfrac{27}{36} = \dfrac{27 \div 9}{36 \div 9} = \dfrac{3}{4}$

7. $\dfrac{20}{10} = \dfrac{20 \div 10}{10 \div 10} = \dfrac{2}{1}$ or 2:1

9. $\dfrac{10}{10+20} = \dfrac{10}{30} = \dfrac{10 \div 10}{30 \div 10} = \dfrac{1}{3}$ or 1:3

11. $\dfrac{24}{x} = \dfrac{12}{7}$

$$12x = 24 \cdot 7$$

$$12x = 168$$

$$\frac{12x}{12} = \frac{168}{12}$$

$$x = 14$$

13. $\dfrac{x}{6} = \dfrac{18}{4}$

$$4x = 6 \cdot 18$$

$$4x = 108$$

$$\frac{4x}{4} = \frac{108}{4}$$

$$x = 27$$

15. $\dfrac{x}{3} = -\dfrac{3}{4}$

$$4x = 3(-3)$$

$$4x = -9$$

$$\frac{4x}{4} = \frac{-9}{4}$$

$$x = -\frac{9}{4}$$

17. $\dfrac{-3}{8} = \dfrac{x}{40}$

$$8x = -3(40)$$

$$8x = -120$$

$$\frac{8x}{8} = \frac{-120}{8}$$

$$x = -15$$

19. $\dfrac{x-2}{5} = \dfrac{3}{10}$

$$10(x-2) = 3 \cdot 5$$

$$10x - 20 = 15$$

$$10x - 20 + 20 = 15 + 20$$

$$10x = 35$$

$$\frac{10x}{10} = \frac{35}{10}$$

$$x = \frac{7}{2}$$

21.
$$\frac{y+10}{10}=\frac{y-2}{4}$$
$$4(y+10)=10(y-2)$$
$$4y+40=10y-20$$
$$4y+40-40=10y-20-40$$
$$4y=10y-60$$
$$4y-10y=10y-60-10y$$
$$-6y=-60$$
$$\frac{-6y}{-6}=\frac{-60}{-6}$$
$$y=10$$

23. Possible Answer:
$$\frac{80}{170}=\frac{80\div10}{170\div10}=\frac{8}{17}\quad\text{or}\quad 8{:}17$$

25. Let x = tax.
$$\frac{725}{65,000}=\frac{x}{100,000}$$
$$65,000x=725\cdot100,000$$
$$65,000x=72,500,000$$
$$\frac{65,000x}{65,000}=\frac{72,500,000}{65,000}$$
$$x\approx1115.38$$
The tax on a property with an assessed value of $100,000 is $1115.38.

27. Let x = total number of fur seal pups in this rookery.
$$\frac{218}{900}=\frac{4963}{x}$$
$$218x=900\cdot4963$$
$$218x=4,466,700$$
$$\frac{218x}{218}=\frac{4,466,700}{218}$$
$$x\approx20,489$$
There were an estimated 20,489 fur seal pups in this rookery.

29. Let x = amount to pay in monthly child support.
$$\frac{1}{40}=\frac{x}{38,000}$$
$$1\cdot38,000=40x$$
$$\frac{38,000}{40}=\frac{40x}{40}$$
$$950=x$$
The father should pay $950 each month.

31. Let x = height of the critter.
$$\frac{67}{10}=\frac{x}{23}$$
$$10x=67\cdot23$$
$$10x=1541$$
$$\frac{10x}{10}=\frac{1541}{10}$$
$$x=154.1$$
The critter was 154.1 inches tall or about 12 ft 10 in.

33. Let x = speed of Blackbird.
$$\frac{1502.2}{2.03}=\frac{x}{3.3}$$
$$2.03x=1502.2(3.3)$$
$$2.03x=4957.26$$
$$\frac{2.03x}{2.03}=\frac{4957.26}{2.03}$$
$$x=2442$$
The Blackbird's speed is 2442 mph.

35. Let x = Mr. Wadlow's weight.
$$\frac{x}{(107)^3}=\frac{170}{(70)^3}$$
$$\frac{x}{1,225,043}=\frac{170}{343,000}$$
$$343,000x=1,225,043\cdot170$$
$$343,000x=208,257,310$$
$$\frac{343,000x}{343,000}=\frac{208,257,310}{343,000}$$
$$x\approx607$$
Robert Wadlow's weight was about 607 pounds.

37. Let x = hours it takes to get to campus averaging 60 mph.
$$\frac{20}{x}=\frac{60}{1.5}$$
$$20\cdot1.5=60x$$
$$30=60x$$
$$\frac{30}{60}=\frac{60x}{60}$$
$$0.5=x$$
At a rate of 60 mph, it takes 0.5 hours, or 30 minutes, to drive to campus.

39. Let x = pressure when the volume is 40 cubic centimeters.

$$\frac{32}{x} = \frac{40}{8}$$

$$40x = 32(8)$$

$$40x = 256$$

$$\frac{40x}{40} = \frac{256}{40}$$

$$x = 6.4$$

The pressure is 6.4 pounds.

41-47. Answers will vary.

49. Let x = age of friend if dog.

$$\frac{7}{56} = \frac{x}{44}$$

$$56x = 7 \cdot 44$$

$$56x = 308$$

$$\frac{56x}{56} = \frac{308}{56}$$

$$x = 5.5$$

The friend would be 5.5 years old if the friend was a dog.

51. Let x_{15} = the light intensity at 15 inches and
Let x_{30} = the light intensity at 30 inches.

$$\frac{x_{15}}{30^2} = \frac{x_{30}}{15^2}$$

$$\frac{x_{15}}{900} = \frac{x_{30}}{225}$$

$$225x_{15} = 900x_{30}$$

$$\frac{225x_{15}}{225} = \frac{900x_{30}}{225}$$

$$x_{15} = 4x_{30}$$

$$\frac{x_{15}}{4} = \frac{4x_{30}}{4}$$

$$\frac{1}{4}x_{15} = x_{30}$$

The illumination at 30 inches is one-fourth of the original intensity.

Check Points 6.5

1. a. $x < 4$

b. $x \geq -2$

c. $-4 \leq x < 1$

2.
$$5x - 3 \geq 17$$
$$5x - 3 + 3 \geq 17 + 3$$
$$5x \geq 20$$
$$\frac{5x}{5} \geq \frac{20}{5}$$
$$x \geq 4$$
$$\{x \mid x \geq 4\}$$

3. a. $\frac{1}{4}x < 2$

$$4 \cdot \frac{1}{4}x < 4 \cdot 2$$
$$x < 8$$
$$\{x \mid x < 8\}$$

b. $-6x < 18$

$$\frac{-6x}{-6} > \frac{18}{-6}$$
$$x > -3$$
$$\{x \mid x > -3\}$$

4.
$$7x - 3 > 13x + 33$$
$$7x - 3 + 3 > 13x + 33 + 3$$
$$7x > 13x + 36$$
$$7x - 13x > 13x + 36 - 13x$$
$$-6x > 36$$
$$\frac{-6x}{-6} < \frac{36}{-6}$$
$$x < -6$$
$$\{x \mid x < -6\}$$

5. $2(x-3)-1 \le 3(x+2)-14$

$2x-6-1 \le 3x+6-14$

$2x-7 \le 3x-8$

$2x-7+7 \le 3x-8+7$

$2x \le 3x-1$

$2x-3x \le 3x-1-3x$

$-x \le -1$

$\dfrac{-x}{-1} \ge \dfrac{-1}{-1}$

$x \ge 1$

$\{x|x \ge 1\}$

6. Let x = your grade on the final exam.

$\dfrac{82+74+78+x+x}{5} \ge 80$

$\dfrac{234+2x}{5} \ge 80$

$5\left(\dfrac{234+2x}{5}\right) \ge 5(80)$

$234+2x \ge 400$

$234+2x-234 \ge 400-234$

$2x \ge 166$

$\dfrac{2x}{2} \ge \dfrac{166}{2}$

$x \ge 83$

You need at least an 83 on the final to get a B in the course.

Exercise Set 6.5

1. $x > 6$

3. $x < -4$

5. $x \ge -3$

7. $x \le 4$

9. $-2 < x \le 5$

11. $-1 < x < 4$

13. $x-3 > 2$

$x-3+3 > 2+3$

$x > 5$

$\{x|x > 5\}$

15. $x+4 \le 9$

$x+4-4 \le 9-4$

$x \le 5$

$\{x|x \le 5\}$

17. $x-3 < 0$

$x-3+3 < 0+3$

$x < 3$

$\{x|x < 3\}$

19. $4x < 20$

$\dfrac{4x}{4} < \dfrac{20}{4}$

$x < 5$

$\{x|x < 5\}$

21. $3x \ge -15$

$\dfrac{3x}{3} \ge \dfrac{-15}{3}$

$x \ge -5$

$\{x|x \ge -5\}$

23. $2x-3 > 7$

$2x-3+3 > 7+3$

$2x > 10$

$\dfrac{2x}{2} > \dfrac{10}{2}$

$x > 5$

$\{x|x > 5\}$

25.
$$3x+3<18$$
$$3x+3-3<18-3$$
$$3x<15$$
$$\frac{3x}{3}<\frac{15}{3}$$
$$x<5$$
$$\{x\,|\,x<5\}$$

27.
$$\frac{1}{2}x<4$$
$$2\cdot\frac{1}{2}x<2\cdot4$$
$$x<8$$
$$\{x\,|\,x<8\}$$

29.
$$\frac{x}{3}>-2$$
$$3\cdot\frac{x}{3}>3\cdot(-2)$$
$$x>-6$$
$$\{x\,|\,x>-6\}$$

31.
$$-3x<15$$
$$\frac{-3x}{-3}>\frac{15}{-3}$$
$$x>-5$$
$$\{x\,|\,x>-5\}$$

33.
$$-3x\geq-15$$
$$\frac{-3x}{-3}\leq\frac{-15}{-3}$$
$$x\leq5$$
$$\{x\,|\,x\leq5\}$$

35.
$$3x+4\leq2x+7$$
$$3x+4-4\leq2x+7-4$$
$$3x\leq2x+3$$
$$3x-2x\leq2x+3-2x$$
$$x\leq3$$
$$\{x\,|\,x\leq3\}$$

37.
$$5x-9<4x+7$$
$$5x-9+9<4x+7+9$$
$$5x<4x+16$$
$$5x-4x<4x+16-4x$$
$$x<16$$
$$\{x\,|\,x<16\}$$

39.
$$-2x-3<3$$
$$-2x-3+3<3+3$$
$$-2x<6$$
$$\frac{-2x}{-2}>\frac{6}{-2}$$
$$x>-3$$
$$\{x\,|\,x>-3\}$$

41.
$$3-7x\leq17$$
$$3-7x-3\leq17-3$$
$$-7x\leq14$$
$$\frac{-7x}{-7}\geq\frac{14}{-7}$$
$$x\geq-2$$
$$\{x\,|\,x\geq-2\}$$

43.
$$-x<4$$
$$\frac{-x}{-1}>\frac{4}{-1}$$
$$x>-4$$
$$\{x\,|\,x>-4\}$$

45.
$$5-x\leq1$$
$$5\quad x\quad 5\leq1\quad 5$$
$$-x\leq-4$$
$$\frac{-x}{-1}\geq\frac{-4}{-1}$$
$$x\geq4$$
$$\{x\,|\,x\geq4\}$$

47.
$$2x - 5 > -x + 6$$
$$2x - 5 + 5 > -x + 6 + 5$$
$$2x > -x + 11$$
$$2x + x > -x + 11 + x$$
$$3x > 11$$
$$\frac{3x}{3} > \frac{11}{3}$$
$$x > \frac{11}{3}$$
$$\left\{ x \middle| x > \frac{11}{3} \right\}$$

49.
$$2x - 5 < 5x - 11$$
$$2x - 5 + 5 < 5x - 11 + 5$$
$$2x < 5x - 6$$
$$2x - 5x < 5x - 6 - 5x$$
$$-3x < -6$$
$$\frac{-3x}{-3} > \frac{-6}{-3}$$
$$x > 2$$
$$\{ x | x > 2 \}$$

51.
$$3(x + 1) - 5 < 2x + 1$$
$$3x + 3 - 5 < 2x + 1$$
$$3x - 2 < 2x + 1$$
$$3x - 2 + 2 < 2x + 1 + 2$$
$$3x < 2x + 3$$
$$3x - 2x < 2x + 3 - 2x$$
$$x < 3$$
$$\{ x | x < 3 \}$$

53.
$$8x + 3 > 3(2x + 1) - x + 5$$
$$8x + 3 > 6x + 3 - x + 5$$
$$8x + 3 > 5x + 8$$
$$8x + 3 - 3 > 5x + 8 - 3$$
$$8x > 5x + 5$$
$$8x - 5x > 5x + 5 - 5x$$
$$3x > 5$$
$$\frac{3x}{3} > \frac{5}{3}$$
$$x > \frac{5}{3}$$
$$\left\{ x \middle| x > \frac{5}{3} \right\}$$

55. $x \geq 34.4\%$
Raleigh, Seattle, San Francisco, and Austin

57. $x < 30\%$
San Diego

59. $30.0\% \leq x < 34.4\%$
Washington, Lexington-Fayette, Minneapolis, Boston, and Arlington

61. $x \leq 40.6\%$

63. $x > 29\%$

65. a. $\dfrac{88+78+86+100}{4} = \dfrac{352}{4} = 88$

The average is 88. An A is not possible.

b. $\dfrac{88+78+86+x}{4} \geq 80$

$\dfrac{252+x}{4} \geq 80$

$4 \cdot \dfrac{252+x}{4} \geq 80 \cdot 4$

$252+x \geq 320$

$252+x-252 \geq 320-252$

$x \geq 68$

You must get at least 68 on the final to earn a B in the course.

67. $N = 550-9x; \quad N < 370$

$550-9x < 370$

$550-9x-550 < 370-550$

$-9x < -180$

$\dfrac{-9x}{-9} > \dfrac{-180}{-9}$

$x > 20$

20 years after 1988 is 2008. According to the model, there will be 370 billion cigarettes consumed in 2008 and less than 370 billion from 2009 onward.

69. Let x = number of miles.

$80 + 0.25x \leq 400$

$80 + 0.25x - 80 \leq 400 - 80$

$0.25x \leq 320$

$\dfrac{0.25x}{0.25} \leq \dfrac{320}{0.25}$

$x \leq 1280$

You can drive at most 1280 miles.

71-75. Answers will vary.

77. Let x = the hours spent working out

$500 + x < 440 + 1.75x$

$500 + x - 440 < 440 + 1.75x - 440$

$60 + x < 1.75x$

$60 + x - x < 1.75x - x$

$60 < 0.75x$

$\dfrac{60}{0.75} < \dfrac{0.75x}{0.75}$

$80 < x \text{ or } x > 80$

If the number of work-out hours exceeds 80 hours, then the first club's membership is cheaper than the second club's.

79.

$126.8 - 9.4x \leq 4.8x + 34.5$

$126.8 - 9.4x - 126.8 \leq 4.8x + 34.5 - 126.8$

$-9.4x \leq 4.8x - 92.3$

$-9.4x - 4.8x \leq 4.8x - 92.3 - 4.8x$

$-14.2x \leq -92.3$

$\dfrac{-14.2x}{-14.2} \geq \dfrac{-92.3}{-14.2}$

$x \geq 6.5$

$\{x \mid x \geq 6.5\}$

Check Points 6.6

1. $(x+5)(x+6) = x \cdot x + x \cdot 6 + 5 \cdot x + 5 \cdot 6$

$= x^2 + 6x + 5x + 30$

$= x^2 + 11x + 30$

2. $(7x+5)(4x-3) = 7x \cdot 4x + 7x(-3) + 5 \cdot 4x + 5(-3)$

$= 28x^2 - 21x + 20x - 15$

$= 28x^2 - x - 15$

3. $x^2 + 5x + 6 = (x+2)(x+3)$

4. $x^2 + 3x - 10 = (x+5)(x-2)$

5. $5x^2 - 14x + 8 = (5x-4)(x-2)$

6. $6y^2 + 19y - 7 = (3y-1)(2y+7)$

7. $(x+6)(x-3) = 0$

$x+6 = 0 \quad \text{or} \quad x-3 = 0$

$x = -6 \qquad\qquad x = 3$

The solution set is $\{-6, \ 3\}$

8.

$x^2 - 6x = 16$

$x^2 - 6x - 16 = 16 - 16$

$x^2 - 6x - 16 = 0$

$(x+2)(x-8) = 0$

$x+2 = 0 \quad \text{or} \quad x-8 = 0$

$x = -2 \qquad\qquad x = 8$

The solution set is $\{-2, \ 8\}$

9.
$$2x^2 + 7x - 4 = 0$$
$$(2x-1)(x+4) = 0$$
$$2x-1 = 0 \quad \text{or} \quad x+4 = 0$$
$$2x = 1 \qquad\qquad x = -4$$
$$x = \frac{1}{2}$$

The solution set is $\left\{-4, \ \frac{1}{2}\right\}$

10. $8x^2 + 2x - 1 = 0$
$$x = \frac{-b \pm \sqrt{b^2 - 4ac}}{2a}$$
$$x = \frac{-(2) \pm \sqrt{(2)^2 - 4(8)(-1)}}{2(8)}$$
$$x = \frac{-2 \pm \sqrt{4 + 32}}{16}$$
$$x = \frac{-2 \pm \sqrt{36}}{16}$$
$$x = \frac{-2 \pm 6}{16}$$
$$x = \frac{-2+6}{16} \quad \text{or} \quad x = \frac{-2-6}{16}$$
$$x = \frac{4}{16} \qquad\qquad x = \frac{-8}{16}$$
$$x = \frac{1}{4} \qquad\qquad x = -\frac{1}{2}$$

The solution set is $\left\{-\frac{1}{2}, \ \frac{1}{4}\right\}$

11.
$$2x^2 = 6x - 1$$
$$2x^2 - 6x + 1 = 0$$
$$x = \frac{-b \pm \sqrt{b^2 - 4ac}}{2a}$$
$$x = \frac{-(-6) \pm \sqrt{(-6)^2 - 4(2)(1)}}{2(2)}$$
$$x = \frac{6 \pm \sqrt{36 - 8}}{4}$$
$$x = \frac{6 \pm \sqrt{28}}{4}$$
$$x = \frac{6 \pm 2\sqrt{7}}{4}$$
$$x = \frac{2\left(3 \pm \sqrt{7}\right)}{4}$$
$$x = \frac{3 \pm \sqrt{7}}{2}$$
$$x = \frac{3 + \sqrt{7}}{2} \quad \text{or} \quad x = \frac{3 - \sqrt{7}}{2}$$

The solution set is $\left\{\frac{3+\sqrt{7}}{2}, \ \frac{3-\sqrt{7}}{2}\right\}$

Exercise Set 6.6

1. $(x+3)(x+5) = x^2 + 5x + 3x + 15$
$$= x^2 + 8x + 15$$

3. $(x-5)(x+3) = x^2 + 3x - 5x - 15$
$$= x^2 - 2x - 15$$

5. $(2x-1)(x+2) = 2x^2 + 4x - 1x - 2$
$$= 2x^2 + 3x - 2$$

7. $(3x-7)(4x-5) = 12x^2 - 15x - 28x + 35$
$$= 12x^2 - 43x + 35$$

9. $x^2 + 5x + 6 = (x+2)(x+3)$
Check: $(x+2)(x+3)$
$$= x^2 + 3x + 2x + 6$$
$$= x^2 + 5x + 6$$

11. $x^2 - 2x - 15 = (x-5)(x+3)$
Check: $(x-5)(x+3)$
$$= x^2 + 3x - 5x - 15$$
$$= x^2 - 2x - 15$$

13. $x^2 - 8x + 15 = (x-3)(x-5)$

Check: $(x-3)(x-5)$

$\quad = x^2 - 5x - 3x + 15$

$\quad = x^2 - 8x + 15$

15. $x^2 - 9x - 36 = (x-12)(x+3)$

Check: $(x-12)(x+3)$

$\quad = x^2 + 3x - 12x - 36$

$\quad = x^2 - 9x - 36$

17. $x^2 - 8x + 32$ is prime.

19. $x^2 + 17x + 16 = (x+16)(x+1)$

Check: $(x+16)(x+1)$

$\quad = x^2 + x + 16x + 16$

$\quad = x^2 + 17x + 16$

21. $2x^2 + 7x + 3 = (2x+1)(x+3)$

Check: $(2x+1)(x+3)$

$\quad = 2x^2 + 6x + x + 3$

$\quad = 2x^2 + 7x + 3$

23. $2x^2 - 17x + 30 = (2x-5)(x-6)$

Check: $(2x-5)(x-6)$

$\quad = 2x^2 - 12x - 5x + 30$

$\quad = 2x^2 - 17x + 30$

25. $3x^2 - x - 2 = (3x+2)(x-1)$

Check: $(3x+2)(x-1)$

$\quad = 3x^2 - 3x + 2x - 2$

$\quad = 3x^2 - x - 2$

27. $3x^2 - 25x - 28 = (3x-28)(x+1)$

Check: $(3x-28)(x+1)$

$\quad = 3x^2 + 3x - 28x - 28$

$\quad = 3x^2 - 25x - 28$

29. $6x^2 - 11x + 4 = (2x-1)(3x-4)$

Check: $(2x-1)(3x-4)$

$\quad = 6x^2 - 8x - 3x + 4$

$\quad = 6x^2 - 11x + 4$

31. $4x^2 + 16x + 15 = (2x+5)(2x+3)$

Check: $(2x+5)(2x+3)$

$\quad = 4x^2 + 6x + 10x + 15$

$\quad = 4x^2 + 16x + 15$

33. $(x-8)(x+3) = 0$

$x - 8 = 0 \quad$ or $\quad x + 3 = 0$

$x = 8 \qquad\qquad x = -3$

The solution set is $\{-3, 8\}$.

35. $(4x+5)(x-2) = 0$

$4x + 5 = 0 \quad$ or $\quad x - 2 = 0$

$4x = -5 \qquad\qquad x = 2$

$x = -\dfrac{5}{4}$

The solution set is $\left\{-\dfrac{5}{4}, 2\right\}$.

37. $x^2 + 8x + 15 = 0$

$(x+5)(x+3) = 0$

$x + 5 = 0 \quad$ or $\quad x + 3 = 0$

$x = -5 \qquad\qquad x = -3$

The solution set is $\{-5, -3\}$.

39. $x^2 - 2x - 15 = 0$

$(x-5)(x+3) = 0$

$x - 5 = 0 \quad$ or $\quad x + 3 = 0$

$x = 5 \qquad\qquad x = -3$

The solution set is $\{-3, 5\}$.

41. $x^2 - 4x = 21$

$x^2 - 4x - 21 = 0$

$(x+3)(x-7) = 0$

$x + 3 = 0 \quad$ or $\quad x - 7 = 0$

$x = -3 \qquad\qquad x = 7$

The solution set is $\{-3, 7\}$.

43. $x^2 + 9x = -8$

$x^2 + 9x + 8 = 0$

$(x+8)(x+1) = 0$

$x + 8 = 0 \quad$ or $\quad x + 1 = 0$

$x = -8 \qquad\qquad x = -1$

The solution set is $\{-8, -1\}$.

45. $x^2 - 12x = -36$

$x^2 - 12x + 36 = 0$

$(x-6)(x-6) = 0$

$x - 6 = 0 \quad$ or $\quad x - 6 = 0$

$x - 6 \qquad\qquad x - 6$

The solution set is $\{6\}$.

47.
$$2x^2 = 7x + 4$$
$$2x^2 - 7x - 4 = 0$$
$$(2x + 1)(x - 4) = 0$$
$$2x + 1 = 0 \quad \text{or} \quad x - 4 = 0$$
$$2x = -1 \qquad\qquad x = 4$$
$$x = -\frac{1}{2}$$

The solution set is $\left\{ -\frac{1}{2}, 4 \right\}$.

49. $5x^2 + x = 18$
$$5x^2 + x - 18 = 0$$
$$(5x - 9)(x + 2) = 0$$
$$5x - 9 = 0 \quad \text{or} \quad x + 2 = 0$$
$$5x = 9 \qquad\qquad x = -2$$
$$x = \frac{9}{5}$$

The solution set is $\left\{ -2, \frac{9}{5} \right\}$.

51. $x(6x + 23) + 7 = 0$
$$6x^2 + 23x + 7 = 0$$
$$(2x + 7)(3x + 1) = 0$$
$$2x + 7 = 0 \quad \text{or} \quad 3x + 1 = 0$$
$$2x = -7 \qquad\qquad 3x = -1$$
$$x = -\frac{7}{2} \qquad\qquad x = -\frac{1}{3}$$

The solution set is $\left\{ -\frac{7}{2}, -\frac{1}{3} \right\}$.

53. $x^2 + 8x + 15 = 0$
$$x = \frac{-b \pm \sqrt{b^2 - 4ac}}{2a}$$
$$x = \frac{-8 \pm \sqrt{8^2 - 4(1)(15)}}{2(1)}$$
$$x = \frac{-8 \pm \sqrt{4}}{2}$$
$$x = \frac{-8 \pm 2}{2}$$
$$x = \frac{-8 - 2}{2} \quad \text{or} \quad x = \frac{-8 + 2}{2}$$
$$x = -5 \qquad\qquad x = -3$$

The solution set is $\{-5, -3\}$.

55. $x^2 + 5x + 3 = 0$
$$x = \frac{-b \pm \sqrt{b^2 - 4ac}}{2a}$$
$$x = \frac{-5 \pm \sqrt{5^2 - 4(1)(3)}}{2(1)}$$
$$x = \frac{-5 \pm \sqrt{13}}{2}$$

The solution set is $\left\{ \dfrac{-5 - \sqrt{13}}{2}, \dfrac{-5 + \sqrt{13}}{2} \right\}$.

57. $x^2 + 4x - 6 = 0$
$$x = \frac{-b \pm \sqrt{b^2 - 4ac}}{2a}$$
$$x = \frac{-4 \pm \sqrt{4^2 - 4(1)(-6)}}{2(1)}$$
$$x = \frac{-4 \pm \sqrt{40}}{2}$$
$$x = \frac{-4 \pm 2\sqrt{10}}{2}$$
$$x = -2 \pm \sqrt{10}$$

The solution set is $\left\{ -2 - \sqrt{10}, \ -2 + \sqrt{10} \right\}$.

59. $x^2 + 4x - 7 = 0$
$$x = \frac{-b \pm \sqrt{b^2 - 4ac}}{2a}$$
$$x = \frac{-4 \pm \sqrt{4^2 - 4(1)(-7)}}{2(1)}$$
$$x = \frac{-4 \pm \sqrt{44}}{2}$$
$$x = \frac{-4 \pm 2\sqrt{11}}{2}$$
$$x = -2 \pm \sqrt{11}$$

The solution set is $\left\{ -2 - \sqrt{11}, \ -2 + \sqrt{11} \right\}$.

61. $x^2 - 3x - 18 = 0$

$$x = \frac{-b \pm \sqrt{b^2 - 4ac}}{2a}$$

$$x = \frac{-(-3) \pm \sqrt{(-3)^2 - 4(1)(-18)}}{2(1)}$$

$$x = \frac{3 \pm \sqrt{81}}{2}$$

$$x = \frac{3 \pm 9}{2}$$

$$x = \frac{3 - 9}{2} \quad \text{or} \quad x = \frac{3 + 9}{2}$$

$$x = -3 \qquad \qquad x = 6$$

The solution sct is $\{-3, 6\}$.

63. $6x^2 - 5x - 6 = 0$

$$x = \frac{-b \pm \sqrt{b^2 - 4ac}}{2a}$$

$$x = \frac{-(-5) \pm \sqrt{(-5)^2 - 4(6)(-6)}}{2(6)}$$

$$x = \frac{5 \pm \sqrt{169}}{12}$$

$$x = \frac{5 \pm 13}{12}$$

$$x = \frac{5 + 13}{12} \quad \text{or} \quad x = \frac{5 - 13}{12}$$

$$x = \frac{18}{12} \qquad \qquad x = \frac{-8}{12}$$

$$x = \frac{3}{2} \qquad \qquad x = -\frac{2}{3}$$

The solution set is $\left\{ \frac{3}{2}, \ -\frac{2}{3} \right\}$.

65. $x^2 - 2x - 10 = 0$

$$x = \frac{-b \pm \sqrt{b^2 - 4ac}}{2a}$$

$$x = \frac{-(-2) \pm \sqrt{(-2)^2 - 4(1)(-10)}}{2(1)}$$

$$x = \frac{2 \pm \sqrt{44}}{2}$$

$$x = \frac{2 \pm 2\sqrt{11}}{2}$$

$$x = 1 \pm \sqrt{11}$$

The solution set is $\left\{ 1 - \sqrt{11}, \ 1 + \sqrt{11} \right\}$.

67. $x^2 - x = 14$

$$x^2 - x - 14 = 0$$

$$x = \frac{-b \pm \sqrt{b^2 - 4ac}}{2a}$$

$$x = \frac{-(-1) \pm \sqrt{(-1)^2 - 4(1)(-14)}}{2(1)}$$

$$x = \frac{1 \pm \sqrt{57}}{2}$$

The solution set is $\left\{ \frac{1 - \sqrt{57}}{2}, \ \frac{1 + \sqrt{57}}{2} \right\}$.

69. $6x^2 + 6x + 1 = 0$

$$x = \frac{-b \pm \sqrt{b^2 - 4ac}}{2a}$$

$$x = \frac{-b \pm \sqrt{b^2 - 4ac}}{2a}$$

$$x = \frac{-6 \pm \sqrt{6^2 - 4(6)(1)}}{2(6)}$$

$$x = \frac{-6 \pm \sqrt{12}}{12}$$

$$x = \frac{-6 \pm 2\sqrt{3}}{12}$$

$$x = \frac{-3 \pm \sqrt{3}}{6}$$

The solution set is $\left\{ \frac{-3 - \sqrt{3}}{6}, \ \frac{-3 + \sqrt{3}}{6} \right\}$.

71. $4x^2 - 12x + 9 = 0$

$$x = \frac{-b \pm \sqrt{b^2 - 4ac}}{2a}$$

$$x = \frac{-(-12) \pm \sqrt{(-12)^2 - 4(4)(9)}}{2(4)}$$

$$x = \frac{12 \pm \sqrt{0}}{8}$$

$$x = \frac{12}{8} = \frac{3}{2}$$

The solution set is $\left\{ \frac{3}{2} \right\}$.

73. $N = 23.4x^2 - 259.1x + 815.8$, for $N = 1000$

$1000 = 23.4x^2 - 259.1x + 815.8$

$0 = 23.4x^2 - 259.1x + 815.8 - 1000$

$0 = 23.4x^2 - 259.1x - 184.2$

$x = \dfrac{-b \pm \sqrt{b^2 - 4ac}}{2a}$

$x = \dfrac{-(-259.1) \pm \sqrt{(-259.1)^2 - 4(23.4)(-184.2)}}{2(23.4)}$

$x = \dfrac{259.1 \pm \sqrt{84373.93}}{46.8}$

$x \approx \dfrac{259.1 \pm 290.47}{46.8}$

$x \approx \dfrac{259.1 + 290.47}{46.8}$ or $x \approx \dfrac{259.1 - 290.47}{46.8}$

$x \approx \dfrac{549.57}{46.8}$ $\qquad$ $x \approx \dfrac{-31.37}{46.8}$

$x \approx 12$ $\qquad\qquad$ $x \approx -1$ (Disregard)

12 years after 1990 is 2002. The formula suggests that in 2002 about 1000 police officers will be convicted of felonies.

75. $N = -0.5x^2 + 4x + 19$, for $N = 20$

$20 = -0.5x^2 + 4x + 19$

$0 = -0.5x^2 + 4x + 19 - 20$

$0 = -0.5x^2 + 4x - 1$

$x = \dfrac{-b \pm \sqrt{b^2 - 4ac}}{2a}$

$x = \dfrac{-(4) \pm \sqrt{(4)^2 - 4(-0.5)(-1)}}{2(-0.5)}$

$x = \dfrac{-4 \pm \sqrt{14}}{-1}$

$x \approx \dfrac{-4 \pm 3.74}{-1}$

$x \approx \dfrac{-4 + 3.74}{-1}$ or $x \approx \dfrac{-4 - 3.74}{-1}$

$x \approx \dfrac{-0.26}{-1}$ $\qquad$ $x \approx \dfrac{-7.74}{-1}$

$x \approx 0$ $\qquad\qquad$ $x \approx 8$

8 years after 1990 is 1998. The formula suggests that in 1990 and 1998 about 20 million people received food stamps.

77. $N = 0.013x^2 - 1.19x + 28.24$, for $N = 3$

$3 = 0.013x^2 - 1.19x + 28.24$

$0 = 0.013x^2 - 1.19x + 28.24 - 3$

$0 = 0.013x^2 - 1.19x + 25.24$

$x = \dfrac{-b \pm \sqrt{b^2 - 4ac}}{2a}$

$x = \dfrac{-(-1.19) \pm \sqrt{(-1.19)^2 - 4(0.013)(25.24)}}{2(0.013)}$

$x = \dfrac{1.19 \pm \sqrt{0.10362}}{0.026}$

$x \approx \dfrac{1.19 \pm 0.3219}{0.026}$

$x \approx \dfrac{1.19 + 0.3219}{0.026}$ or $x \approx \dfrac{1.19 - 0.3219}{0.026}$

$x \approx \dfrac{1.5119}{0.026}$ $\qquad$ $x \approx \dfrac{0.8681}{0.026}$

$x \approx 58$ $\qquad\qquad$ $x \approx 33$

33-year-olds and 58-year-olds are expected to be involved in 3 fatal crashes per 100 million miles driven. The formula models the data well.

79-83. Answers will vary.

85. $x^2 + bx + 15$

$(x + 3)(x + 5) = x^2 + 8x + 15$

$(x + 1)(x + 15) = x^2 + 16x + 15$

Therefore, $b = 8, 16$.

87. $x^{2n} + 20x^n + 99 = (x^n + 11)(x^n + 9)$

Chapter 6 Review Exercises

1. $6x + 9 = 6 \cdot 4 + 9 = 24 + 9 = 33$

2. $4x^2 - 3x + 2 = 4 \cdot 5^2 - 3 \cdot 5 + 2$
$\qquad\qquad\quad = 4 \cdot 25 - 3 \cdot 5 + 2$
$\qquad\qquad\quad = 100 - 15 + 2$
$\qquad\qquad\quad = 87$

3. $7x^2 + 4x - 5 = 7(-2)^2 + 4(-2) - 5$
$\qquad\qquad\qquad = 7(4) + 4(-2) - 5$
$\qquad\qquad\qquad = 28 - 8 - 5$
$\qquad\qquad\qquad = 15$

4. $7x + 9 - 12 - x = 7x - x + 9 - 12$
$\qquad\qquad\qquad\quad = 6x - 3$

5. $6(5x+3)-20 = 6\cdot5x+6\cdot3-20$
$$= 30x+18-20$$
$$= 30x-2$$

6. $4(7x-1)+11x = 28x-4+11x$
$$= 39x-4$$

7. $9x = 9\cdot15 = 135$
You can stay in the sun 135 minutes without burning with a number 9 spf lotion.

8. $x-0.25x = 2400-0.25(2400)$
$$= 2400-600$$
$$= 1800$$
The computer's discount price is $1800.

9. $x = 2005-1995 = 10$
$$N = 1.2x^2+15.2x+181.4$$
$$N = 1.2(10)^2+15.2(10)+181.4$$
$$N = 120+152+181.4$$
$$N = 453.4$$
The formula suggests that $453.4 billion will be spent in 2005. The formula models the graph quite well.

10. $4x+9 = 33$
$$4x+9-9 = 33-9$$
$$4x = 24$$
$$\frac{4x}{4} = \frac{24}{4}$$
$$x = 6$$
The solution set is $\{6\}$.

11. $5x-3 = x+5$
$$5x-3+3 = x+5+3$$
$$5x - x+8$$
$$5x-x = x+8-x$$
$$4x = 8$$
$$\frac{4x}{4} = \frac{8}{4}$$
$$x = 2$$
The solution set is $\{2\}$.

12. $3(x+4) = 5x-12$
$$3x+12 = 5x-12$$
$$3x+12-12 = 5x-12-12$$
$$3x = 5x-24$$
$$3x-5x = 5x-24-5x$$
$$-2x = -24$$
$$\frac{-2x}{-2} = \frac{-24}{-2}$$
$$x = 12$$
The solution set is $\{12\}$.

13. $2(x-2)+3(x+5) = 2x-2$
$$2x-4+3x+15 = 2x-2$$
$$5x+11-2x-2$$
$$5x+11-11 = 2x-2-11$$
$$5x = 2x-13$$
$$5x-2x = 2x-13-2x$$
$$3x = -13$$
$$\frac{3x}{3} = \frac{-13}{3}$$
$$x = -\frac{13}{3}$$
The solution set is $\left\{-\dfrac{13}{3}\right\}$.

14. $\dfrac{2x}{3} = \dfrac{x}{6}+1$
$$6\left(\frac{2x}{3}\right) = 6\left(\frac{x}{6}+1\right)$$
$$4x = x+6$$
$$4x-x = x+6-x$$
$$3x = 6$$
$$\frac{3x}{3} = \frac{6}{3}$$
$$x = 2$$
The solution set is $\{2\}$.

15. $3x+y = 9$
$$3x+y-3x = -3x+9$$
$$y = -3x+9$$

16. $4x + 2y = 16$

$4x + 2y - 4x = -4x + 16$

$2y = -4x + 16$

$\dfrac{2y}{2} = \dfrac{-4x + 16}{2}$

$y = \dfrac{-4x}{2} + \dfrac{16}{2}$

$y = -2x + 8$

17. $D = RT$, for T

$\dfrac{D}{R} = \dfrac{RT}{R}$

$\dfrac{D}{R} = T$ or $T = \dfrac{D}{R}$

18. $P = 2l + 2w$ for w

$P - 2l = 2l + 2w - 2l$

$P - 2l = 2w$

$\dfrac{P - 2l}{2} = \dfrac{2w}{2}$

$\dfrac{P - 2l}{2} = w$ or $w = \dfrac{P - 2l}{2}$

19. $A = \dfrac{1}{2}bh$ for h

$2 \cdot A = 2 \cdot \dfrac{1}{2}bh$

$2A = bh$

$\dfrac{2A}{b} = \dfrac{bh}{b}$

$\dfrac{2A}{b} = h$ or $h = \dfrac{2A}{b}$

20. $A = \dfrac{B + C}{2}$ for B

$2 \cdot A = 2 \cdot \dfrac{B + C}{2}$

$2A = B + C$

$2A - C = B + C - C$

$2A - C = B$ or $B = 2A - C$

21. $y = 420x + 720$

$4080 = 420x + 720$

$4080 - 720 = 420x + 720 - 720$

$3360 = 420x$

$\dfrac{3360}{420} = \dfrac{420x}{420}$

$8 = x$

Losses amounted to $4080 million in 1997 (8 years after 1989).

22. $17 - 9x$

23. $3x - 7$

24. $5x + 8$

25. $\dfrac{6}{x} + 3x$

26. $17 - 4x = 5$

$17 - 4x - 17 = 5 - 17$

$-4x = -12$

$\dfrac{-4x}{-4} = \dfrac{-12}{-4}$

$x = 3$

The number is 3.

27. $5x + 2 = x + 22$

$5x + 2 - 2 = x + 22 - 2$

$5x = x + 20$

$5x - x = x + 20 - x$

$4x = 20$

$\dfrac{4x}{4} = \dfrac{20}{4}$

$x = 5$

The number is 5.

28. $7x - 1 = 5x + 9$

$7x - 1 + 1 = 5x + 9 + 1$

$7x = 5x + 10$

$7x - 5x = 5x + 10 - 5x$

$2x = 10$

$\dfrac{2x}{2} = \dfrac{10}{2}$

$x = 5$

The number is 5.

29.
$$8(x-5)=56$$
$$8x-40=56$$
$$8x-40+40=56+40$$
$$8x=96$$
$$\frac{8x}{8}=\frac{96}{8}$$
$$x=12$$
The number is 12.

30. Let x = the years after 2000.
$$567+15x=702$$
$$567+15x-567=702-567$$
$$15x=135$$
$$\frac{15x}{15}=\frac{135}{15}$$
$$x=9$$
The average weekly salary will reach $702 nine years after 2000 or in 2009.

31. Let x = the number of Madonna's platinum albums, then
$x+2$ = the number of Barbra Streisand's platinum albums.
$$x+x+2=96$$
$$2x+2=96$$
$$2x+2-2=96-2$$
$$2x=94$$
$$\frac{2x}{2}=\frac{94}{2}$$
$$x=47$$
$$x+2=49$$
Madonna has had 47 platinum albums and Barbra Streisand has had 49.

32. Let x = the number of unhealthy air days in New York City, then
$3x+48$ = the number of unhealthy air days in Los Angeles.
$$x+3x+48=268$$
$$4x+48=268$$
$$4x+48-48=268-48$$
$$4x=220$$
$$\frac{4x}{4}=\frac{220}{4}$$
$$x=55$$
$$3x+48=213$$
New York City averages 55 unhealthy air days and Los Angeles averages 213.

33. Let x = the number of minutes at which the costs of the two plans are the same.
$$\overbrace{15+0.05x}^{\text{first plan}}=\overbrace{5+0.07x}^{\text{other plan}}$$
$$15+0.05x-15=5+0.07x-15$$
$$0.05x=0.07x-10$$
$$0.05x-0.07x=0.07x-10-0.07x$$
$$-0.02x=-10$$
$$\frac{-0.02x}{-0.02}=\frac{-10}{-0.02}$$
$$x=500$$
The two plans are the same at 500 minutes.

34. a. $\frac{30}{22}=\frac{30\div2}{22\div2}=\frac{15}{11}$ or $15:11$

b. $\frac{16+10}{18}=\frac{26}{18}=\frac{26\div2}{18\div2}=\frac{13}{9}$ or $13:9$

35.
$$\frac{3}{x}=\frac{15}{25}$$
$$3\cdot25=x\cdot15$$
$$75=15x$$
$$\frac{75}{15}=\frac{15x}{15}$$
$$5=x$$
The solution set is $\{5\}$.

36.
$$\frac{-7}{5}=\frac{91}{x}$$
$$-7\cdot x=5\cdot91$$
$$-7x=455$$
$$\frac{-7x}{-7}=\frac{455}{-7}$$
$$x=-65$$
The solution set is $\{-65\}$.

37.
$$\frac{x+2}{3} = \frac{4}{5}$$
$$5(x+2) = 3 \cdot 4$$
$$5x + 10 = 12$$
$$5x + 10 - 10 = 12 - 10$$
$$5x = 2$$
$$\frac{5x}{5} = \frac{2}{5}$$
$$x = \frac{2}{5}$$

The solution set is $\left\{\dfrac{2}{5}\right\}$.

38.
$$\frac{5}{x+7} = \frac{3}{x+3}$$
$$5(x+3) = 3(x+7)$$
$$5x + 15 = 3x + 21$$
$$5x + 15 - 15 = 3x + 21 - 15$$
$$5x = 3x + 6$$
$$5x - 3x = 3x + 6 - 3x$$
$$2x = 6$$
$$\frac{2x}{2} = \frac{6}{2}$$
$$x = 3$$

The solution set is {3}.

39. Let x = number of teachers
$$\frac{3}{50} = \frac{x}{5400}$$
$$50 \cdot x = 3 \cdot 5400$$
$$50x = 16,200$$
$$\frac{50x}{50} = \frac{16,200}{50}$$
$$x = 324$$
There should be 324 teachers for 5400 students.

40. Let x = number of trout in lake
$$\frac{32}{82} = \frac{112}{x}$$
$$32x = 82 \cdot 112$$
$$32x = 9184$$
$$\frac{32x}{32} = \frac{9184}{32}$$
$$x = 287$$
There are 287 trout in the lake.

41. Let x = the dollar amount of the electric bill.
$$\frac{1400}{98} = \frac{2200}{x}$$
$$1400 \cdot x = 98 \cdot 2200$$
$$1400x = 215,600$$
$$\frac{1400x}{1400} = \frac{215,600}{1400}$$
$$x = 154$$
The electric bill is $154.

42. Let x = feet the object will fall in 10 seconds.
$$\frac{144}{3^2} = \frac{x}{10^2}$$
$$\frac{144}{9} = \frac{x}{100}$$
$$144 \cdot 100 = 9x$$
$$14,400 = 9x$$
$$\frac{14,400}{9} = \frac{9x}{9}$$
$$1600 = x$$
The object will fall 1600 feet in 10 seconds.

43. Let x = hours needed when driving 40 mph.
$$\frac{50}{x} = \frac{40}{4}$$
$$50 \cdot 4 = 40x$$
$$200 = 40x$$
$$\frac{200}{40} = \frac{40x}{40}$$
$$5 = x$$
At a rate of 40 mph, it will take 5 hours.

44.
$$2x - 5 < 3$$
$$2x - 5 + 5 < 3 + 5$$
$$2x < 8$$
$$\frac{2x}{2} < \frac{8}{2}$$
$$x < 4$$
$$\{x \mid x < 4\}$$

45. $\dfrac{x}{2} > -4$

$2 \cdot \dfrac{x}{2} > 2(-4)$

$x > -8$

$\{x \mid x > -8\}$

46. $3 - 5x \le 18$

$3 - 5x - 3 \le 18 - 3$

$-5x \le 15$

$\dfrac{-5x}{-5} \ge \dfrac{15}{-5}$

$x \ge -3$

$\{x \mid x \ge -3\}$

47. $4x + 6 < 5x$

$4x + 6 - 6 < 5x - 6$

$4x < 5x - 6$

$4x - 5x < 5x - 6 - 5x$

$-x < -6$

$\dfrac{-x}{-1} > \dfrac{-6}{-1}$

$x > 6$

$\{x \mid x > 6\}$

48. $6x - 10 \ge 2(x + 3)$

$6x - 10 + 10 \ge 2x + 6 + 10$

$6x \ge 2x + 16$

$6x - 2x \ge 2x + 16 - 2x$

$4x \ge 16$

$\dfrac{4x}{4} \ge \dfrac{16}{4}$

$x \ge 4$

$\{x \mid x \ge 4\}$

49. $4x + 3(2x - 7) \le x - 3$

$4x + 6x - 21 \le x - 3$

$10x - 21 \le x - 3$

$10x - 21 + 21 \le x - 3 + 21$

$10x \le x + 18$

$10x - x \le x + 18 - x$

$9x \le 18$

$\dfrac{9x}{9} \le \dfrac{18}{9}$

$x \le 2$

$\{x \mid x \le 2\}$

50. Let x = score on third test.

$\dfrac{42 + 74 + x}{3} \ge 60$

$\dfrac{116 + x}{3} \ge 60$

$3 \cdot \dfrac{116 + x}{3} \ge 3 \cdot 60$

$116 + x \ge 180$

$116 + x - 116 \ge 180 - 116$

$x \ge 64$

The score on the third test must be at least 64.

51. $(x + 9)(x - 5) = x^2 - 5x + 9x - 45$

$\qquad\qquad\qquad = x^2 + 4x - 45$

52. $(4x - 7)(3x + 2) = 12x^2 + 8x - 21x - 14$

$\qquad\qquad\qquad = 12x^2 - 13x - 14$

53. $x^2 - x - 12 = (x - 4)(x + 3)$

54. $x^2 - 8x + 15 = (x - 5)(x - 3)$

55. $x^2 + 2x + 3$ is prime.

56. $3x^2 - 17x + 10 = (3x - 2)(x - 5)$

57. $6x^2 - 11x - 10 = (3x + 2)(2x - 5)$

58. $3x^2 - 6x - 5$ is prime.

59. $x^2 + 5x - 14 = 0$

$(x + 7)(x - 2) = 0$

$x + 7 = 0 \quad$ or $\quad x - 2 = 0$

$x = -7 \qquad\qquad x = 2$

The solution set is $\{-7, 2\}$.

60. $x^2 - 4x = 32$
$x^2 - 4x - 32 = 0$
$(x-8)(x+4) = 0$
$x - 8 = 0$ or $x + 4 = 0$
   $x = 8$ $x = -4$
The solution set is $\{-4, 8\}$.

61. $2x^2 + 15x - 8 = 0$
$(2x-1)(x+8) = 0$
$2x - 1 = 0$ or $x + 8 = 0$
  $2x = 1$ $x = -8$
   $x = \dfrac{1}{2}$
The solution set is $\left\{-8, \dfrac{1}{2}\right\}$.

62. $3x^2 = -21x - 30$
$3x^2 + 21x + 30 = 0$
$(3x+6)(x+5) = 0$
$3x + 6 = 0$ or $x + 5 = 0$
  $3x = -6$ $x = -5$
   $x = -2$
The solution set is $\{-5, -2\}$.

63. $x^2 - 4x + 3 = 0$
$x = \dfrac{-b \pm \sqrt{b^2 - 4ac}}{2a}$
$x = \dfrac{-(-4) \pm \sqrt{(-4)^2 - 4(1)(3)}}{2(1)}$
$x = \dfrac{4 \pm \sqrt{4}}{2}$
$x = \dfrac{4 \pm 2}{2}$
$x = \dfrac{4-2}{2}$ or $x = \dfrac{4+2}{2}$
$x = 1$ $x = 3$
The solution set is $\{1, 3\}$.

64. $x^2 - 5x = 4$
$x^2 - 5x - 4 = 0$
$x = \dfrac{-b \pm \sqrt{b^2 - 4ac}}{2a}$
$x = \dfrac{-(-5) \pm \sqrt{(-5)^2 - 4(1)(-4)}}{2(1)}$
$x = \dfrac{5 \pm \sqrt{41}}{2}$
The solution set is $\left\{\dfrac{5 - \sqrt{41}}{2}, \dfrac{5 + \sqrt{41}}{2}\right\}$.

65. $2x^2 + 5x - 3 = 0$
$x = \dfrac{-b \pm \sqrt{b^2 - 4ac}}{2a}$
$x = \dfrac{-5 \pm \sqrt{5^2 - 4(2)(-3)}}{2(2)}$
$x = \dfrac{-5 \pm \sqrt{49}}{4}$
$x = \dfrac{-5 \pm 7}{4}$
$x = \dfrac{-5+7}{4}$ or $x = \dfrac{-5-7}{4}$
$x = \dfrac{1}{2}$ $x = -3$
The solution set is $\left\{-3, \dfrac{1}{2}\right\}$.

66. $3x^2 - 6x = 5$
$3x^2 - 6x - 5 = 0$
$x = \dfrac{-b \pm \sqrt{b^2 - 4ac}}{2a}$
$x = \dfrac{-(-6) \pm \sqrt{(-6)^2 - 4(3)(-5)}}{2(3)}$
$x = \dfrac{6 \pm \sqrt{96}}{6}$
$x = \dfrac{6 \pm 4\sqrt{6}}{6}$
$x = \dfrac{3 \pm 2\sqrt{6}}{3}$
The solution set is $\left\{\dfrac{3 - 2\sqrt{6}}{3}, \dfrac{3 + 2\sqrt{6}}{3}\right\}$.

67. $x - 1990 - 1980 = 10$

$N = 2x^2 + 22x + 320$

$N = 2(10)^2 + 22(10) + 320$

$N = 200 + 220 + 320$

$N = 740$

The prison population was 740 thousand in 1990. The formula models the data quite well.

68. Solve when $N = 480$.

$$N = 2x^2 + 22x + 320$$

$$480 = 2x^2 + 22x + 320$$

$$480 - 480 = 2x^2 + 22x + 320 - 480$$

$$0 = 2x^2 + 22x - 160$$

$$x = \frac{-b \pm \sqrt{b^2 - 4ac}}{2a}$$

$$x = \frac{-(22) \pm \sqrt{(22)^2 - 4(2)(-160)}}{2(2)}$$

$$x = \frac{-22 \pm \sqrt{1764}}{4}$$

$$x = \frac{-22 \pm 42}{4}$$

$$x = \frac{-22 + 42}{4} \quad \text{or} \quad x = \frac{-22 - 42}{4}$$

$$x = 5 \qquad\qquad x = -16 \text{ (Disregard)}$$

The prison population was 480 thousand in 1985.

Chapter 6 Test

1. Evaluate $5x^2 - 7x - 2$ when $x = -3$.

$$5x^2 - 7x - 2 = 5(-3)^2 - 7(-3) - 2$$

$$= 5(9) - 7(-3) - 2$$

$$= 45 + 21 - 2$$

$$= 64$$

2. $5(3x - 2) + 7x = 5 \cdot 3x + 5(-2) + 7x$

$$-15x - 10 + 7x$$

$$-15x + 7x - 10$$

$$= 22x - 10$$

3. $t - 1994 - 1984 = 10$

$S = 91t + 164$

$S = 91(10) + 164$

$S = 910 + 164$

$S = 1074$

The annual salary in 1994 was $1074 thousand or $1,074,000.

4. $8x - 5(x - 2) = x + 26$

$$8x - 5x + 10 = x + 26$$

$$3x + 10 = x + 26$$

$$3x + 10 - 10 = x + 26 - 10$$

$$3x = x + 16$$

$$3x - x = x + 16 - x$$

$$2x = 16$$

$$\frac{2x}{2} = \frac{16}{2}$$

$$x = 8$$

The solution set is $\{8\}$.

5. $3(2x - 4) = 9 - 3(x + 1)$

$$6x - 12 = 9 - 3x - 3$$

$$6x - 12 = 6 - 3x$$

$$6x - 12 + 12 = 6 - 3x + 12$$

$$6x = -3x + 18$$

$$6x + 3x = -3x + 18 + 3x$$

$$9x - 18$$

$$\frac{9x}{9} = \frac{18}{9}$$

$$x = 2$$

The solution set is $\{2\}$.

6. Solve when $N = 142$

$N = 3.5x + 58$

$$142 = 3.5x + 58$$

$$142 - 58 = 3.5x + 58 - 58$$

$$84 = 3.5x$$

$$\frac{84}{3.5} = \frac{3.5x}{3.5}$$

$$24 = x$$

The average mortgage loan will be $142 thousand in 2004.

7. $2x + 4y = 8$

$2x + 4y - 2x = -2x + 8$

$$4y = -2x + 8$$

$$\frac{4y}{4} = \frac{-2x + 8}{4}$$

$$y = \frac{-2x}{4} + \frac{8}{4}$$

$$y = -\frac{1}{2}x + 2$$

8.
$$L = \frac{P - 2W}{2}$$
$$2 \cdot L = 2 \cdot \frac{P - 2W}{2}$$
$$2L = P - 2W$$
$$2L + 2W = P - 2W + 2W$$
$$2L + 2W = P \text{ or } P = 2L + 2W$$

9. Let x = the number.
$$5x - 9 = 310$$
$$5x - 9 + 9 = 310 + 9$$
$$5x = 319$$
$$\frac{5x}{5} = \frac{319}{5}$$
$$x = 63.8$$
The number is 63.8.

10. Let x = Buchanan's age the time he took office, then
$x + 4$ = Reagan's age the time he took office.
$$x + x + 4 = 134$$
$$2x + 4 = 134$$
$$2x + 4 - 4 = 134 - 4$$
$$2x = 130$$
$$\frac{2x}{2} = \frac{130}{2}$$
$$x = 65$$
$$x + 4 = 69$$
At the time they took office, Buchanan was 65 and Reagan was 69.

11. Let x = the number of minutes
$$15 + 0.05x = 45$$
$$15 + 0.05x - 15 = 45 - 15$$
$$0.05x = 30$$
$$\frac{0.05x}{0.05} = \frac{30}{0.05}$$
$$x = 600$$
You can talk for 600 minutes, or 10 hours.

12. $\dfrac{10}{15 + 10} = \dfrac{10}{25} = \dfrac{10 \div 5}{25 \div 5} = \dfrac{2}{5}$ or $2:5$

13. $\dfrac{5}{8} = \dfrac{x}{12}$
$$8 \cdot x = 5 \cdot 12$$
$$8x = 60$$
$$\frac{8x}{8} = \frac{60}{8}$$
$$x = 7.5$$
The solution set is $\{7.5\}$.

14.
$$\frac{x + 5}{8} = \frac{x + 2}{5}$$
$$5(x + 5) = 8(x + 2)$$
$$5x + 25 = 8x + 16$$
$$5x + 25 - 25 = 8x + 16 - 25$$
$$5x = 8x - 9$$
$$5x - 8x = 8x - 9 - 8x$$
$$-3x = -9$$
$$\frac{-3x}{-3} = \frac{-9}{-3}$$
$$x = 3$$
The solution set is $\{3\}$.

15. Let x = number of elk in the park.
$$\frac{5}{150} = \frac{200}{x}$$
$$5x = 150 \cdot 200$$
$$5x = 30,000$$
$$\frac{5x}{5} = \frac{30,000}{5}$$
$$x = 6000$$
There are 6000 elk in the park.

16. Let x = pressure.
$$\frac{25}{60} = \frac{x}{330}$$
$$60x = 25 \cdot 330$$
$$60x = 8250$$
$$\frac{60x}{60} = \frac{8250}{60}$$
$$x = 137.5$$
The pressure will be 137.5 pounds per square inch.

17. Let x = the current when the resistance is 5 ohms.

$$\frac{5}{x} = \frac{4}{42}$$

$$4x = 5(42)$$

$$4x = 210$$

$$\frac{4x}{4} = \frac{210}{4}$$

$$x = 52.5$$

The current will be 52.5 amperes when the resistance is 5 ohms.

18.
$$6 - 9x \geq 33$$
$$6 - 9x - 6 \geq 33 - 6$$
$$-9x \geq 27$$
$$\frac{-9x}{-9} \leq \frac{27}{-9}$$
$$x \leq -3$$
$$\{x \mid x \leq -3\}$$

19.
$$4x - 2 > 2(x + 6)$$
$$4x - 2 > 2x + 12$$
$$4x - 2 + 2 > 2x + 12 + 2$$
$$4x > 2x + 14$$
$$4x - 2x > 2x + 14 - 2x$$
$$2x > 14$$
$$\frac{2x}{2} > \frac{14}{2}$$
$$x > 7$$
$$\{x \mid x > 7\}$$

20. Let x = grade on 4th examination.

$$\frac{76 + 80 + 72 + x}{4} \geq 80$$

$$\frac{228 + x}{4} \geq 80$$

$$4 \cdot \frac{228 + x}{4} \geq 80 \cdot 4$$

$$228 + x \geq 320$$

$$228 + x - 228 \geq 320 - 228$$

$$x \geq 92$$

The student must earn at least a 92 to receive a B.

21. $(2x - 5)(3x + 4) = 6x^2 + 8x - 15x - 20$
$$= 6x^2 - 7x - 20$$

22. $2x^2 - 9x + 10 = (2x - 5)(x - 2)$

23.
$$x^2 + 5x = 36$$
$$x^2 + 5x - 36 = 0$$
$$(x + 9)(x - 4) = 0$$
$$x + 9 = 0 \quad \text{or} \quad x - 4 = 0$$
$$x = -9 \qquad x = 4$$

The solution set is $\{-9, 4\}$.

24. $2x^2 + 4x = -1$
$$2x^2 + 4x + 1 = 0$$
$$x = \frac{-b \pm \sqrt{b^2 - 4ac}}{2a}$$
$$x = \frac{-4 \pm \sqrt{4^2 - 4(2)(1)}}{2(2)}$$
$$x = \frac{-4 \pm \sqrt{8}}{4}$$
$$x = \frac{-4 \pm 2\sqrt{2}}{4}$$
$$x = \frac{-2 \pm \sqrt{2}}{2}$$

The solution set is $\left\{ \frac{-2 - \sqrt{2}}{2}, \ \frac{-2 + \sqrt{2}}{2} \right\}$.

Check Points 7.1

1.

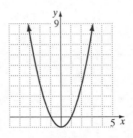

A graph with points plotted: $A(-2, 4)$, $C(-3, 0)$, $D(0, -3)$, $B(4, -2)$

2.

x	$y = x^2 - 1$	(x, y)
-3	$y = (-3)^2 - 1 = 9 - 1 = 8$	$(-3, 8)$
-2	$y = (-2)^2 - 1 = 4 - 1 = 3$	$(-2, 3)$
-1	$y = (-1)^2 - 1 = 1 - 1 = 0$	$(-1, 0)$
0	$y = (0)^2 - 1 = 0 - 1 = -1$	$(0, -1)$
1	$y = (1)^2 - 1 = 1 - 1 = 0$	$(1, 0)$
2	$y = (2)^2 - 1 = 4 - 1 = 3$	$(2, 3)$
3	$y = (3)^2 - 1 = 9 - 1 = 8$	$(3, 8)$

3. a.

Without the coupon book

x	$y = 2x$	(x, y)
0	$y = 2(0) = 0$	$(0, 0)$
2	$y = 2(2) = 4$	$(2, 4)$
4	$y = 2(4) = 8$	$(4, 8)$
6	$y = 2(6) = 12$	$(6, 12)$
8	$y = 2(8) = 16$	$(8, 16)$
10	$y = 2(10) = 20$	$(10, 20)$
12	$y = 2(12) = 24$	$(12, 24)$

With the coupon book

x	$y = 10 + x$	(x, y)
0	$y = 10 + 0 = 10$	$(0, 10)$
2	$y = 10 + 2 = 12$	$(2, 12)$
4	$y = 10 + 4 = 14$	$(4, 14)$
6	$y = 10 + 6 = 16$	$(6, 16)$
8	$y = 10 + 8 = 18$	$(8, 18)$
10	$y = 10 + 10 = 20$	$(10, 20)$
12	$y = 10 + 12 = 22$	$(12, 22)$

b.

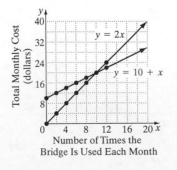

c. The graphs intersect at $(10, 20)$. This means that if the bridge is used ten times in a month, the total monthly cost is $20 with or without the coupon book.

4. a. We can express $y = -0.05x^2 + 4.2x - 26$ in function notation by replacing y by $f(x)$.

The formula in function notation is $f(x) = -0.05x^2 + 4.2x - 26$

b. $f(x) = -0.05x^2 + 4.2x - 26$

$f(20) = -0.05(20)^2 + 4.2(20) - 26$

$\quad\quad = -20 + 84 - 26$

$\quad\quad = 38$

c. We see that $f(20) = 38$, so 38% of 20-year-old coffee drinkers become irritable if they do not have coffee at their regular time.

5.

x	$f(x) = 2x$	(x, y) or $(x, f(x))$
-2	$f(-2) = 2(-2) = -4$	$(-2, -4)$
-1	$f(-1) = 2(-1) = -2$	$(-1, -2)$
0	$f(0) = 2(0) = 0$	$(0, 0)$
1	$f(1) = 2(1) = 2$	$(1, 2)$
2	$f(2) = 2(2) = 4$	$(2, 4)$

x	$g(x) = 2x - 3$	(x, y) or $(x, f(x))$
-2	$g(-2) = 2(-2) - 3 = -7$	$(-2, -7)$
-1	$g(-1) = 2(-1) - 3 = -5$	$(-1, -5)$
0	$g(0) = 2(0) - 3 = -3$	$(0, -3)$
1	$g(1) = 2(1) - 3 = -1$	$(1, -1)$
2	$g(2) = 2(2) - 3 = 1$	$(2, 1)$

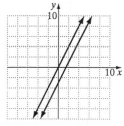

The graph of g is the graph of f shifted vertically down 3 units.

6. a. y is a function of x.

b. y is a function of x.

c. y is not a function of x. Two values of y correspond to an x-value.

7. To find $f(10)$, locate 10 on the x-axis. Follow that value up to the graph, then look to the y-axis to find the corresponding y-coordinate. A reasonable estimate of the y-coordinate is 16. Thus $f(10) \approx 16$.

Exercise Set 7.1

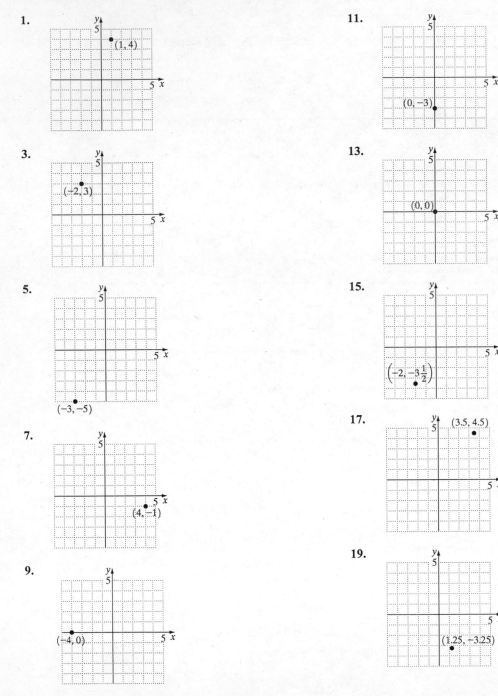

1.

3.

5.

7.

9.

11.

13.

15.

17.

19.

21.

x	−3	−2	−1	0	1	2	3
$y = x^2 - 2$	7	2	−1	−2	−1	2	7

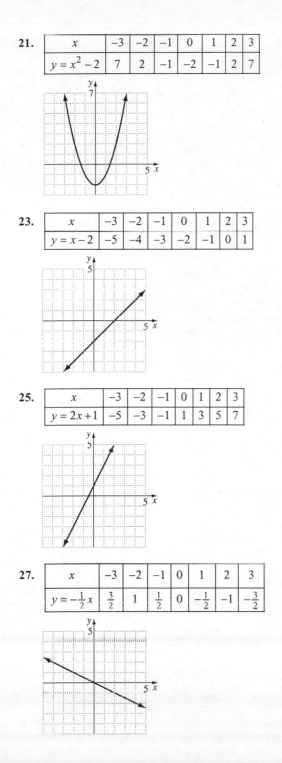

23.

x	−3	−2	−1	0	1	2	3
$y = x - 2$	−5	−4	−3	−2	−1	0	1

25.

x	−3	−2	−1	0	1	2	3
$y = 2x + 1$	−5	−3	−1	1	3	5	7

27.

x	−3	−2	−1	0	1	2	3
$y = -\frac{1}{2}x$	$\frac{3}{2}$	1	$\frac{1}{2}$	0	$-\frac{1}{2}$	−1	$-\frac{3}{2}$

29.

x	−3	−2	−1	0	1	2	3
$y = x^3$	−27	−8	−1	0	1	8	27

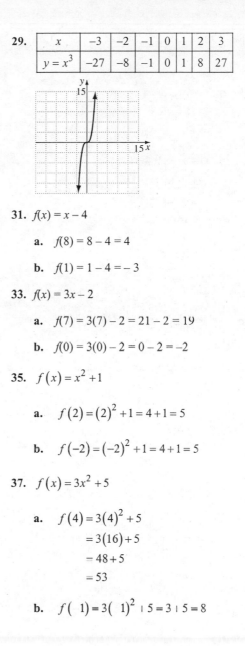

31. $f(x) = x - 4$

    a. $f(8) = 8 - 4 = 4$

    b. $f(1) = 1 - 4 = -3$

33. $f(x) = 3x - 2$

    a. $f(7) = 3(7) - 2 = 21 - 2 = 19$

    b. $f(0) = 3(0) - 2 = 0 - 2 = -2$

35. $f(x) = x^2 + 1$

    a. $f(2) = (2)^2 + 1 = 4 + 1 = 5$

    b. $f(-2) = (-2)^2 + 1 = 4 + 1 = 5$

37. $f(x) = 3x^2 + 5$

    a. $f(4) = 3(4)^2 + 5$
            $= 3(16) + 5$
            $= 48 + 5$
            $= 53$

    b. $f(1) = 3(1)^2 + 5 = 3 + 5 = 8$

133

39. $f(x) = 2x^2 + 3x - 1$

   a. $f(3) = 2(3)^2 + 3(3) - 1$
$$= 2(9) + 9 - 1$$
$$= 18 + 9 - 1$$
$$= 26$$

   b. $f(-4) = 2(-4)^2 + 3(-4) - 1$
$$= 2(16) - 12 - 1$$
$$= 32 - 12 - 1$$
$$= 19$$

41.

x	$f(x) = x^2 - 1$
-2	3
-1	0
0	-1
1	0
2	3

43.

x	$f(x) = x - 1$
-2	-3
-1	-2
0	-1
1	0
2	1

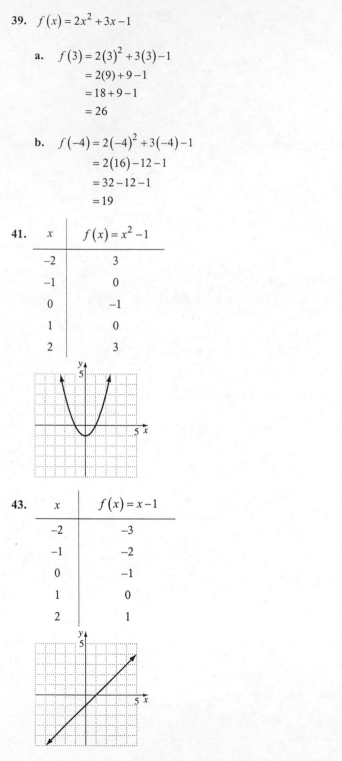

45.

x	$f(x) = (x-2)^2$
0	4
1	1
2	0
3	1
4	4

47.

x	$f(x) = x^3 + 1$
-2	-7
-1	0
0	1
1	2
2	9

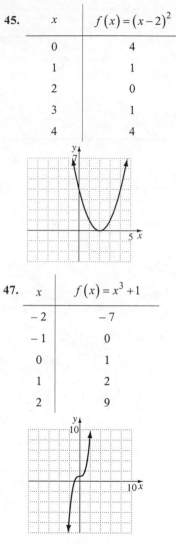

49. y is a function of x.

51. y is a function of x.

53. y is not a function of x.
Two values of y correspond to an x-value.

55. y is a function of x.

57. A (2, 7); The ball is 7 feet above the ground when it is 2 yards from the quarterback.

59. C (6, 9.25)

61. The maximum height is 12 feet. It reaches this maximum when it is 15 yards from the quarterback.

63. $f(20) = 0.76(20) + 171.4 = 186.6$

At age 20 an average American man's cholesterol level is expected to be 186.6.

65. $f(4) = -\dfrac{1}{2}(4)^2 + 4(4) + 19 = -8 + 16 + 19 = 27$

In 1994, 27 million people received food stamps.

67. (4, 27); This is the highest point on the graph. It represents when the most food stamps were given out after 1990.

69. $f(60) \approx 3.1$; In 1960, Jewish Americans comprised about 3.1% of the U.S. Population.

71. The percentage of Jewish American in the U.S. population reached a maximum of 3.7% in 1940 .

73. f is a function of x because it's graph passes the vertical line test.

75. a.

x	$f(x) = 0.1x^2 - 0.4 + 0.6$
0	0.6
1	0.3
2	0.2
3	0.3
4	0.6
5	1.1

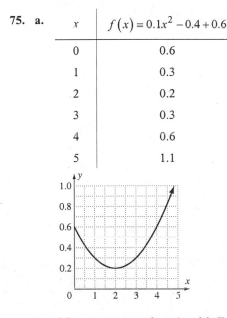

b. 0.3 ppm corresponds to 1 and 3. To avoid unsafe air, runners should exercise between 1 and 3 hours after 9 A.M., which means between 10 A.M. and 12 P.M.

77-81. Answers will vary.

83. a. $f(x) = -\dfrac{x}{2}$ if x is even.

b. $f(x) = \dfrac{x+1}{2}$ if x is odd.

c. $f(20) + f(40) + f(65)$

$= \left(-\dfrac{20}{2}\right) + \left(-\dfrac{40}{2}\right) + \left(\dfrac{65+1}{2}\right)$

$= 3$

Check Points 7.2

1. Find the x-intercept by setting $y = 0$

$2x + 3(0) = 6$

$2x = 6$

$x = 3$; resulting point (3, 0)

Find the y-intercept by setting $x = 0$

$2(0) + 3y = 6$

$3y = 6$

$y = 2$; resulting point (0, 2)

Find a checkpoint by substituting any value.

$2(1) + 3y = 6$

$2 + 3y = 6$

$3y = 4$

$y = \dfrac{4}{3}$; resulting point $\left(1, \dfrac{4}{3}\right)$

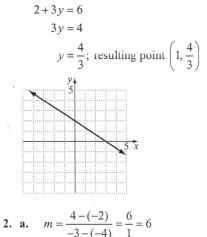

2. a. $m = \dfrac{4 - (-2)}{-3 - (-4)} = \dfrac{6}{1} = 6$

b. $m = \dfrac{5 - (-2)}{-1 - 4} = \dfrac{7}{-5} = -\dfrac{7}{5}$

3. Step 1. Plot the y-intercept of $(0, 1)$

Step 2. Obtain a second point using the slope m.

$$m = \frac{3}{5} = \frac{\text{Rise}}{\text{Run}}$$

Starting from the y-intercept move up 3 units and move 5 units to the right. This puts the second point at $(3, 6)$.

Step 3. Draw the line through the two points.

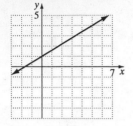

4. Solve for y.

$$3x + 4y = 0$$

$$4y = -3x + 0$$

$$\frac{4y}{4} = \frac{-3x}{4} + \frac{0}{4}$$

$$y = \frac{-3}{4}x + 0$$

$m = \dfrac{-3}{4}$ and the y-intercept is $(0, 0)$

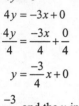

5. Draw horizontal line that intersects the y-axis at 3.

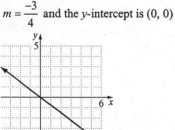

6. Draw vertical line that intersects the x-axis at -2.

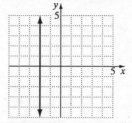

Exercise Set 7.2

1. Find the x-intercept by setting $y = 0$

$$x - y = 3$$

$$x - 0 = 3$$

$$x = 3; \text{ resulting point } (3, 0)$$

Find the y-intercept by setting $x = 0$

$$0 - y = 3$$

$$-y = 3$$

$$y = -3; \text{ resulting point } (0, -3)$$

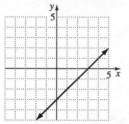

3. Find the x-intercept by setting $y = 0$

$$3x - 4(0) = 12$$

$$3x = 12$$

$$x = 4; \text{ resulting point } (4, 0)$$

Find the y-intercept by setting $x = 0$

$$3(0) - 4y = 12$$

$$-4y = 12$$

$$y = -3; \text{ resulting point } (0, -3)$$

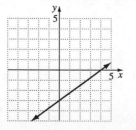

5. Find the x-intercept by setting $y = 0$
$$2x + 0 = 6$$
$$2x = 6$$
$$x = 3; \text{ resulting point } (3, 0)$$
Find the y-intercept by setting $x = 0$
$$2(0) + y = 6$$
$$y = 6; \text{ resulting point } (0, 6)$$

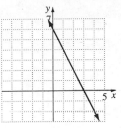

7. Find the x-intercept by setting $y = 0$
$$5x = 3(0) - 15$$
$$5x = -15$$
$$x = -3; \text{ resulting point } (-3, 0)$$
Find the y-intercept by setting $x = 0$
$$5(0) = 3y - 15$$
$$0 = 3y - 15$$
$$-3y = -15$$
$$y = 5; \text{ resulting point } (0, 5)$$

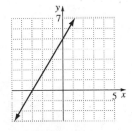

9. $m = \dfrac{5-6}{3-2} = \dfrac{-1}{1} = -1$; line falls.

11. $m = \dfrac{2-1}{2-(-2)} = \dfrac{1}{4}$; line rises.

13. $m = \dfrac{-1-4}{-1-(-2)} = \dfrac{-5}{1} = -5$; line falls.

15. $m = \dfrac{-2-3}{5-5} = \dfrac{-5}{0}$;
Slope undefined. Line is vertical.

17. $m = \dfrac{8-0}{0-2} = \dfrac{8}{-2} = -4$; line falls.

19. $m = \dfrac{1-1}{-2-5} = \dfrac{0}{-7} = 0$; line is horizontal.

21. $y = 2x + 3$
Slope: 2, y-intercept: 3
Plot point $(0, 3)$ and second point using
$$m = \frac{2}{1} = \frac{\text{rise}}{\text{run}}$$

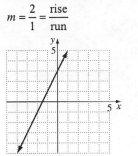

23. $y = -2x + 4$
Slope: -2, y-intercept: 4
Plot point $(0, 4)$ and second point using
$$m = \frac{-2}{1} = \frac{\text{rise}}{\text{run}}$$

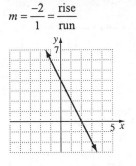

25. $y = \dfrac{1}{2}x + 3$

Slope: $\dfrac{1}{2}$, y-intercept: 3

Plot point $(0, 3)$ and second point using
$$m = \frac{1}{2} = \frac{\text{rise}}{\text{run}}.$$

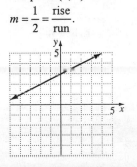

27. $f(x) = \dfrac{2}{3}x - 4$

Slope: $\dfrac{2}{3}$, y-intercept: -4

Plot point $(0, -4)$ and second point using

$m = \dfrac{2}{3} = \dfrac{\text{rise}}{\text{run}}$.

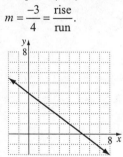

29. $y = -\dfrac{3}{4}x + 4$

Slope: $-\dfrac{3}{4}$, y-intercept: 4

Plot point $(0, 4)$ and second point using

$m = \dfrac{-3}{4} = \dfrac{\text{rise}}{\text{run}}$.

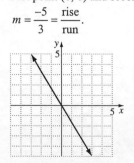

31. $f(x) = -\dfrac{5}{3}x$ or $f(x) = -\dfrac{5}{3}x + 0$

Slope: $-\dfrac{5}{3}$, y-intercept: 0

Plot point $(0, 0)$ and second point using

$m = \dfrac{-5}{3} = \dfrac{\text{rise}}{\text{run}}$.

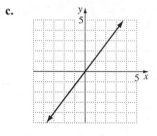

33. a. $3x + y = 0$

$y = -3x$ or $y = -3x + 0$

b. Slope $= -3$
y-intercept $= 0$

c.

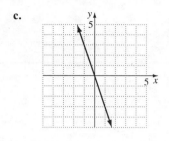

35. a. $3y = 4x$

$y = \dfrac{4}{3}x$ or $y = \dfrac{4}{3}x + 0$

b. Slope $= \dfrac{4}{3}$

y-intercept $= 0$

c.

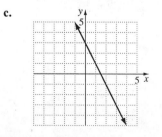

37. a. $2x + y = 3$

$y = -2x + 3$

b. Slope $= -2$
y-intercept $= 3$

c.

39. a.
$$7x + 2y = 14$$
$$2y = -7x + 14$$
$$y = -\frac{7}{2}x + 7$$

b. Slope $= -\frac{7}{2}$

y-intercept $= 7$

c.

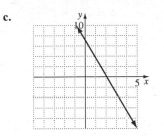

41. $y = 4$

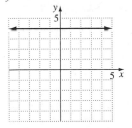

43. $y = -2$

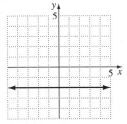

45. $x = 2$

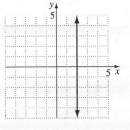

47. $x + 1 = 0$ or $x = -1$

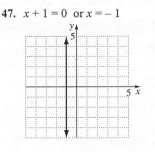

49. $m = \dfrac{1500 - 1000}{2001 - 1999} = \dfrac{500}{2} = 250$

The amount spent online per U.S. online household increased \$250 per year from 1999 to 2001.

51. $m = \dfrac{-286 - 50}{2010 - 2001} = \dfrac{-336}{9} \approx -37.33$

The federal budget surplus is expected to decrease \$37.33 billion per year from 2001 through 2010.

53. a. The y-intercept is 24. This represents an average retail prescription cost of \$24 in 1991.

b. $m = \dfrac{38 - 24}{7 - 0} = \dfrac{14}{7} = 2$; The cost increased at a rate of \$2 per year from 1991 to 2000.

c. $y = 2x + 24$

d. Since 2005 is 14 years after 1991, let $x = 14$.
$y = 2(14) + 24 = 28 + 24 = 52$; The expected average cost will be \$52 in 2005.

55-63. Answers will vary.

65. Let A be the coefficient of x and let B be the coefficient of y, giving the equation $Ax + By = 10$.
The x-intercept is (5, 0). The y-intercept is (0, 2)
Substitute these coordinates into the equation:

x-intercept: (5, 0)	y-intercept: (0, 2)
$Ax + By = 10$	$Ax + By = 10$
$A(5) + B(0) = 10$	$A(0) + B(2) = 10$
$5A = 10$	$2B = 10$
$A = 2$	$B = 5$

Thus the equation is $2x + 5y = 10$

67. $m = 2$

69. $m = -\dfrac{1}{2}$

Check Points 7.3

1. $f(x) = 0.4x^2 - 36x + 1000$

$$f(30) = 0.4(30)^2 - 36(30) + 1000$$
$$= 0.4(900) - 36(30) + 1000$$
$$= 360 - 1080 + 1000$$
$$= 280$$

Thus $f(30) = 280$ which indicates that 30-year-olds have 280 accidents per 50 million miles driven.

2. **a.** The coefficient, a, of x^2 is 1. Since $a > 0$, the parabola opens up.

b.

x	$y = x^2 - 6x + 8$	(x, y)
0	$y = (0)^2 - 6(0) + 8 = 8$	$(0, 8)$
1	$y = (1)^2 - 6(1) + 8 = 3$	$(1, 3)$
2	$y = (2)^2 - 6(2) + 8 = 0$	$(2, 0)$
3	$y = (3)^2 - 6(3) + 8 = -1$	$(3, -1)$
4	$y = (4)^2 - 6(4) + 8 = 0$	$(4, 0)$
5	$y = (5)^2 - 6(5) + 8 = 3$	$(5, 3)$
6	$y = (6)^2 - 6(6) + 8 = 8$	$(6, 8)$

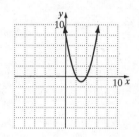

3. Replace y with 0 and solve the equation for x by factoring.

$$x^2 - 6x + 8 = 0$$
$$(x - 4)(x - 2) = 0$$
$$x - 4 = 0 \quad \text{or} \quad x - 2 = 0$$
$$x = 4 \qquad\qquad x = 2$$

Thus the x-intercepts are 4 and 2, which are located at the points (4, 0) and (2, 0).

4. Replace x with 0 and solve the equation for y.

$$y = x^2 - 6x + 8$$
$$y = (0)^2 - 6(0) + 8$$
$$y = 8$$

Thus the y-intercept is 8, which is located at the point (0, 8).

5. In the equation $y = x^2 - 6x + 8$, $a = 1$ and $b = -6$.

$$x\text{-coordinate of vertex } = \frac{-b}{2a} = \frac{-(-6)}{2(1)} = \frac{6}{2} = 3$$

To find the y-coordinate of vertex, substitute 3 for x in $y = x^2 - 6x + 8$, and evaluate.

$$y\text{-coordinate of vertex } = (3)^2 - 6(3) + 8 = 9 - 18 + 8 = -1$$

Thus, the vertex is the point $(3, -1)$.

6. Step 1. Since $a > 0$, the parabola opens upward ($a = 1$).

Step 2. Find the vertex given $a = 1$ and $b = 6$.

x-coordinate of vertex $= \dfrac{-b}{2a} = \dfrac{-6}{2(1)} = \dfrac{-6}{2} = -3$

y-coordinate of vertex $= (-3)^2 + 6(-3) + 5 = 9 - 18 + 5 = -4$

Thus, the vertex is the point $(-3, -4)$.

Step 3. Replace y with 0 and solve the equation for x by factoring.

$$x^2 + 6x + 5 = 0$$
$$(x + 5)(x + 1) = 0$$
$$x + 5 = 0 \quad \text{or} \quad x + 1 = 0$$
$$x = -5 \qquad\qquad x = -1$$

Thus the x-intercepts are -5 and -1, , which are located at the points $(-5, 0)$ and $(-1, 0)$.

Step 4. Replace x with 0 and solve the equation for y.

$$y = x^2 + 6x + 5$$
$$y = (0)^2 + 6(0) + 5$$
$$y = 5$$

Thus the y-intercept is 5, which is located at the point $(0, 5)$.

Steps 5 and 6. Plot the intercepts at the vertex. Connect these points with a smooth curve.

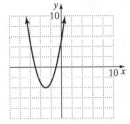

7. Step 1. Since $a < 0$, the parabola opens downward ($a = -1$).

Step 2. Find the vertex given $a = -1$ and $b = -2$.

x-coordinate of vertex $= \dfrac{-b}{2a} = \dfrac{-(-2)}{2(-1)} = \dfrac{2}{-2} = -1$

y-coordinate of vertex $= -(-1)^2 - 2(-1) + 5 = -1 + 2 + 5 = 6$

Thus, the vertex is the point $(-1, 6)$.

Step 3. Replace y with 0 and solve for x by using the quadratic formula ($a = -1$, $b = -2$, and $c = 5$).

$$x = \frac{-b \pm \sqrt{b^2 - 4ac}}{2a} = \frac{-(-2) \pm \sqrt{(-2)^2 - 4(-1)(5)}}{2(-1)} = \frac{2 \pm \sqrt{24}}{-2}$$

$$x = \frac{2 + \sqrt{24}}{-2} \approx -3.4 \quad \text{or} \quad x = \frac{2 - \sqrt{24}}{?} \approx 1.4$$

Thus the points are $(-3.4, 0)$ and $(1.4, 0)$.

Step 4.   Replace x with 0 and solve the equation for y.

$$y = -x^2 - 2x + 5 = -(0)^2 - 2(0) + 5 = 5$$

Thus the y-intercept is 5, which is located at the point (0, 5).

Steps 5 and 6.   Plot the intercepts at the vertex. Connect these points with a smooth curve.

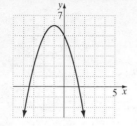

8. In the equation $y = -0.5x^2 + 4x + 19$, $a = -0.5$ and $b = 4$.

x-coordinate of vertex $= \dfrac{-b}{2a} = \dfrac{-4}{2(-0.5)} = 4$

To find the y-coordinate of vertex, substitute 4 for x in $y = -0.5x^2 + 4x + 19$, and evaluate.

y-coordinate of vertex $= -0.5(4)^2 + 4(4) + 19 = -8 + 16 + 19 = 27$

Thus, the vertex is the point $(4, 27)$. This represents that the number of people receiving food reached a maximum of 27 million in 1994, 4 years after 1990.

Exercise Set 7.3

1. $y = x^2 - 4x + 3$

    $a = 1 > 0$; the parabola opens upward.

3. $y = -2x^2 + x + 6$

    $a = -2 < 0$; the parabola opens downward.

5. $y = x^2 - 4x + 3$

    Replace y with 0.

    $0 = x^2 - 4x + 3$

    $0 = (x - 3)(x - 1)$

    $x - 3 = 0$ or $x - 1 = 0$

    $x = 3 \qquad x = 1$

    The x-intercepts are 3 and 1.
    The parabola passes through (3, 0) and (1, 0).

7. $y = -x^2 + 8x - 12$

    Replace y with 0.

    $0 = -x^2 + 8x - 12$ or $x^2 - 8x + 12 = 0$

    $(x - 6)(x - 2) = 0$

    $x - 6 = 0$ or $x - 2 = 0$

    $x = 6 \qquad x = 2$

    The x-intercepts are 6 and 2.
    The parabola passes through (6, 0) and (2, 0).

9. $y = x^2 + 2x - 4$

    Replace y with 0.

    $0 = x^2 + 2x - 4$

    $x = \dfrac{-b \pm \sqrt{b^2 - 4ac}}{2a}$

    $= \dfrac{-2 \pm \sqrt{2^2 - 4(1)(-4)}}{2(1)}$

    $= \dfrac{-2 \pm \sqrt{20}}{2}$

    $= -1 \pm \sqrt{5}$

    $x \approx -3.2 \quad x \approx 1.2$

    The x-intercepts are approximately -3.2 and 1.2.
    The parabola passes through
    $(-3.2, 0)$ and $(1.2, 0)$.

11. $y = x^2 - 4x + 3$

Replace x with 0.

$y = 0^2 - 4(0) + 3 = 3$

The y-intercept is 3.

The parabola passes through $(0, 3)$.

13. $y = -x^2 + 8x - 12$

Replace x with 0.

$y = -(0)^2 + 8(0) - 12 = -12$

The y-intercept is -12.

The parabola passes through $(0, -12)$.

15. $y = x^2 + 2x - 4$

Replace x with 0.

$y = (0)^2 + 2(0) - 4 = -4$

The y-intercept is -4.

The parabola passes through $(0, -4)$.

17. $y = x^2 + 6x$

Replace x with 0.

$y = (0)^2 + 6(0) = 0.$

The y-intercept is 0.

The parabola passes through $(0, 0)$.

19. Because $y = x^2 - 4x + 3$, a (the coefficient of x^2) is 1 and b (the coefficient of x) is -4. The x-coordinate of the vertex is:

$$x = \frac{-b}{2a} = \frac{-(-4)}{2(1)} = \frac{4}{2} = 2$$

Substitute 2 for x in the equation of the function to find the y-coordinate of the vertex.

$y = (2)^2 - 4(2) + 3 = 4 - 8 + 3 = -1$

The y-coordinate of the vertex is -1.

The vertex is at $(2, -1)$.

21. Because $y = 2x^2 + 4x - 6$, a (the coefficient of x^2) is 2 and b (the coefficient of x) is 4. The x-coordinate of the vertex is:

$$x = \frac{-b}{2a} = \frac{-4}{2(2)} = \frac{-4}{4} = -1$$

Substitute -1 for x in the equation of the function to find the y-coordinate of the vertex.

$y = 2(-1)^2 + 4(-1) - 6 = 2 - 4 - 6$

$\quad = -8$

The y-coordinate of the vertex is -8.

The vertex is at $(-1, -8)$.

23. Because $y = x^2 + 6x$, a, (the coefficient of x^2) is 1 and b (the coefficient of x) is 6. The x-coordinate of the vertex is:

$$x = \frac{-b}{2a} = \frac{-6}{2(1)} = -3$$

Substitute -3 for x in the equation of the function to find the y-coordinate of the vertex.

$y = (-3)^2 + 6(-3) = 9 - 18 = -9$

The y-coordinate of the vertex is -9.

The vertex is at $(-3, -9)$.

25. $y = x^2 + 8x + 7$

The parabola opens upward because $a > 0$.

$$x = \frac{-b}{2a} = \frac{-8}{2(1)} = -4$$

$y = (-4)^2 + 6(-4) + 5 = -3$

The vertex is at $(-4, -3)$.

Find the x-intercepts.

$0 = x^2 + 8x + 7$

$0 = (x + 7)(x + 1)$

$x + 7 = 0 \quad$ or $\quad x + 1 = 0$

$\quad x = -7 \qquad\qquad x = -1$

The parabola passes through $(-7, 0)$, and $(-1, 0)$.

Find the y-intercept.

$y = 0^2 + 8(0) + 7$

$y = 7$

The parabola passes through $(0, 7)$.

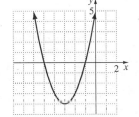

27. $y = x^2 - 2x - 8$

The parabola opens upward because $a > 0$.

$x = \dfrac{-b}{2a} = \dfrac{-(-2)}{2(1)} = 1$

$y = (1)^2 - 2(1) - 8 = -9$

The vertex is at $(1, -9)$.

Find the x-intercepts.

$0 = x^2 - 2x - 8$

$0 = (x + 2)(x - 4)$

$x + 2 = 0 \quad$ or $\quad x - 4 = 0$

$\quad x = -2 \qquad\qquad x = 4$

The parabola passes through $(-2, 0)$, and $(4, 0)$.

Find the y-intercept.

$y = 0^2 - 2 \cdot 0 - 8$

$y = -8$

The parabola passes through $(0, -8)$.

29. $y = -x^2 + 4x - 3$

The parabola opens downward because $a < 0$.

$x = \dfrac{-b}{2a} = \dfrac{-4}{2(-1)} = 2$

$y = -2^2 + 4 \cdot 2 - 3 = 1$

The vertex is at $(2, 1)$.

Find the x-intercepts.

$0 = -x^2 + 4x - 3$ or $x^2 - 4x + 3 = 0$

$(x - 1)(x - 3) = 0$

$x - 1 = 0$ or $x - 3 = 0$

$\quad x = 1 \qquad\qquad x = 3$

The parabola passes through $(1, 0)$ and $(3, 0)$.

Find the y-intercept.

$y = -0^2 + 4 \cdot 0 - 3$

$y = -3$

The parabola passes through $(0, -3)$.

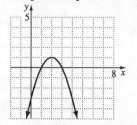

31. $y = x^2 - 1$

The parabola opens upward because $a > 0$.

$x = \dfrac{-b}{2a} = \dfrac{-0}{2(1)} = 0$

$y = 0^2 - 1 = -1$

The vertex is at $(0, -1)$.

Find the x-intercepts.

$0 = x^2 - 1$

$0 = (x + 1)(x - 1)$

$x + 1 = 0$ or $x - 1 = 0$

$\quad x = -1 \qquad\qquad x = 1$

The parabola passes through $(-1, 0)$, and $(1, 0)$.

Find the y-intercept.

$y = 0^2 - 1$

$y = -1$

The parabola passes through $(0, -1)$.

33. $y = x^2 + 2x + 1$

The parabola opens upward because $a > 0$.

$x = \dfrac{-b}{2a} = \dfrac{-2}{2(1)} = -1$

$y = (-1)^2 + 2(-1) + 1 = 0$

The vertex is $(-1, 0)$.

Find the x-intercept.

$\quad 0 = x^2 + 2x + 1$

$\quad 0 = (x + 1)(x + 1)$

$x + 1 = 0$

$\quad x = -1$

The parabola passes through $(-1, 0)$.
Find the y-intercept.
$y = 0^2 + 2 \cdot 0 + 1$
$y = 1$
The parabola passes through $(0, 1)$.

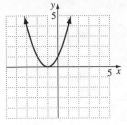

35. $f(x) = -2x^2 + 4x + 5$
The parabola opens downward because $a < 0$.
$x = \dfrac{-b}{2a} = \dfrac{-4}{2(-2)} = 1$

$y = f(1) = -2(1)^2 + 4(1) + 5 = 7$
The vertex is at $(1, 7)$.
Find the x-intercepts.
$0 = -2x^2 + 4x + 5$

$x = \dfrac{-4 \pm \sqrt{4^2 - 4(-2)(5)}}{2(-2)}$

$x \approx -0.9$ or $x \approx 2.9$
The parabola passes through the approximate points $(-0.9, 0)$ and $(2.9, 0)$.
Find the y-intercept.
$y = f(0) = -2(0)^2 + 4(0) + 5$
$y = f(0) = 5$
The parabola passes through $(0, 5)$.

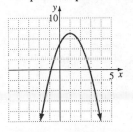

37. $f(50) = 0.036(50)^2 - 2.8(50) + 58.14$
$\qquad = 90 - 140 + 58.14$
$\qquad = 8.14$
There are 8.14 deaths per year per thousand people among 50-year-olds.

39. $(50, 8.14)$

41. 57-year-olds

43. In the function $y = -16x^2 + 200x + 4$,
$\quad a = -16$ and $b = 200$.
x-coordinate of vertex
$\quad = \dfrac{-b}{2a} = \dfrac{-200}{2(-16)} = 6.25$
y-coordinate of vertex
$\quad = -16(6.25)^2 + 200(6.25) + 4 = 629$
The vertex $(6.25, 629)$ represents that after 6.25 seconds the firework reaches it's greatest height of 629 feet.

45. In the function $y = 0.022x^2 - 0.4x + 60.07$,
$\quad a = 0.022$ and $b = -0.4$.
x-coordinate of vertex
$\quad = \dfrac{-b}{2a} = \dfrac{-(0.4)}{2(0.022)} \approx 9$
y-coordinate of vertex
$\quad = 0.022(9)^2 - 0.4(9) + 60.07 \approx 58$
The vertex $(9, 58)$ represents that 9 years after 1960, or 1969, women's earnings were 58% of men's earnings.

47-53. Answers will vary.

55. There are two x-intercepts because the vertex is above the x-axis and the parabola opens downward $(a < 0)$.

57. There are no x-intercepts because the vertex is above the x-axis and the parabola opens upward $(a > 0)$.

Check Points 7.4

1. Let $x = -3$: $y = 3^{-3} = \dfrac{1}{3^3} = \dfrac{1}{27}$

   Let $x = -2$: $y = 3^{-2} = \dfrac{1}{3^2} = \dfrac{1}{9}$

   Let $x = -1$: $y = 3^{-1} = \dfrac{1}{3^1} = \dfrac{1}{3}$

   Let $x = 0$: $y = 3^0 = 1$

   Let $x = 1$: $y = 3^1 = 3$

   Let $x = 2$: $y = 3^2 = 9$

   Let $x = 3$: $y = 3^3 = 27$

x	-3	-2	-1	0	1	2	3
$y = 3^x$	$\frac{1}{27}$	$\frac{1}{9}$	$\frac{1}{3}$	1	3	9	27

   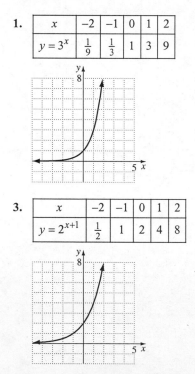

2. Let $x = -3$: $y = \left(\dfrac{1}{3}\right)^{-3} = \dfrac{1}{\left(\frac{1}{3}\right)^3} = \dfrac{1}{\frac{1}{27}} = 27$

   Let $x = -2$: $y = \left(\dfrac{1}{3}\right)^{-2} = \dfrac{1}{\left(\frac{1}{3}\right)^2} = \dfrac{1}{\frac{1}{9}} = 9$

   Let $x = -1$: $y = \left(\dfrac{1}{3}\right)^{-1} = \dfrac{1}{\left(\frac{1}{3}\right)^1} = \dfrac{1}{\frac{1}{3}} = 3$

   Let $x = 0$: $y = \left(\dfrac{1}{3}\right)^0 = 1$

   Let $x = 1$: $y = \left(\dfrac{1}{3}\right)^1 = \dfrac{1}{3}$

   Let $x = 2$: $y = \left(\dfrac{1}{3}\right)^2 = \dfrac{1}{9}$

   Let $x = 3$: $y = \left(\dfrac{1}{3}\right)^3 = \dfrac{1}{27}$

x	-3	-2	-1	0	1	2	3
$y = \left(\frac{1}{3}\right)^x$	27	9	3	1	$\frac{1}{3}$	$\frac{1}{9}$	$\frac{1}{27}$

   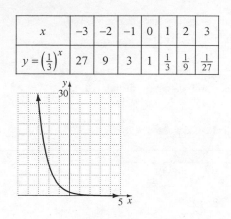

3. $f(60) = 13.49(0.967)^{60} - 1 \approx 0.8 \approx 1$. Thus one 0-ring is expected to fail at $60°\text{F}$.

4. $f(90) = 1000(0.5)^{\left(\frac{90}{30}\right)} = 125$. After 90 years, 125 kilograms of cesium-137 will be in the atmosphere. Because this exceeds 100, the Chernobyl area will still be unsafe in 2076.

5. $f(50) = 6e^{0.013(50)} \approx 11.49$. This indicates that world population in 2050 will be approximately 11.49 billion.

Exercise Set 7.4

1.

x	-2	-1	0	1	2
$y = 3^x$	$\frac{1}{9}$	$\frac{1}{3}$	1	3	9

3.

x	-2	-1	0	1	2
$y = 2^{x+1}$	$\frac{1}{2}$	1	2	4	8

5.

x	-2	-1	0	1	2
$y = 3^{x-1}$	$\frac{1}{27}$	$\frac{1}{9}$	$\frac{1}{3}$	1	3

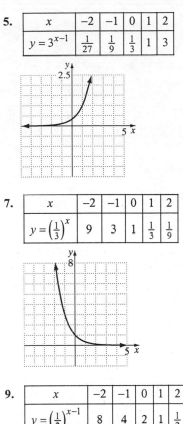

7.

x	-2	-1	0	1	2
$y = \left(\frac{1}{3}\right)^x$	9	3	1	$\frac{1}{3}$	$\frac{1}{9}$

9.

x	-2	-1	0	1	2
$y = \left(\frac{1}{2}\right)^{x-1}$	8	4	2	1	$\frac{1}{2}$

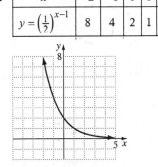

11. $y = 67.38(1.026)^x$

   a. $y = 67.38(1.026)^0 = 67.38$ million

   b. $y = 67.38(1.026)^{27} \approx 134.74$ million

   c. $2034 - 1980 = 54$
     $y = 67.38(1.026)^{54} \approx 269.46$ million

   d. $2061 - 1980 = 81$
     $y = 67.38(1.026)^{81} \approx 538.85$ million.

13. $S = 65,000(1+0.06)^{10} \approx \$116,405.10$

15. $f(40) = 1.6e^{0.039(40)} \approx \$7.61$

17. $f(30) = \dfrac{90}{1 + 270e^{-0.122(30)}} \approx 11.3\%$
About 11.3% of 30-year-olds have some coronary heart disease.

19-23. Answers will vary.

Check Points 7.5

  1. Replace x with 4 and y with -1.

$$x + 2y = 2 \qquad x - 2y = 6$$
$$4 + 2(-1) = 2 \qquad 4 - 2(-1) = 6$$
$$4 - 2 = 2 \qquad 4 + 2 = 6$$
$$2 = 2 \text{ true} \qquad 6 = 6 \text{ true}$$

The pair $(4, -1)$ is a solution of the system.

  2.

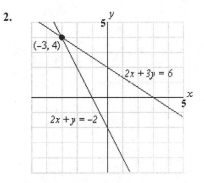

Check coordinates of intersection:

$$2x + 3y = 6 \qquad 2x + y = -2$$
$$2(-3) + 3(4) = 6 \qquad 2(-3) + (4) = -2$$
$$-6 + 12 = 6 \qquad -6 + 2 = -2$$
$$6 = 6 \text{ true} \qquad -2 = -2 \text{ true}$$

The solution set is $\{(-3, 4)\}$.

3. Step 1. Solve one of the equations for one variable: $y = 3x - 7$

Step 2. Substitute into the other equation:
$$5x - 2y = 8$$
$$5x - 2(\overset{y}{\overbrace{3x - 7}}) = 8$$

Step 3. Solve: $5x - 2(3x - 7) = 8$
$$5x - 6x + 14 = 8$$
$$-x + 14 = 8$$
$$-x = -6$$
$$x = 6$$

Step 4. Back-substitute the obtained value into the equation from step 1:
$$y = 3x - 7$$
$$y = 3(6) - 7$$
$$y = 11$$

Step 5. Check (6, 11) in both equations:
$$y = 3x - 7 \qquad 5x - 2y = 8$$
$$11 = 3(6) - 7 \quad 5(6) - 2(11) = 8$$
$$11 = 11, \text{ true} \qquad 8 = 8, \text{ true}$$
The solution set is $\{(6, 11)\}$.

4. Step 1. Solve one of the equations for one variable: $x - y = 3$
$$x = y + 3$$

Step 2. Substitute into the other equation:
$$3x + 2y = -1$$
$$3(\overset{x}{\overbrace{y + 3}}) + 2y = -1$$

Step 3. Solve: $3(y + 3) + 2y = -1$
$$3y + 9 + 2y = -1$$
$$5y + 9 = -1$$
$$5y = -10$$
$$y = -2$$

Step 4. Back-substitute the obtained value into the equation from step 1:
$$x = y + 3$$
$$x = -2 + 3$$
$$x = 1$$

Step 5. Check $(1, -2)$ in both equations:
$$x - y = 3 \qquad 3x + 2y = -1$$
$$1 - (-2) = 3 \qquad 3(1) + 2(-2) = -1$$
$$3 = 3, \text{ true} \qquad -1 = -1, \text{ true}$$
The solution set is $\{(1, -2)\}$.

5. $$2x - 5y = 26$$
$$\underline{-2x + 9y = -42}$$
$$4y = -16$$
$$y = -4$$

Substitute into either equation:

$$2x - 5\overset{y}{(-4)} = 26$$
$$2x + 20 = 26$$
$$2x = 6$$
$$x = 3$$

Check:

$$2(3) - 5(-4) = 26 \qquad -2(3) + 9(-4) = -42$$
$$6 + 20 = 26 \qquad\qquad -6 - 36 = -42$$
$$26 = 26 \text{ true} \qquad\qquad -42 = -42 \text{ true}$$

The solution set is $\{(3, -4)\}$.

6. Rewrite one or both equations:

$$4x + 5y = 3 \xrightarrow{\text{No change}} 4x + 5y = 3$$
$$2x - 3y = 7 \xrightarrow{\text{Mult. by } -2} \underline{-4x + 6y = -14}$$
$$11y = -11$$
$$y = -1$$

Back-substitute into either equation:

$$4x + 5y = 3$$
$$4x + 5(-1) = 3$$
$$4x - 5 = 3$$
$$4x = 8$$
$$x = 2$$

Checking confirms the solution set is $\{(2, -1\}$.

7. Rewrite both equations in the form $Ax + By = C$:

$$3x = 2 - 4y \quad \rightarrow \quad 3x + 4y = 2$$
$$5y = -1 - 2x \quad \rightarrow \quad 2x + 5y = -1$$

Rewrite with opposite coefficients, then add and solve:

$$3x + 4y = 2 \xrightarrow{\text{Mult. by } 2} 6x + 8y = 4$$
$$2x + 5y = -1 \xrightarrow{\text{Mult. by } -3} \underline{-6x - 15y - 3}$$
$$-7y = 7$$
$$y = -1$$

Back-substitute into either equation:

$$3x = 2 - 4y$$
$$3x = 2 - 4(-1)$$
$$3x = 6$$
$$x = 2$$

Checking confirms the solution set is $\{(2, -1\}$.

8. Rewrite with a pair of opposite coefficients, then add:

$x + 2y = 4$ $\xrightarrow{\text{Mult. by } -3}$ $-3x - 6y = -12$

$3x + 6y = 13$ $\xrightarrow{\text{No change}}$ $\underline{3x + 6y = 13}$

$\phantom{3x + 6y = 13 \xrightarrow{\text{No change}}}0 = 1$

The statement $0 = 1$ is false which indicates that the system has no solution. The solution set is the empty set, $\varnothing$.

9. Substitute $4x - 4$ for y in the other equation:

$$8x - 2\overbrace{(4x - 4)}^{y} = 8$$
$$8x - 8x + 8 = 8$$
$$8 = 8$$

The statement $8 = 8$ is true which indicates that the system has infinitely many solutions. The solution set is $\left\{(x, y) \mid y = 4x - 4\right\}$ or $\left\{(x, y) \mid 8x - 2y = 8\right\}$.

10. Step 1. Use variables to represent the unknown quantities:
Let x represent calories in a Quarter Pounder.
Let y represent calories in a Whopper with cheese.

Step 2. Write a system of equations:
$2x + 3y = 2607$
$x + y = 1009$

Step 3. Solve and answer:

$2x + 3y = 2607$ $\xrightarrow{\text{No change}}$ $2x + 3y = 2607$

$x + y = 1009$ $\xrightarrow{\text{Mult. by } -2}$ $\underline{-2x - 2y = -2018}$

$\phantom{x + y = 1009 \xrightarrow{\text{Mult. by } -2}}y = 589$

$x + y = 1009$
$x + 589 = 1009$
$x = 420$

Therefore a Quarter Pounder contains 420 calories and a Whopper with cheese contains 589 calories.

Step 4. Checking confirms the proposed answers:

$2x + 3y = 2607$ $\qquad$ $x + y = 1009$

$2(420) + 3(589) = 2607$ $\qquad$ $420 + 589 = 1009$

$840 + 1767 = 2607$ $\qquad$ $1009 = 1009, \text{ true}$

$2607 = 2607, \text{ true}$

Exercise Set 7.5

1. Replace x with 2 and y with 3.

$x + 3y = 11$ $\qquad$ $x - 5y = -13$

$2 + 3(3) = 11$ $\qquad$ $2 - 5(3) = 13$

$2 + 9 = 11$ $\qquad$ $2 - 15 = 13$

$11 = 11, \text{ true}$ $\qquad$ $13 = 13, \text{ true}$

The pair (2, 3) is a solution of the system.

3. Replace x with 2 and y with 5.
$$2x + 3y = 17$$
$$2(2) + 3(5) = 17$$
$$4 + 15 = 17$$
$$19 = 17, \text{ false.}$$
The pair (2, 5) is not a solution of the system.

5.

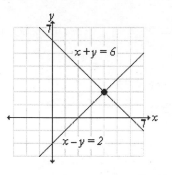

Check coordinates of intersection:
$$x + y = 6 \qquad x - y = 2$$
$$4 + 2 = 6 \qquad 4 - 2 = 2$$
$$6 = 6, \text{ true} \qquad 2 = 2, \text{ true}$$
The solution set is $\{(4, 2)\}$.

7.

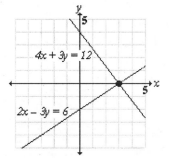

Check coordinates of intersection:
$$2x - 3y = 6 \qquad 4x + 3y = 12$$
$$2(3) - 3(0) = 6 \qquad 4(3) + (0) = 12$$
$$6 = 6, \text{ true} \qquad 12 = 12, \text{ true}$$
The solution set is $\{(3, 0)\}$.

9.

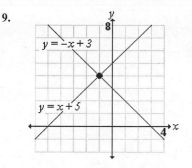

Check coordinates of intersection:
$$y = x + 5 \qquad y = -x + 3$$
$$4 = -1 + 5 \qquad 4 = -(-1) + 3$$
$$4 = 4, \text{ true} \qquad 4 = 4, \text{ true}$$
The solution set is $\{(-1, 4)\}$.

11.

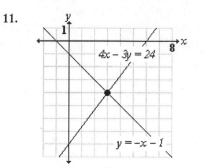

Check coordinates of intersection:
$$y = -x - 1 \qquad 4x - 3y = 24$$
$$-4 = -(3) - 1 \qquad 4(3) - 3(-4) = 24$$
$$-4 = -4, \text{ true} \qquad 24 = 24, \text{ true}$$
The solution set is $\{(3, -4)\}$.

13. $y = 3x \quad x + y = 4$
$$x + 3x = 4$$
$$4x = 4$$
$$x = 1$$
$$y = 3(1) = 3$$
The proposed solution is (1, 3)
Check: $3 - 3(1) \qquad 1 + 3 = 4$
$$3 = 3, \text{ true} \qquad 4 = 4, \text{ true}$$
The pair (1, 3) satisfies both equations.
The system's solution set is $\{(1, 3)\}$.

15. $y = 2x - 9 \quad x + 3y = 8$

$x + 3(2x - 9) = 8$

$x + 6x - 27 = 8$

$7x = 35$

$x = 5$

$y = 2(5) - 9 = 1$

The proposed solution is (5, 1).

Check:

$1 = 2(5) - 9 \quad 5 + 3(1) = 8$

$1 = 10 - 9 \qquad 5 + 3 = 8$

$1 = 1, \text{true} \qquad 8 = 8, \text{true}$

The pair (5, 1) satisfies both equations.

The system's solution set is $\{(5, 1)\}$.

17. $x + 3y = 5$

$x = 5 - 3y \quad 4x + 5y = 13$

$4(5 - 3y) + 5y = 13$

$20 - 12y + 5y = 13$

$20 - 7y = 13$

$-7y = -7$

$y = 1$

$x = 5 - 3(1) = 2$

The proposed solution is (2, 1).

Check:

$2 + 3(1) = 5 \qquad 4(2) + 5(1) = 13$

$5 = 5, \text{true} \qquad 8 + 5 = 13$

$\qquad\qquad\qquad 13 = 13, \text{true}$

The pair (2, 1) satisfies both equations.

The system's solution set is $\{(2, 1)\}$.

19. $2x - y = -5$

$y = 2x + 5 \quad x + 5y = 14$

$x + 5(2x + 5) = 14$

$x + 10x + 25 = 14$

$11x = -11$

$x = -1$

$y = 2(-1) + 5 = -2 + 5 = 3$

The proposed solution is (–1, 3).

Check:

$2(-1) - 3 = -5 \qquad -1 + 5(3) = 14$

$-2 - 3 = -5 \qquad -1 + 15 = 14$

$-5 = -5, \text{true} \qquad 14 = 14, \text{true}$

The pair (– 1, 3) satisfies both equations.

The system's solution set is $\{(-1, 3)\}$.

21. $2x - y = 3$

$y = 2x - 3 \quad 5x - 2y = 10$

$5x - 2(2x - 3) = 10$

$5x - 4x + 6 = 10$

$x = 4$

$y = 2(4) - 3 = 8 - 3 = 5$

The proposed solution is (4, 5).

Check:

$2(4) - 5 = 3 \qquad 5(4) - 2(5) = 10$

$8 - 5 = 3 \qquad 20 - 10 = 10$

$3 = 3, \text{true} \qquad 10 = 10, \text{true}$

The pair (4, 5) satisfies both equations.

The system's solution set is $\{(4, 5)\}$.

23. $x + 8y = 6$

$x = 6 - 8y \quad 2x + 4y = -3$

$2(6 - 8y) + 4y = -3$

$12 - 16y + 4y = -3$

$-12y = -15$

$\dfrac{-12y}{-12} = \dfrac{-15}{-12}$

$y = \dfrac{15}{12} = \dfrac{5}{4}$

$x = 6 - 8\left(\dfrac{5}{4}\right) = 6 - 10 = -4$

The proposed solution is $\left(-4, \dfrac{5}{4}\right)$

Check:

$-4 + 8\left(\dfrac{5}{4}\right) = 6 \quad 2(-4) + 4\left(\dfrac{5}{4}\right) = -3$

$-4 + 10 = 6 \qquad\qquad -8 + 5 = -3$

$6 = 6, \text{true} \qquad\qquad -3 = -3, \text{ true}$

The pair $\left(-4, \dfrac{5}{4}\right)$ satisfies both equations.

The system's solution set is $\left\{\left(-4, \dfrac{5}{4}\right)\right\}$.

25. $x + y = 1$

$\dfrac{x - y = 3}{2x = 4}$

$x = 2$

$x + y = 1$

$2 + y = 1$

$y = -1$

Check: $2 + (-1) = 1 \quad 2 - (-1) = 3$

$\qquad\qquad 1 = 1, \text{true} \qquad 3 = 3, \text{ true}$

The solution set is $\{(2, -1)\}$.

27.
$$2x + 3y = 6$$
$$\underline{2x - 3y = 6}$$
$$4x = 12$$
$$x = 3$$
$$2x + 3y = 6$$
$$2 \cdot 3 + 3y = 6$$
$$6 + 3y = 6$$
$$3y = 0$$
$$y = 0$$
Check:

$$2(3) + 3(0) = 6 \qquad 2(3) - 3(0) = 6$$
$$6 + 0 = 6 \qquad\qquad 6 - 0 = 6$$
$$6 = 6, \text{truc} \qquad\quad 6 = 6, \text{truc}$$

The solution set is $\{(3, 0)\}$.

29.
$$x + 2y = 2 \quad \text{Mult. by 3.} \quad 3x + 6y = 6$$
$$-4x + 3y = 25 \quad \text{Mult. by} -2. \quad \underline{8x - 6y = -50}$$
$$11x = -44$$
$$x = -4$$

$$x + 2y = 2$$
$$-4 + 2y = 2$$
$$2y = 6$$
$$y = 3$$
Check:

$$-4 + 2(3) = 2 \qquad -4(-4) + 3(3) = 25$$
$$-4 + 6 = 2 \qquad\qquad 16 + 9 = 25$$
$$2 = 2, \text{true} \qquad\quad 25 = 25, \text{ true}$$

The solution set is $\{(-4, 3)\}$.

31.
$$4x + 3y = 15 \quad \text{Mult. by 5.} \quad 20x + 15y = 75$$
$$2x - 5y = 1 \quad \text{Mult. by 3.} \quad \underline{6x - 15y = 3}$$
$$26x = 78$$
$$x = 3$$

$$4x + 3y = 15$$
$$4 \cdot 3 + 3y = 15$$
$$12 + 3y = 15$$
$$3y = 3$$
$$y = 1$$
Check:

$$4(3) + 3(1) = 15 \qquad 2(3) - 5(1) = 1$$
$$12 + 3 = 15 \qquad\qquad 6 - 5 = 1$$
$$15 = 15, \text{true} \qquad\quad 1 = 1, \text{true}$$

The solution set is $\{(3, 1)\}$.

33.
$$3x - 4y = 11 \quad \text{Mult. by 3.} \quad 9x - 12y = 33$$
$$2x + 3y = -4 \quad \text{Mult. by 4.} \quad \underline{8x + 12y = -16}$$
$$17x = 17$$
$$x = 1$$

$$2x + 3y = -4$$
$$2 \cdot 1 + 3y = -4$$
$$2 + 3y = -4$$
$$3y = -6$$
$$y = -2$$
Check:

$$3(1) - 4(-2) = 11$$
$$3 + 8 = 11$$
$$11 = 11, \text{true}$$
$$2(1) + 3(-2) = -4$$
$$2 - 6 = -4$$
$$-4 = -4, \text{true}$$

The solution set is $\{(1, -2)\}$.

35.
$$2x = 3y - 4 \quad \begin{matrix}\text{Rearrange and}\\\text{Mult. by 3.}\end{matrix} \quad 6x - 9y = -12$$
$$-6x + 12y = 6 \quad \text{No change.} \quad \underline{-6x + 12y = 6}$$
$$3y = -6$$
$$y = -2$$

$$2x = 3y - 4$$
$$2x = 3(-2) - 4$$
$$2x = -6 - 4$$
$$2x = -10$$
$$x = -5$$
Check:

$$2(-5) = 3(-2) - 4 \qquad -6(-5) + 12(-2) = 6$$
$$-10 = -6 - 4 \qquad\qquad 30 - 24 = 6$$
$$-10 = -10, \text{true} \qquad\quad 6 = 6, \text{true}$$

The solution set is $\{(-5, -2)\}$.

37.
$$x = 9 - 2y \quad x + 2y = 13$$
$$(9 - 2y) + 2y = 13$$
$$9 = 13 \quad \text{false}$$

The system has no solution.

The solution set is the empty set, $\varnothing$.

39. $y = 3x - 5 \quad 21x - 35 = 7y$

$\quad\quad 21x - 35 = 7(3x - 5)$

$\quad\quad 21x - 35 = 21x - 35$

$\quad\quad 21x - 21x = 35 - 35$

$\quad\quad\quad\quad\quad 0 = 0$, true

The system has infinitely many solutions. The solution set is $\{(x, y) | y = 3x - 5\}$.

41. $3x - 2y = -5 \quad$ No change. $\quad 3x - 2y = -5$

$\quad 4x + y = 8 \quad\quad$ Mult. by 2. $\quad\underline{8x + 2y = 16}$

$\quad\quad\quad\quad\quad\quad\quad\quad\quad\quad\quad\quad 11x = 11$

$\quad\quad\quad\quad\quad\quad\quad\quad\quad\quad\quad\quad x = 1$

$\quad 4x + y = 8$

$\quad 4(1) + y = 8$

$\quad\quad\quad\quad y = 4$

Check:

$\quad 3(1) - 2(4) = -5 \quad\quad 4(1) + (4) = 8$

$\quad\quad\quad 3 - 8 = -5 \quad\quad\quad\quad 4 + 4 = 8$

$\quad\quad\quad\quad -5 = -5$, true $\quad\quad\quad 8 = 8$, true

The solution set is $\{(1, 4)\}$.

43. $x + 3y = 2$

$\quad x = 2 - 3y \quad 3x + 9y = 6$

$\quad 3(2 - 3y) + 9y = 6$

$\quad\quad 6 - 9y + 9y = 6$

$\quad\quad\quad\quad\quad\quad 6 = 6$ true

The system has infinitely many solutions. The solution set is $\{(x, y) | x + 3y = 2\}$.

45. $x + y = 7$

$\quad \underline{x - y = -1}$

$\quad\quad 2x = 6$

$\quad\quad\quad x = 3$

$\quad y = 7 - x = 7 - 3 = 4$

The numbers are 3 and 4.

47. $3x - y = 1 \quad\quad$ Multiply by 2. $\quad 6x - 2y = 2$

$\quad x + 2y = 12 \quad$ No change. $\quad\quad\underline{x + 2y = 12}$

$\quad\quad\quad\quad\quad\quad\quad\quad\quad\quad\quad\quad 7x = 14$

$\quad\quad\quad\quad\quad\quad\quad\quad\quad\quad\quad\quad x = 2$

$\quad 3x - y = 1$

$\quad 3(2) - y = 1$

$\quad\quad\quad -y = -5$

$\quad\quad\quad\quad y = 5$

The numbers are 2 and 5.

49. Let x represent calories in a pan pizza.
Let y represent calories in a beef burrito.

$$x + 2y = 1980 \xrightarrow{\text{Mult. by } -2} -2x - 4y = -3960$$
$$2x + y = 2670 \xrightarrow{\text{No change}} \underline{\quad 2x + y = 2670 \quad}$$
$$-3y = -1290$$
$$y = 430$$

$$x + 2y = 1980$$
$$x + 2(430) = 1980$$
$$x + 860 = 1980$$
$$x = 1120$$

A pan pizza has 1120 calories and a beef burrito has 430 calories.

51. Let x represent the number of years after 1985.
Let y represent the average weekly earnings of a high school graduate.

This question seeks to find the year in which the average college graduate earns twice as much as a high school graduate. Thus the average college graduate's earnings will be $2y$.

College graduate's earnings will be: $2y = 508 + 25x$

High school graduate's earnings will be: $y = 345 + 9x$

Substitute $345 + 9x$ for y in the equation $2y = 508 + 25x$:

$$2\overbrace{(345 + 9x)}^{y} = 508 + 25x$$
$$690 + 18x = 508 + 25x$$
$$182 = 7x$$
$$x = 26 \text{ years}$$

When $x = 26$, $\quad y = 345 + 9(26) \quad$ and $\quad 2y = 508 + 25(26)$
$$y = \$579 \qquad\qquad 2y = \$1158$$

In 26 years after 1985, or 2011, average college graduate's earnings will be \$1158 per week and the average high school graduate will earn \$579 per week.

53. Substitute $-0.4x + 28$ for y in the equation $0.07x + y = 15$:

$$0.07x + \overbrace{(-0.4x + 28)}^{y} = 15$$
$$0.07x - 0.4x + 28 = 15$$
$$-0.33x + 28 = 15$$
$$-0.33x = -13$$
$$x \approx 39.39 \text{ years}$$

When $x \approx 39.39$, $\quad y \approx -0.4(39.39) + 28$
$$y \approx 12.24 \text{ car accident deaths per } 100{,}000$$

Using the other equation, $\quad y \approx 15 - 0.07(39.39)$
$$y \approx 12.24 \text{ gunfire deaths per } 100{,}000$$

In 39 years after 1965, or 2004, there will be about 12.24 deaths per 100,000 from car accidents and from gunfire. This is represented by the point (2004, 12.24).

55-65. Answers will vary

67. Let x be the number of people downstairs and y the number of people upstairs.

$$x + 1 = y - 1 \quad x = y - 2$$
$$2x = y + 1 \quad 2(y - 2) = y + 1$$
$$2y - 4 = y + 1$$
$$y = 5$$
$$x = 5 - 2$$
$$x = 3$$

There are 3 people downstairs and 5 people upstairs.

Check Points 7.6

1. To graph $2x - 4y \geq 8$, begin by graphing $2x - 4y = 8$ with a solid line because $\geq$ includes equality.

x-intercept:	y-intercept:	test point $(0, 0)$:
$2x - 4(0) = 8$	$2(0) - 4y = 8$	$2x - 4y \geq 8$
$2x = 8$	$-4y = 8$	$2(0) - 4(0) \geq 8$
$x = 4$	$y = -2$	$0 \geq 8$, false

   Since the test point makes the inequality <u>false</u>, shade the half-plane not containing test point $(0, 0)$.

   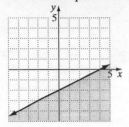

2. To graph $y > -\frac{3}{4}x$, begin by graphing $y = -\frac{3}{4}x$ with a dashed line because $>$ does not include equality.

   test point $(1, 1)$:

   $y > -\frac{3}{4}x$

   $1 > -\frac{3}{4} \cdot 1$

   $1 > -\frac{3}{4}$, true

   Since the test point makes the inequality <u>true</u>, shade the half-plane <u>containing</u> test point $(1, 1)$.

   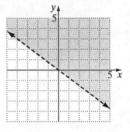

3. To graph $x + 2y > 4$, begin by graphing $x + 2y = 4$ with a dashed line because $>$ does not include equality.

   test point $(0, 0)$:

   $x + 2y > 4$

   $0 > 4$

   $0 + 2(0) > 4$, false

   Since the test point makes the inequality <u>false</u>, shade the half-plane <u>not containing</u> test point $(0, 0)$.

   To graph $2x - 3y \leq -6$, begin by graphing $2x - 3y = -6$ with a solid line because $\leq$ does include equality.

   test point $(0, 0)$:

   $2x - 3y \leq -6$

   $0 - 0 \leq -6$

   $0 \leq -6$, false

   Since the test point makes the inequality <u>false</u>, shade the half-plane <u>not containing</u> test point $(0, 0)$.

   The graph shows the intersection of the two half-planes.

   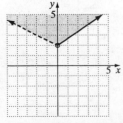

4. To graph $x < 3$, begin by graphing $x = 3$ with a dashed line because $<$ does not include equality.
The graph of $x < 3$ is the half-plane to the left of the dashed line.

To graph $y \geq -1$, begin by graphing $y = -1$ with a solid line because $\geq$ does include equality.
The graph of $y \geq -1$ is the half-plane above the solid line.
The graph shows the intersection of the two half-planes.

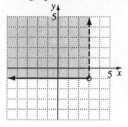

5. a. Another point in the target zone is (30, 140). This means that a pulse rate of 140 beats per minute is within the target zone for a 30-year-old person engaged in aerobic exercise.

   b. The pair (30, 140) makes each inequality of the system true.

   $$2a + 3p \geq 450 \qquad\qquad a + p \leq 190$$
   $$2(30) + 3(140) \geq 450 \qquad 30 + 140 \leq 190$$
   $$60 + 420 \geq 450 \qquad\qquad 170 \leq 190, \text{ true}$$
   $$480 \geq 450, \text{ true}$$

Exercise Set 7.6

1. To graph $x + y \geq 2$, begin by graphing $x + y = 2$ with a solid line because $\geq$ includes equality.
   test point (0, 0):

   $$x + y \geq 2$$
   $$0 + 0 \geq 2$$
   $$0 \geq 2, \text{ false}$$

   Since the test point makes the inequality <u>false</u>, shade the half-plane <u>not containing</u> test point (0, 0).

   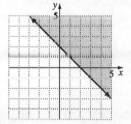

3. To graph $3x - y \geq 6$, begin by graphing $3x - y = 6$ with a solid line because $\geq$ includes equality.
   test point (0, 0):

   $$3x - y \geq 6$$
   $$3(0) - 0 \geq 6$$
   $$0 \geq 6, \text{ false}$$

   Since the test point makes the inequality <u>false</u>, shade the half-plane <u>not containing</u> test point (0, 0).

   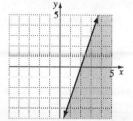

5. To graph $2x + 3y > 12$, begin by graphing $2x + 3y = 12$ with a dashed line because $>$ does not include equality.

   test point (0, 0):

$$2x + 3y > 12$$
$$2(0) + 3(0) > 12$$
$$0 > 12, \text{ false}$$

Since the test point makes the inequality <u>false</u>, shade the half-plane <u>not containing</u> test point (0, 0).

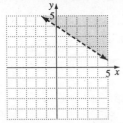

7. To graph $5x + 3y \leq -15$, begin by graphing $5x + 3y = -15$ with a solid line because $\leq$ includes equality.

   test point (0, 0):

$$5x + 3y \leq -15$$
$$5(0) + 3(0) \leq -15$$
$$0 \leq -15, \text{ false}$$

Since the test point makes the inequality <u>false</u>, shade the half-plane <u>not containing</u> test point (0, 0).

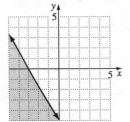

9. To graph $2y - 3x > 6$, begin by graphing $2y - 3x = 6$ with a dashed line because $>$ does not include equality.

   test point (0, 0):

$$2y - 3x > 6$$
$$2(0) - 3(0) > 6$$
$$0 > 6, \text{ false}$$

Since the test point makes the inequality <u>false</u>, shade the half-plane <u>not containing</u> test point (0, 0).

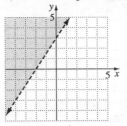

11. To graph $y > \frac{1}{3}x$, begin by graphing $y = \frac{1}{3}x$ as a dashed line passing through the origin with a slope of $\frac{1}{3}$, then shade above the line.

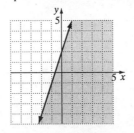

13. To graph $y \leq 3x + 2$, begin by graphing $y = 3x + 2$ as a solid line passing through (0, 2) with a slope of $\frac{3}{1}$, then shade below the line.

15. To graph $y < -\frac{1}{4}x$, begin by graphing $y = -\frac{1}{4}x$ as a dashed line passing through the origin with a slope of $-\frac{1}{4}$, then shade below the line.

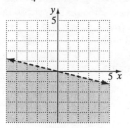

17. To graph $x \le 2$, begin by graphing $x = 2$ as a solid vertical line passing through $x = 2$, then shade to the left of the line.

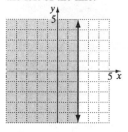

19. To graph $y > -4$, begin by graphing $y = -4$ as a dashed horizontal line passing through $y = -4$, then shade above the line.

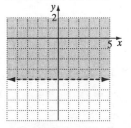

21. To graph $y \ge 0$, begin by graphing $y = 0$ as a solid horizontal line passing through $y = 0$, then shade above the line.

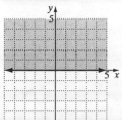

23. $3x + 6y \le 6$

$2x + y \le 8$

Graph $3x + 6y = 6$ as a solid line.

If $x = 0$, then $y = 1$ and if $y = 0$, then $x = 2$.
Because $(0, 0)$ makes the inequality true, shade the half-plane containing $(0, 0)$.
Graph $2x + y = 8$ as a solid line.
If $x = 0$, then $y = 8$ and if $y = 0$, then $x = 4$.
Because $(0, 0)$ makes the inequality true, shade the half-plane containing $(0, 0)$.

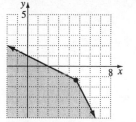

25. $2x + y < 3$

$x - y > 2$

Graph $2x + y = 3$ as a dashed line.

If $x = 0$, then $y = 3$ and if $y = 0$, then $x = \frac{3}{2}$.

Because $(0, 0)$ makes the inequality true, shade the half-plane containing $(0, 0)$.
Graph $x - y = 2$ as a dashed line.
If $x = 0$, then $y = -2$ and if $y = 0$, then $x = 2$.
Because $(0, 0)$ makes the inequality false, shade the half-plane not containing $(0, 0)$.

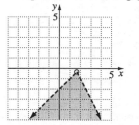

27. $2x + y < 4$

$x - y > 4$

Graph $2x + y = 4$ as a dashed line.

If $x = 0$, then $y = 4$ and if $y = 0$, then $x = 2$.

Because (0, 0) makes the inequality true, shade the half-plane containing (0, 0)

Graph $x - y = 4$ as a dashed line.

If $x = 0$, then $y = -4$. and if $y = 0$, then $x = 4$.

Because (0, 0) makes the inequality false, shade the half-plane not containing (0, 0).

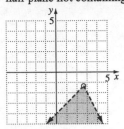

29. $x \geq 2$

$y \leq 3$

Graph $x = 2$ as a solid line.

The points in the half-plane to the right of the line satisfy $x > 2$.

Graph $y = 3$ as a solid line.

The points in the half-plane below the line satisfy $y < 3$.

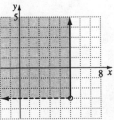

31. $x \leq 5$

$y > -3$

Graph $x = 5$ as a solid line.

The points in the half-plane to the left of the line satisfy $x < 5$.

Graph $y = -3$ as a dashed line.

The points in the half-plane above the line satisfy $y > -3$.

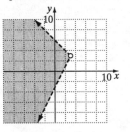

33. $x - y \leq 1$

$x \geq 2$

Graph $x - y = 1$ as a solid line.

If $x = 0$, then $y = -1$ and if $y = 0$, then $x = 1$.

Because (0, 0) satisfies the inequality, shade the half-plane containing (0, 0).

Graph $x = 2$ as a solid line.

The points in the half-plane to the right of $x = 2$ satisfy the inequality $x \geq 2$.

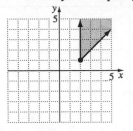

35. To graph $y > 2x - 3$, begin by graphing $y = 2x - 3$ as a dashed line with a y-intercept of -3 and a slope of $\frac{2}{1}$, then shade above the line. To graph $y < -x + 6$, begin by graphing $y = -x + 6$ as a dashed line with a y-intercept of 6 and a slope of $\frac{-1}{1}$, then shade below the line.

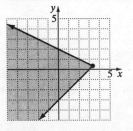

37. To graph $x + 2y \leq 4$, begin by graphing $x + 2y = 4$ as a solid line with a y-intercept of 2 and an x-intercept of 4. Because (0, 0) satisfies the inequality, shade the half-plane containing (0, 0). To graph $y \geq x - 3$, begin by graphing $y = x - 3$ as a solid line with a y-intercept of -3 and a slope of $\frac{1}{1}$, then shade above this line.

39. Point A is (50, 30). Substitution shows that point A satisfies each inequality for forests.

$T \geq 35$ $5T - 7P < 70$

$50 \geq 35$, true $5(50) - 7(30) < 70$

                            $40 < 70$, true

41. a. $50x + 150y > 2000$

   b.

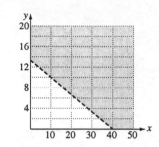

   c. An example of a solution is (30, 8). This represents that 30 children and 8 adults would exceed the 2000 pound weight limit.

43. a. $\text{BMI} = \dfrac{703W}{H^2}$

         $= \dfrac{703(200)}{(72)^2}$

         ≈ 27.1

   b. Locating the point (20, 27.1) on the male chart indicates the man is borderline overweight.

45-49. Answers will vary.

51. $x + 2y \geq 8$

      $x - y \leq 2$

        $y \leq 4$

Graph $x + 2y = 8$ as a solid line.
If $x = 0$, then $y = 4$ and if $y = 0$, then $x = 8$.
Because (0, 0) makes the inequality false, shade the half-plane not containing (0, 0).
Graph $x - y = 2$ as a solid line.
If $x = 0$, then $y = -2$ and if $y = 0$, then $x = 2$.
Because (0, 0) makes the inequality true, shade the half-plane containing (0, 0).
Graph $y = 4$ as a solid line.
The points in the half-plane below the line satisfy $y < 4$.

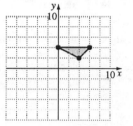

53.

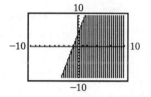

Check Points 7.7

1. The objective function is $z = 25x + 55y$

2. Not more than a total of 80 bookshelves and desks can be manufactured per day. This is represented by the inequality $x + y \leq 80$.

3. Objective function: $z = 25x + 55y$
Constraints: $x + y \leq 80$
                $30 \leq x \leq 80$
                $10 \leq y \leq 30$

4. Graph the constraints and find the corners, or vertices, of the region of intersection.

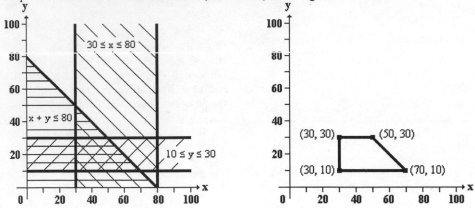

Find the value of the objective function at each corner of the graphed region.

Corner (x, y)	Objective Function $z = 25x + 55y$ z
(30, 10)	$z = 25(30) + 55(10)$ $= 750 + 550 = 1300$
(30, 30)	$z = 25(30) + 55(30)$ $= 750 + 1650 = 2400$
(50, 30)	$z = 25(50) + 55(30)$ $= 1250 + 1650 = 2900 \leftarrow$ Maximum
(70, 10)	$z = 25(70) + 55(10)$ $= 1750 + 550 = 2300$

The maximum value of z is 2900 and it occurs at the point (50, 30).
In order to maximize profit, 50 bookshelves and 30 desks must be produced each day for a profit of $2900.

Exercise Set 7.7

1. Objective function: $z = 5x + 6y$
   at (2, 10) $z = 5(2) + 6(10) = 70$
   at (7, 5) $z = 5(7) + 6(5) = 65$
   at (8, 3) $z = 5(8) + 6(3) = 58$
   at (1, 2) $z = 5(1) + 6(2) = 17$
   Maximum value = 70
   Minimum value = 17

3. Objective function: $z = 40x + 50y$
   at (0, 8) $z = 40(0) + 50(8) = 400$
   at (4, 9) $z = 40(4) + 50(9) = 610$
   at (8, 0) $z = 40(8) + 50(0) = 320$
   at (0, 0) $z = 40(0) + 50(0) = 0$
   Maximum value = 610
   Minimum value = 0

5. a.

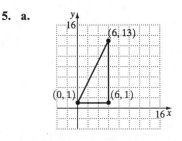

   b.   at (0, 1) $z = 0 + 1 = 1$
        at (6, 13) $z = 6 + 13 = 19$
        at (6, 1) $z = 6 + 1 = 7$

   c.   Maximum = 19
        occurs at $x = 6$ and $y = 13$
        Minimum = 1
        occurs at $x = 0$ and $y = 1$

7. a.

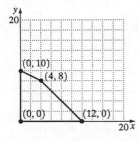

b. at (0, 10) $z = 6(0) + 10(10) = 100$
at (4, 8) $z = 6(4) + 10(8) = 104$
at (12, 0) $z = 6(12) + 10(0) = 72$
at (0, 0) $z = 6(0) + 10(0) = 0$

c. Maximum = 104
occurs at $x = 4$ and $y = 8$
Minimum = 0
occurs at $x = 0$ and $y = 0$

9. a. $z = 10x + 7y$

b. Constraints
$x + y \le 20$
$x \ge 3$
$x \le 8$

c.

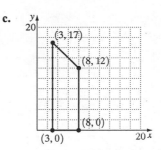

d. at (3, 0) $z = 10(3) + 7(0) = 30$
at (8, 0) $z = 10(8) + 7(0) = 80$
at (3, 17) $z = 10(3) + 7(17) = 149$
at (8, 12) $z = 10(8) + 7(12) = 164$

e. The student can earn the maximum amount per week by tutoring for <u>8</u> hours per week and working as a teacher's aid for <u>12</u> hours per week. The maximum amount that the student can earn each week is <u>$164</u>.

11. Let x represent the number of cartons of food and let y represent the number of cartons of clothing.
Objective function: $z = 12x + 5y$ Constraints: $50x + 20y \le 19,000$
$20x + 10y \le 8000$

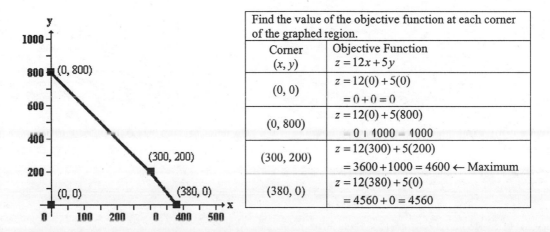

Find the value of the objective function at each corner of the graphed region.	
Corner (x, y)	Objective Function $z = 12x + 5y$
(0, 0)	$z = 12(0) + 5(0)$ $= 0 + 0 = 0$
(0, 800)	$z = 12(0) + 5(800)$ $= 0 + 1000 = 1000$
(300, 200)	$z = 12(300) + 5(200)$ $= 3600 + 1000 = 4600 \leftarrow$ Maximum
(380, 0)	$z = 12(380) + 5(0)$ $= 4560 + 0 = 4560$

The maximum the number of people that can be helped is 4600.
This occurs if 300 cartons of food and 200 cartons of clothing are shipped.

13. Let x and y be the number of parents and students, respectively.

Objective function: $z = 2x + y$ Constraints: $x + y \leq 150$

$$2y \geq x$$

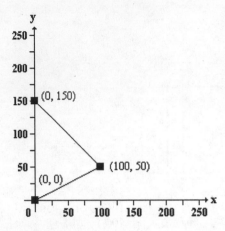

Find the value of the objective function at each corner of the graphed region.	
Corner (x, y)	Objective Function $z = 2x + y$
$(0, 0)$	$z = 2(0) + 0$ $= 0 + 0 = 0$
$(0, 150)$	$z = 2(0) + 150$ $= 0 + 150 = 150$
$(100, 50)$	$z = 2(100) + 50$ $= 200 + 50 = 250 \leftarrow$ Maximum

To raise the maximum amount of money, 100 parents and 50 students should attend.

15-19. Answers will vary.

Chapter 7 Review Exercises

1.

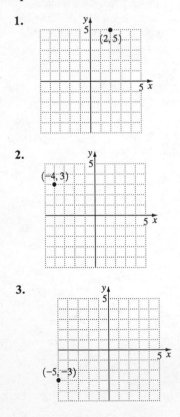

4.

2.

3.

5.

x	$y = 2x - 2$
-3	-8
-2	-6
-1	-4
0	-2
1	0
2	2
3	4

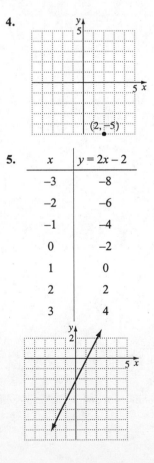

6.

x	$y = x^2 - 3$
−3	6
−2	1
−1	−2
0	−3
1	−2
2	1
3	6

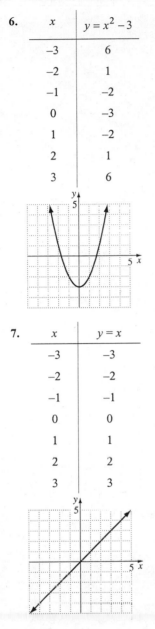

7.

x	$y = x$
−3	−3
−2	−2
−1	−1
0	0
1	1
2	2
3	3

8. $f(x) = 4x + 11$

$f(-2) = 4(-2) + 11 = -8 + 11 = 3$

9. $f(x) = -7x + 5$

$f(-3) = -7(-3) + 5 = 21 + 5 = 26$

10. $f(x) = 3x^2 - 5x + 2$

$f(4) = 3(4)^2 - 5(4) + 2 = 48 - 20 + 2 = 30$

11. $f(x) = -3x^2 + 6x + 8$

$f(-4) = -3(-4)^2 + 6(-4) + 8$

$= -48 - 24 + 8 = -64$

12.

x	$f(x) = \dfrac{1}{2}x$
−6	−3
−4	−2
−2	−1
0	0
2	1
4	2
6	3

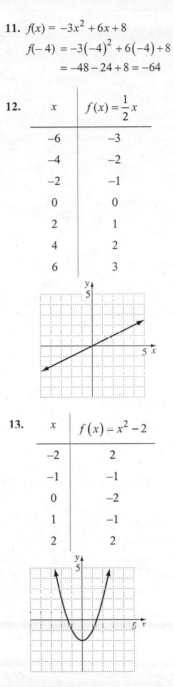

13.

x	$f(x) = x^2 - 2$
−2	2
−1	−1
0	−2
1	−1
2	2

14. y is not a function of x because it fails the vertical line test.

15. y is a function of x because it passes the vertical line test.

16. y is not a function of x because it fails the vertical line test.

17. $2x + y = 4$

x-intercept is 2; y-intercept is 4.

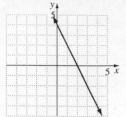

18. $2x - 3y = 6$

x-intercept is 3; y-intercept is –2.

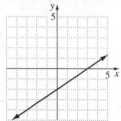

19. $5x - 3y = 15$

x-intercept is 3; y-intercept is –5.

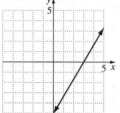

20. Slope $= \dfrac{1-2}{5-3} = -\dfrac{1}{2}$; line falls

21. Slope $= \dfrac{-4-2}{-3-(-1)} = \dfrac{-6}{-2} = 3$; line rises

22. Slope $= \dfrac{4-4}{6-(-3)} = 0$; line horizontal

23. Slope $= \dfrac{-3-3}{5-5} = \dfrac{-6}{0}$ is undefined, vertical line

24. $y = 2x - 4$; Slope: 2, y-intercept: –4

Plot point (0, –4) and second point using

$m = \dfrac{2}{1} = \dfrac{\text{rise}}{\text{run}}$.

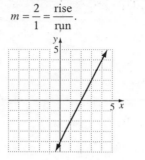

25. $y = -\dfrac{2}{3}x + 5$; Slope: $-\dfrac{2}{3}$, y-intercept: 5

Plot point (0, 5) and second point using

$m = \dfrac{-2}{3} = \dfrac{\text{rise}}{\text{run}}$.

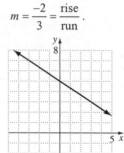

26. $y = \dfrac{3}{4}x - 2$; Slope: $\dfrac{3}{4}$, y-intercept: –2

Plot point (0, –2) and second point using

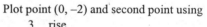

$m = \dfrac{3}{4} = \dfrac{\text{rise}}{\text{run}}$.

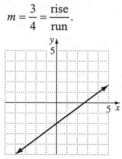

27. $y = \frac{1}{2}x + 0$; Slope: $\frac{1}{2}$, y-intercept: 0

Plot point (0, 0) and second point using

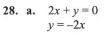

$m = \frac{1}{2} = \frac{\text{rise}}{\text{run}}$.

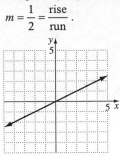

28. **a.** $2x + y = 0$
    $y = -2x$

    b. Slope $= -2$
    y-intercept $= 0$

    c.
    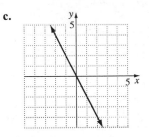

29. **a.** $3y = 5x$
    $y = \frac{5}{3}x$

    b. Slope $= \frac{5}{3}$
    y-intercept $= 0$

    c.
    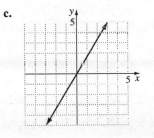

30. **a.** $3x + 2y = 4$
    $2y = -3x + 4$
    $y = -\frac{3}{2}x + 2$

    b. Slope $= -\frac{3}{2}$
    y-intercept $= 2$

    c.
    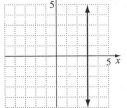

31. $x = 3$

32. $y = -4$

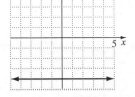

33. $x + 2 = 0$ or $x = -2$

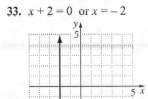

34. a. $m = \dfrac{74,300 - 21,800}{2000 - 1955} = \dfrac{52,500}{45} = \dfrac{3500}{3}$

b. $\dfrac{3500}{3} \approx \$1167$

35. a. The parabola opens upward because $a > 0$ $(a = 1)$.

b. x-intercepts: Set $y = 0$

$0 = x^2 - 6x - 7$

$0 = (x - 7)(x + 1)$

$x - 7 = 0$ or $x + 1 = 0$

$x = 7$ $\qquad x = -1$

c. y-intercept: Set $x = 0$

$y = 0^2 - 6(0) - 7 = -7$

d. Vertex: $x = \dfrac{-b}{2a} = \dfrac{-(-6)}{2(1)} = 3$

$y = (3)^2 - 6(3) - 7 = 9 - 18 - 7 = -16$

Vertex is at $(3, -16)$.

e.

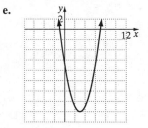

36. a. The parabola opens downward because $a < 0$ $(a = -1)$.

b. x-intercepts: Set $y = 0$

$0 = -x^2 - 2x + 3$

$0 = x^2 + 2x - 3$

$0 = (x + 3)(x - 1)$

$x + 3 = 0$ or $x - 1 = 0$

$x = -3$ $\qquad x = 1$

c. y-intercept: Set $x = 0$

$y = -0^2 - 2(0) + 3 = 3$

d. Vertex: $x = \dfrac{-b}{2a} = \dfrac{-(-2)}{2(-1)} = -1$

$y = -(-1)^2 - 2(-1) + 3 = -1 + 2 + 3 = 4$

Vertex is at $(-1, 4)$.

e.

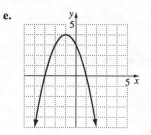

37. a. The parabola opens downward because $a < 0$ $(a = -3)$.

b. x-intercepts: Set $y = 0$

$x = \dfrac{-b \pm \sqrt{b^2 - 4ac}}{2a}$

$= \dfrac{-6 \pm \sqrt{6^2 - 4(-3)(1)}}{2(-3)}$

$= \dfrac{-6 \pm \sqrt{48}}{-6}$

Thus, $x = \dfrac{-6 + \sqrt{48}}{-6} \approx -0.2$

or $x = \dfrac{-6 - \sqrt{48}}{-6} \approx 2.2$

c. y-intercept: Set $x = 0$

$y = -3(0)^2 + 6(0) + 1 = 1$

d. Vertex: $x = \dfrac{-b}{2a} = \dfrac{-6}{2(-3)} = 1$

$y = -3(1)^2 + 6(1) + 1 = -3 + 6 + 1 = 4$

Vertex is at $(1, 4)$.

e.

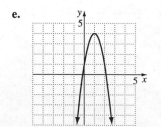

38. Vertex: $x = \dfrac{-b}{2a} = \dfrac{-4.2}{2(-0.05)} = 42$

$y = -0.05(42)^2 + 4.2(42) - 26$

$= -88.2 + 176.4 - 26$

$= 62.2$

The vertex is (42, 62.2) shows 42-year-olds have the maximum percentage of irritable coffee drinkers at 62.2%.

39.

x	$y = 2^x$
–2	$\frac{1}{4}$
–1	$\frac{1}{2}$
0	1
1	2
2	4

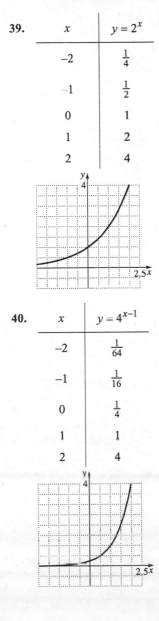

40.

x	$y = 4^{x-1}$
–2	$\frac{1}{64}$
–1	$\frac{1}{16}$
0	$\frac{1}{4}$
1	1
2	4

41.

x	$y = \left(\frac{1}{2}\right)^{2x}$
–2	16
–1	4
0	1
1	$\frac{1}{4}$
2	$\frac{1}{16}$

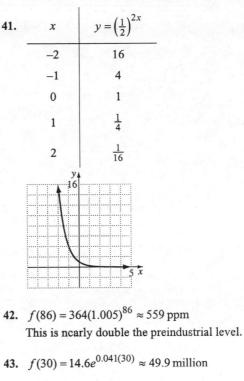

42. $f(86) = 364(1.005)^{86} \approx 559$ ppm

This is nearly double the preindustrial level.

43. $f(30) = 14.6e^{0.041(30)} \approx 49.9$ million

44. Answers will vary.

45. The intersection is (2, 3).

Check: $2 + 3 = 5$ $\qquad$ $3(2) - 3 = 3$

$\qquad\qquad 5 = 5$ true $\qquad$ $6 - 3 = 3$

$\qquad\qquad\qquad\qquad\qquad 3 = 3$ true

The solution set is {(2, 3)}.

46. The intersection is (–2, –3).

Check: $2(-2) - (-3) = -1$ $\qquad$ $-2 - 3 = -5$

$\qquad\qquad -4 + 3 = -1$ $\qquad\qquad$ $-5 = -5$ true

$\qquad\qquad\quad -1 = -1$ true

The solution set is {(–2, –3)}.

47. The intersection is (3, 2).

Check: $2 = -3 + 5$ $\qquad$ $2(3) - 2 = 4$

$\qquad 2 = 2$ true $\qquad$ $6 - 2 = 4$

$\qquad\qquad\qquad\qquad 4 = 4$ true

The solution set is {(3, 2)}.

169

48. $x = 3y + 10 \qquad 2x + 3y = 2$

$\qquad 2(3y + 10) + 3y = 2$

$\qquad 6y + 20 + 3y = 2$

$\qquad\qquad 9y = -18$

$\qquad\qquad y = -2$

$\qquad x = 3(-2) + 10 = -6 + 10 = 4$

The solution set is $\{(4, -2)\}$.

49. $y = 4x + 1 \qquad\qquad 3x + 2y = 13$

$\qquad 3x + 2(4x + 1) = 13$

$\qquad 3x + 8x + 2 = 13$

$\qquad\qquad 11x = 11$

$\qquad\qquad x = 1$

$y = 4(1) + 1 = 5$

The solution set is $\{(1, 5)\}$.

50. $x + 4y = 14$

$\quad x = 14 - 4y \qquad 2x - y = 1$

$\quad 2(14 - 4y) - y = 1$

$\qquad 28 - 8y - y = 1$

$\qquad\qquad -9y = -27$

$\qquad\qquad y = 3$

$x = 14 - 4(3) = 2$

The solution set is $\{(2, 3)\}$.

51. $x + 2y = -3 \quad$ No change. $\quad x + 2y = -3$

$x - y = -12 \quad$ Multiply by -1. $\quad \underline{-x + \; y = 12}$

$\qquad\qquad\qquad\qquad\qquad\qquad 3y = 9$

$\qquad\qquad\qquad\qquad\qquad\qquad y = 3$

$x - y = -12$

$x - 3 = -12$

$\quad x = -9$

The solution set is $\{(-9, 3)\}$.

52. $2x - y = 2 \quad$ Mult. by 2. $\quad 4x - 2y = 4$

$x + 2y = 11 \quad$ No change $\quad \underline{x + 2y = 11}$

$\qquad\qquad\qquad\qquad\qquad\qquad 5x = 15$

$\qquad\qquad\qquad\qquad\qquad\qquad x = 3$

$x + 2y = 11$

$3 + 2y = 11$

$\quad 2y = 8$

$\quad y = 4$

The solution set is $\{(3, 4)\}$.

53. $5x + 3y = 1 \quad$ Mult. by 3. $\qquad 15x + 9y = 3$

$3x + 4y = -6 \quad$ Mult. by -5. $\quad \underline{-15x - 20y = 30}$

$\qquad\qquad\qquad\qquad\qquad\qquad\qquad -11y = 33$

$\qquad\qquad\qquad\qquad\qquad\qquad\qquad y = -3$

$\qquad 5x + 3y = 1$

$\qquad 5x + 3(-3) = 1$

$\qquad\qquad 5x = 10$

$\qquad\qquad x = 2$

The solution set is $\{(2, -3)\}$.

54. $y = -x + 4 \qquad\qquad 3x + 3y = -6$

$\quad 3x + 3(-x + 4) = -6$

$\quad 3x - 3x + 12 = -6$

$\qquad\qquad 12 = -6$, false

There is no solution or $\{ \}$.

55. $3x + y = 8$

$\quad y = 8 - 3x \qquad\qquad 2x - 5y = 11$

$\quad 2x - 5(8 - 3x) = 11$

$\quad 2x - 40 + 15x = 11$

$\qquad\qquad 17x = 51$

$\qquad\qquad x = 3$

$y = 8 - 3(3) = -1$

The solution set is $\{(3, -1)\}$.

56. $3x - 2y = 6 \quad$ Mult. by -2. $\quad -6x + 4y = -12$

$6x - 4y = 12 \quad$ No change. $\quad \underline{6x - 4y = 12}$

$\qquad\qquad\qquad\qquad\qquad\qquad\qquad 0 = 0$

The system has infinitely many solutions.

The solution set is $\{(x, y) \mid 3x - 2y = 6\}$.

57. $3x + 8y = -1$

$\quad x - 2y = -5$

$\quad x = -5 + 2y \qquad\qquad 3x + 8y = -1$

$\qquad\qquad\qquad\qquad 3(-5 + 2y) + 8y = -1$

$\qquad\qquad\qquad\qquad -15 + 6y + 8y = -1$

$\qquad\qquad\qquad\qquad\qquad 14y = 14$

$\qquad\qquad\qquad\qquad\qquad y = 1$

$x = -5 + 2y = -5 + 2 = -3$

The numbers are -3 and 1.

58. Let x and y be the cholesterol intake of 1 ounce of shrimp and scallops, respectively.

$3x + 2y = 156$

$5x + 3y = 300 - 45 = 255$

$15x + 10y = 780$　　Mult. 1st equation by 5.

$\underline{-15x - 9y = -765}$　　Mult. 2nd equation by -3.

$\qquad y = 15$

$3x + 2y = 156$

$3x + 2(15) = 156$

$3x = 126$

$x = 42$

The cholesterol content in an ounce of shrimp is 42 mg. And the cholesterol content in an ounce of scallops is 15 mg.

59. Substitute $0.19x + 54.91$ for E in the equation

$E = 0.11x + 68.41$

$\overbrace{(0.19x + 54.91)}^{E} = 0.11x + 68.41$

$0.08x = 13.5$

$x \approx 169$ years

$E = 0.11(169) + 68.41 \approx 87$

Thus 169 years after 1900, or 2069, life expectancies for men and women will both be about 87 years.

60. To graph $x - 3y \leq 6$, begin by graphing $x - 3y = 6$ with a solid line because $\leq$ includes equality.

test point (0, 0):

$x - 3y \leq 6$

$0 - 3(0) \leq 6$

$0 \leq 6$, true

Since the test point makes the inequality <u>true</u>, shade the half-plane <u>containing</u> test point (0, 0).

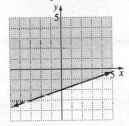

61. To graph $2x + 3y \geq 12$, begin by graphing $2x + 3y = 12$ with a solid line because $\geq$ includes equality.

test point (0, 0):

$2x + 3y \geq 12$

$2(0) + 3(0) \geq 12$

$0 \geq 12$, false

Since the test point makes the inequality <u>false</u>, shade the half-plane <u>not containing</u> test point (0, 0).

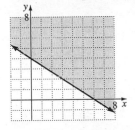

62. To graph $2x - 7y > 14$, begin by graphing $2x - 7y = 14$ with a dashed line because $>$ does not include equality.

test point (0, 0):

$2x - 7y > 14$

$2(0) - 7(0) > 14$

$0 > 14$, false

Since the test point makes the inequality <u>false</u>, shade the half-plane <u>not containing</u> test point (0, 0).

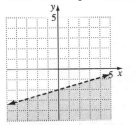

63. To graph $y > \frac{3}{5}x$, begin by graphing $y - \frac{3}{5}x$ as a dashed line passing through the origin with a slope of $\frac{3}{5}$, then shade above the line.

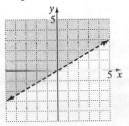

64. To graph $y \le -\frac{1}{2}x + 2$, begin by graphing $y = -\frac{1}{2}x + 2$ as a solid line passing through $(0, 2)$ with a slope of $\frac{-1}{2}$, then shade below the line.

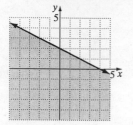

65. To graph $x \le 2$, begin by graphing $x = 2$ as a solid vertical line passing through $x = 2$, then shade to the left of the line.

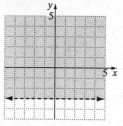

66. To graph $y > -3$, begin by graphing $y = -3$ as a dashed horizontal line passing through $y = -3$, then shade above the line.

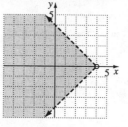

67. $3x - y \le 6$
    $x + y \ge 2$
Graph $3x - y = 6$ as a solid line.
Because $(0, 0)$ makes the inequality true, shade the half-plane containing $(0,0)$.
Graph $x + y = 2$ as a solid line.
Because $(0, 0)$ makes the inequality false, shade the half-plane not containing $(0, 0)$.

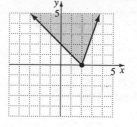

68. $x + y < 4$
    $x - y < 4$
Graph $x + y = 4$ as a dashed line.
Because $(0, 0)$ makes the inequality true, shade the half-plane containing $(0, 0)$.
Graph $x - y = 4$ as a dashed line.
Because $(0, 0)$ makes the inequality true, shade the half-plane containing $(0, 0)$.

69. $x \le 3$
    $y > -2$
Graph $x = 3$ as a solid line.
The points in the half-plane to the left of the line satisfy $x < 3$.
Graph $y = -2$ as a dashed line.
The points in the half-plane above the line satisfy $y > -2$.

172

70. $4x + 6y = 24$

$y > 2$

Graph $4x + 6y = 24$ as a solid line.

Because (0, 0) makes the inequality true, shade the half-plane containing (0, 0).

Graph $y = 2$ as a dashed line.

The points in the half-plane above the line satisfy $y > 2$.

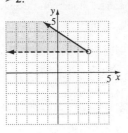

71. $x + y \le 6$

$y \ge 2x - 3$

Graph $x + y = 6$ as a solid line.

Because (0, 0) makes the inequality true, shade the half-plane containing (0, 0).

Graph $y = 2x - 3$ as a solid line.

Because (0, 0) makes the inequality true, shade the half-plane containing (0, 0).

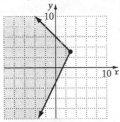

72. $y < -x + 4$

$y > x - 4$

Graph $y < -x + 4$ as a dashed line.

Because (0, 0) makes the inequality true, shade the half-plane containing (0, 0).

Graph $y = x - 4$ as a dashed line.

Because (0, 0) makes the inequality true, shade the half-plane containing (0, 0).

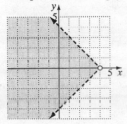

73. $z - 2x + 3y$

at (1, 0) $z = 2(1) + 3(0) = 2$

at $\left(\dfrac{1}{2}, \dfrac{1}{2}\right)$ $z = 2\left(\dfrac{1}{2}\right) + 3\left(\dfrac{1}{2}\right) = \dfrac{5}{2}$

at (2, 2) $z = 2(2) + 3(2) = 10$

at (4, 0) $z = 2(4) + 3(0) = 8$

Maximum value of the objective function is 10.

Minimum value of the objective function is 2.

74. $z = 2x + 3y$

Constraints: $x \le 6$

$y \le 5$

$x + y \ge 2$

$x > 0$

$y \ge 0$

a.

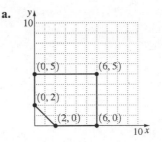

b. at (0, 2) $z = 2(0) + 3(2) = 6$

at (0, 5) $z = 2(0) + 3(5) = 15$

at (6, 5) $z = 2(6) + 3(5) = 27$

at (6, 0) $z = 2(6) + 3(0) = 12$

at (2, 0) $z = 2(2) + 3(0) = 4$

c. The maximum value of the objective function is 27. It occurs at $x = 6$ and $y = 5$.

The minimum value of the objective function is 4. It occurs at $x = 2$ and $y = 0$.

Chapter 7 Test

1.

x	$y = (x-1)^2$
-1	4
0	1
1	0
2	1
3	4

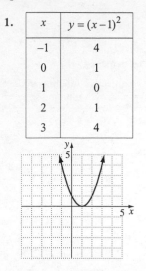

2. $f(-2) = 3(-2)^2 - 7(-2) - 5 = 12 + 14 - 5 = 21$

3. Set $y = 0$ Set $x = 0$
   $4x - 2 \cdot 0 = -8$ $4 \cdot 0 - 2y = -8$
   $\quad 4x = -8$ $\quad -2y = -8$
   $\quad\quad x = -2$ $\quad\quad y = 4$
   x-intercept is -2. y-intercept is 4.

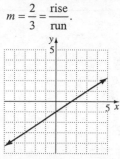

4. Slope $= \dfrac{-2-4}{-5-(-3)} = \dfrac{-6}{-2} = 3$

5. Slope $= \dfrac{530-320}{5-0} = \dfrac{210}{5} = 42$

   This indicates that one-way fares are increasing at $42 per year.

6. $y = \dfrac{2}{3}x - 1$

   Slope: $\dfrac{2}{3}$, y-intercept: -1

   Plot the point $(0, -1)$ and a second point using

   $m = \dfrac{2}{3} = \dfrac{\text{rise}}{\text{run}}$.

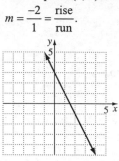

7. $y = -2x + 3$
   Slope: -2, y-intercept: 3
   Plot the point $(0, 3)$ and a second point using

   $m = \dfrac{-2}{1} = \dfrac{\text{rise}}{\text{run}}$.

8. a. $a - 1 > 0$; parabola opens upward.

b. Vertex: $x = \dfrac{-b}{2a} = \dfrac{-(-2)}{2(1)} = 1$

$y = (1)^2 - 2(1) - 8 = -9$

Vertex is at $(1, -9)$.

c. x-intercepts: set $y = 0$

$0 = x^2 - 2x - 8$

$0 = (x - 4)(x + 2)$

$x - 4 = 0$ or $x + 2 = 0$

$x = 4$ $x = -2$

x-intercepts are -2 and 4.

d. y-intercept: set $x = 0$

$y = (0)^2 - 2(0) - 8 = -8$

y-intercept is -8.

e.

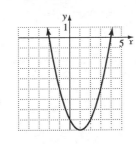

9. Vertex: $x = \dfrac{-b}{2a} = \dfrac{-96}{2(-16)} = 3$

$y = -16(3)^2 + 96(3) + 3 = 147$

The vertex $(3, 147)$ represents that after 3 seconds the baseball will reach its maximum height of 147 feet.

10.

x	$y = 3^x$
-2	$\frac{1}{9}$
-1	$\frac{1}{3}$
0	1
1	3
2	9

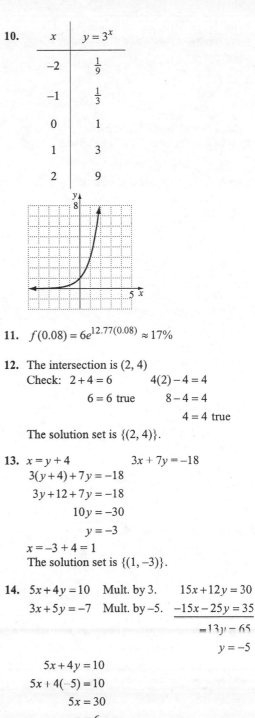

11. $f(0.08) = 6e^{12.77(0.08)} \approx 17\%$

12. The intersection is $(2, 4)$

Check: $2 + 4 = 6$ $4(2) - 4 = 4$

           $6 = 6$ true $8 - 4 = 4$

                                $4 = 4$ true

The solution set is $\{(2, 4)\}$.

13. $x = y + 4$ $3x + 7y = -18$

      $3(y + 4) + 7y = -18$

        $3y + 12 + 7y = -18$

              $10y = -30$

                 $y = -3$

$x = -3 + 4 = 1$

The solution set is $\{(1, -3)\}$.

14. $5x + 4y = 10$ Mult. by 3. $15x + 12y = 30$

$3x + 5y = -7$ Mult. by -5. $\underline{-15x - 25y = 35}$

                                      $-13y = 65$

                                        $y = -5$

              $5x + 4y = 10$

            $5x + 4(-5) = 10$

                 $5x = 30$

                  $x = 6$

The solution set is $\{(6, -5)\}$.

15. Let x represent the cost of World War II.
Let y represent the cost of the Vietnam War.

$x + y = 500$

$\dfrac{x - y = 120}{2x = 620}$

$x = 310$

$x + y = 500$

$310 + y = 500$

$y = 190$

Cost of World War II: $310 billion.
Cost of the Vietnam War: $190 billion.

16. $3x - 2y < 6$

Graph $3x - 2y = 6$ as a dashed line.

x-intercept:

$3x - 2 \cdot 0 = 6$

$3x = 6$

$x = 2$

y-intercept:

$3 \cdot 0 - 2y = 6$

$-2y = 6$

$y = -3$

Test point: (0, 0).

Is $3 \cdot 0 - 2 \cdot 0 < 6$?

$0 < 6$, true

Shade the half-plane containing (0, 0).

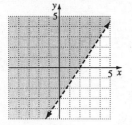

17. Graph $y = \frac{1}{2}x - 1$ as a solid line.

Use y-intercept of -1 and slope of $\frac{1}{2}$

Shade below this line.

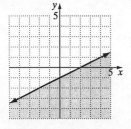

18. $2x - y \leq 4$

$2x - y > -1$

Graph $2x - y = 4$ as a solid line.

Because (0, 0) makes the inequality true, shade the half-plane containing (0, 0).

Graph $2x - y = -1$ as a dashed line.

Because (0, 0) makes the inequality true, shade the half-plane containing (0, 0).

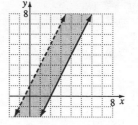

19. $z = 3x + 2y$

at (2, 0) $\quad z = 3(2) + 2(0) = 6$

at (2, 6) $\quad z = 3(2) + 2(6) = 18$

at (6, 3) $\quad z = 3(6) + 2(3) = 24$

at (8, 0) $\quad z = 3(8) + 2(0) = 24$

The maximum value of the objective function is 24.
The minimum value of the objective function is 6.

20.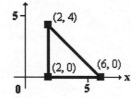

Objective function: $z = 3x + 5y$

at (2, 0) $\quad z = 3(2) + 5(0) = 6$

at (6, 0) $\quad z = 3(6) + 5(0) = 18$

at (2, 4) $\quad z = 3(2) + 5(4) = 26$

The maximum value of the objective function is 26.

Chapter 8
Consumer Mathematics and Financial Management

Check Points 8.1

1. Step 1: $\frac{1}{8} = 1 \div 8 = 0.125$

   Step 2: $0.125 \cdot 100 = 12.5$
   Step 3: 12.5%

2. $0.023 = 2.3\%$

3. **a.** $67\% = 0.67$

   b. $250\% = 2.50 = 2.5$

4. $\dfrac{\$3.44}{\$4.30} = \dfrac{3.44}{4.30} = 0.8 = 80\%$

   In England, 80% of fuel cost is for taxes.

5. $\dfrac{539,533}{2,314,690} \approx 0.23 = 23\%$

   23% of deaths are caused by cancer.

6. 13% of $8000 = 0.13 \times 8000 = 1040$
   Thus, 1040 Americans suffer spinal cord injuries due to sports injuries each year.

7. **a.** 6% of $\$1260 = 0.06 \times \$1260 = \$75.60$
   The tax paid is $75.60

   b. $\$1260.00 + \$75.60 = \$1335.60$
   The total cost is $1335.60

8. **a.** 35% of $\$380 = 0.35 \times \$380 = \$133$
   The discount is $133

   b. $\$380 - \$133 = \$247$
   The sale price is $247

9. Amount of increase: $2.2 - 1.7 = 0.5$

   $\dfrac{\text{amount of increase}}{\text{original amount}} = \dfrac{0.5}{1.7} \approx 0.29 = 29\%$

   There was a 29% increase from 1998 to 1999.

10. Amount of decrease: $\$940 - \$611 = \$329$

    $\dfrac{\text{amount of decrease}}{\text{original amount}} = \dfrac{\$329}{\$940} = 0.35 = 35\%$

    There was a 35% decrease from 1998 to 1999.

11. Amount of increase: $12\% - 10\% = 2\%$

    $\dfrac{\text{amount of increase}}{\text{original amount}} = \dfrac{2\%}{10\%} = 0.2 = 20\%$

    There was a 20% increase for this episode.

12. **a.** 20% of $\$1200 = 0.20 \times \$1200 = \$240$
    Taxes for year 1 are $\$1200 - \$240 = \$960$
    20% of $\$960 = 0.20 \times \$960 = \$192$
    Taxes for year 2 are $\$960 + \$192 = \$1152$

    b. $\dfrac{\$1200 - \$1152}{\$1200} = \dfrac{\$48}{\$1200} = 0.04 = 4\%$

    Taxes for year 2 are 4% less than the original amount.

Exercise Set 8.1

1. $\dfrac{2}{5} = 2 \div 5 = 0.4 = 40\%$

3. $\dfrac{1}{4} = 1 \div 4 = 0.25 = 25\%$

5. $\dfrac{3}{8} = 3 \div 8 = 0.375 = 37.5\%$

7. $\dfrac{1}{40} = 1 \div 40 = 0.025 = 2.5\%$

9. $\dfrac{9}{80} = 9 \div 80 = 0.1125 = 11.25\%$

11. $0.59 = 59\%$

13. $0.3844 = 38.44\%$

15. $2.87 = 287\%$

17. $14.87 = 1487\%$

19. $100 = 10,000\%$

21. $72\% = 0.72$

23. $43.6\% = 0.436$

25. $130\% = 1.3$

27. $2\% = 0.02$

29. $\dfrac{1}{2}\% = 0.5\% = 0.005$

31. $\dfrac{5}{8}\% = 0.625\% = 0.00625$

33. $62\dfrac{1}{2}\% = 62.5\% = .625$

35. $\dfrac{7500}{60,000} = 0.125 = 12.5\%$

37. $\dfrac{214}{418 + 214} = \dfrac{214}{632} = 0.34 = 34\%$

39. $(0.34)(1200) = 408$ women

41. $(0.15)(60) = \$9$

43. a. $(0.06)(16,800) = \$1,008$

b. $16,800 + 1008 = \$17,808$

45. a. $(0.12)(860) = \$103.20$

b. $860 - 103.20 = \$756.80$

47. $\dfrac{315 - 288}{288} = \dfrac{27}{288} \approx 0.09 = 9\%$

49. $\dfrac{840 - 714}{840} = 0.15 = 15\%$

51. $\dfrac{81 - 26}{26} \approx 2.12 = 212\%$

53. Amount after first year
$= 10,000 - (0.3)(10,000)$
$= \$7000$
Amount after second year
$= 7000 + (0.4)(7000)$
$= \$9800$
Your adviser is not using percentages properly.

Actual change:
$\dfrac{10,000 - 9800}{10,000} = 0.02 = 2\%$ decrease.

55-59. Answers will vary.

61. c is true.

$1 - \left(\dfrac{1}{10} + \dfrac{1}{3} + \dfrac{1}{5}\right) = 1 - \dfrac{19}{30} = \dfrac{11}{30} \approx 0.36\overline{6} = 36\dfrac{2}{3}\%$

63. January sales $= 60 \cdot 200 = \$12,000$
Number of customers in February
$= 60 - (0.10)(60) = 60 - 6 = 54$
Price of washing machine in February
$= 200 + (0.20)(200) = 200 + 40 = \$240$
February sales $= 54 \cdot \$240 = \$12,960$
$12,960 - 12,000 = \$960$ increase.

65. $\$18.71$

Check Points 8.2

1. $I = (\$3000)(0.05)(1) = \$150$

2. $I = (\$2400)(0.07)(2) = \$336$

3. a. $I = (\$2040)(0.075)\left(\dfrac{4}{12}\right) = \$51$

b. $\$2040 + \$51 = \$2091$

4. a. $A = \$1000(1 + 0.04)^5 = \$1216.65$

b. $\$1216.65 - \$1000 = \$216.65$

5. a. $A = \$4200\left(1 + \dfrac{0.04}{4}\right)^{4 \cdot 10} = \$6253.23$

b. $\$6253.23 - \$4200 = \$2053.23$

6. $A = \$1000\left(1 + \dfrac{0.0425}{2}\right)^{2 \cdot 1} \approx \$1042.95$ versus

$A = \$1000\left(1 + \dfrac{0.043}{1}\right)^{1 \cdot 1} = \$1043.00$

A 4.3% rate compounded annually is better.

7. $P = \dfrac{A}{\left(1 + \dfrac{r}{n}\right)^{nt}}$

$A = \$10,000, \ r = 0.06, \ n = 52, \ t = 8$

$P = \dfrac{10,000}{\left(1 + \dfrac{0.06}{52}\right)^{52 \cdot 8}} = \dfrac{10,000}{1.6156273} = \$6189.55$

8. $Y = \left(1 + \dfrac{r}{n}\right)^{n} - 1$

$Y = \left(1 + \dfrac{0.08}{4}\right)^{4} - 1 \approx 0.0824 = 8.24\%$

9. $A = P\dfrac{\left(1 + \dfrac{r}{n}\right)^{nt} - 1}{\dfrac{r}{n}}$

$= 3000\dfrac{\left(1 + \dfrac{0.10}{1}\right)^{1 \cdot 40} - 1}{\dfrac{0.10}{1}} \approx \$1,327,778$

Exercise Set 8.2

1. a. $I = (\$4000)(0.06)(1) = \$240$

   b. $\$4000 + \$240 = \$4240$

3. a. $I = (\$180)(0.03)(2) = \$10.80$

   b. $\$180 + \$10.80 = \$190.80$

5. a. $I = (\$900)(0.04)\left(\dfrac{9}{12}\right) = \$27$

   b. $\$900 + \$27 = \$927$

7. a. $I - (\$9200)(0.065)\left(\dfrac{18}{12}\right) - \$897$

   b. $\$9200 + \$897 = \$10,097$

9. a. $I = (\$20,000)(0.0525)\left(\dfrac{3}{12}\right) = \$262.50$

   b. $\$20,000 + \$262.50 = \$20,262.50$

11. a. $I = (\$15,500)(0.11)\left(\dfrac{90}{360}\right) = \$426.25$

   b. $\$15,500 + \$426.25 = \$15,926.25$

13. a. $A = \$10,000(1+0.04)^2 = \$10,816$

   b. $\$10,816 - \$10,000 = \$816$

15. a. $A = \$3000\left(1+\dfrac{0.05}{2}\right)^{2\cdot4}$

     $= \$3000(1.025)^8$

     $= \$3655.21$

   b. $\$3655.21 - \$3000 = \$655.21$

17. a. $A = \$9500\left(1+\dfrac{0.06}{4}\right)^{4\cdot5}$

     $= \$9500(1.015)^{20}$

     $= \$12,795.12$

   b. $\$12,795.12 - \$9500 = \$3295.12$

19. a. $A = \$4500\left(1+\dfrac{0.045}{12}\right)^{12\cdot3}$

     $= \$4500(1.0038)^{36}$

     $= \$5149.12$

   b. $\$5149.12 - \$4500 = \$649.12$

21. a. $A = \$1500\left(1+\dfrac{0.085}{360}\right)^{360\cdot2.5}$

     $= \$1500(1.000236)^{900}$

     $= \$1855.10$

   b. $\$1855.10 - \$1500 = \$355.10$

23. a. $A = \$20,000\left(1+\dfrac{0.045}{360}\right)^{360\cdot20}$

     $= \$20,000(1.000125)^{7200}$

     $= \$49,189.30$

   b. $\$49,189.30 - \$20,000 = \$29,189.30$

25. $A = \$1000\left(1+\dfrac{0.055}{2}\right)^{2\cdot1} = \$1055.76$ versus

   $A = \$1000\left(1+\dfrac{0.054}{12}\right)^{12\cdot1} = \$1055.36$

   A 5.5% rate compounded semiannually is better.

27. $A = \$16,000\left(1+\dfrac{0.07}{12}\right)^{12\cdot3} = \$19,726.81$ versus

   $A = \$16,000\left(1+\dfrac{0.0685}{360}\right)^{360\cdot3} = \$19,649.84$

   A 7% rate compounded monthly is better.

29. $A = \$10,000, r = 0.06, n = 2, t = 3$

   $P = \dfrac{10,000}{\left(1+\frac{0.06}{2}\right)^{2\cdot3}} = \dfrac{10,000}{(1.03)^6} = \$8374.84$

31. $A = \$10,000, r = 0.095, n = 12, t = 3$

   $P = \dfrac{10,000}{\left(1+\frac{0.095}{12}\right)^{12\cdot3}}$

   $= \dfrac{10,000}{(1.00791667)^{36}}$

   $= \$7528.59$

33. $Y - \left(1+\dfrac{0.06}{2}\right)^2 - 1 - 0.0609 - 6.09\%$

35. $Y = \left(1 + \dfrac{0.06}{12}\right)^{12} - 1 \approx 0.0617 = 6.17\%$

37. $Y = \left(1 + \dfrac{0.06}{1000}\right)^{1000} - 1 \approx 0.0618 = 6.18\%$

39. $A = \$2500 \dfrac{\left(1 + \dfrac{0.09}{1}\right)^{1\cdot40} - 1}{\dfrac{0.09}{1}} \approx \$844,706$

41. a. $I = Prt$

$= (\$4000)(0.0825)\left(\dfrac{9}{12}\right)$

$= \$247.50$

b. $\$4000 + \$247.50 = \$4247.50$

43. $A = P\left(1 + \dfrac{r}{n}\right)^{nt}$

a. $A = \$24\left(1 + \dfrac{0.05}{12}\right)^{12\cdot374}$

$= \$24(1.00416667)^{4488}$

$\approx \$3,052,400,000$

b. $A = \$24\left(1 + \dfrac{0.05}{360}\right)^{360\cdot374}$

$= \$24(1.00013889)^{134,640}$

$\approx \$3,169,200,000$

45. $A = P\left(1 + \dfrac{r}{n}\right)^{nt}$

$A = \$10,000\left(1 + \dfrac{0.09}{12}\right)^{12\cdot21}$

$= \$10,000(1.0075)^{252}$

$= \$65,728.51$

47. $P = \dfrac{A}{\left(1 + \dfrac{r}{n}\right)^{nt}}$

$A = \$500,000,\ r = 0.09,\ n = 12,\ t = 65 - 30 = 35$

$P = \dfrac{500,000}{\left(1 + \dfrac{0.09}{12}\right)^{12\cdot35}} = \dfrac{500,000}{(1.0075)^{420}} = \$21,679.39$

49-53. Answers will vary.

55. $A = P\left(1 + \dfrac{r}{n}\right)^{nt}$

Start with $5000 in account for 2 years:

$A = \$5000\left(1 + \dfrac{0.08}{12}\right)^{12\cdot2} = \$5864.44$

Then withdraw $1500, which leaves
$5864.44 − $1500 = $4364.44

Leave $4364.44 in the account for 1 year:

$A = \$4364.44\left(1 + \dfrac{0.08}{12}\right)^{12\cdot1} = \$4726.69$

Then add $2000, so now have
$4726.69 + $2000 = $6726.69
Have $6726.69 in account for 3 years:

$A = 6726.69\left(1 + \dfrac{0.08}{12}\right)^{12\cdot3} = \$8544.49$

57. The effective annual yield is approximately 6.18%.

Check Points 8.3

1. a. Amount financed
$= \$14,000 - \$280 = \$13,720$

b. Total installment price
$= 60 \cdot \$315 + \$280 = \$19,180$

c. Finance charge
$\$19,180 - \$14,000 = \$5180$

2. Find the finance charge per $100 financed

$\dfrac{\$5180}{\$13,720} \cdot \$100 = \$37.76$

For 60 monthly payments, $37.76 is closest to
$38.06 in Table 8.3. Therefore the APR is 13.5%.

3. a. $u = \dfrac{kRV}{100 + V}$, where $k = 60 - 24 = 36$,
$R = \$315,\ V = \$22.17$
V is found by looking up the APR from Check
Point 2 in the "36-payments" row of table 8.3
Interest saved =

$u = \dfrac{36(\$315)(\$22.17)}{100 + 22.17} \approx \$2057.85$

b. Payoff amount =
(Payment number 24) + (Total of remaining
payments after payment 24) − (Interest saved)
$= \$315 + 36 \cdot \$315 - \$2057.85 = \$9597.15$

4. a. $u = \dfrac{k(k+1)}{n(n+1)} \cdot F$, where $k = 36$, $n = 60$, $F = \$5180$ (Computed in Check Point 1)

Interest saved $= u = \dfrac{36(36+1)}{60(60+1)} \cdot 5180 \approx \$1885.18$

  b.   Payoff amount = (Payment number 24) + (Total of remaining payments after payment 24) – (Interest saved)
$= \$315 + 36 \cdot \$315 - \$1885.18 = \$9769.82$

5. a.   The interest is $I = Prt = (\$4720 - \$1000) \times 0.016 \times 1 = \$3720 \times 0.016 \times 1 = \$59.52$

  b.   New balance $= \$3720 + \$59.52 + \$1025 + \$45 = \$4849.52$

  c.   Minimum monthly payment $= \dfrac{\text{balance owed}}{36} = \dfrac{\$4849.52}{36} \approx \$135$.

6. a.   Unpaid Balance Method: $I = Prt = (\$6800 - \$500) \times 0.018 \times 1 = \$6300 \times 0.018 \times 1 = \$113.40$

  b.   Previous Balance Method: $I = Prt = \$6800 \times 0.018 \times 1 = \$122.40$

  c.   Average daily balance $= \dfrac{(\$6800)(7) + (\$6300)(24)}{31} \approx \$6412.90$

Average Daily Balance Method: $I = Prt = \$6412.90 \times 0.018 \times 1 \approx \$115.43$.

Exercise Set 8.3

1. a.   Amount financed $= \$27,000 - \$5000 = \$22,000$

  b.   Total installment price $= 60 \cdot \$410 + \$5000 = \$29,600$

  c.   Finance charge $= \$29,600 - \$27,000 = \$2,600$

3. a.   Amount financed $= \$1100 - 100 = \$1000$

  b.   Total installment price $= 12 \cdot \$110 + \$100 = \$1420$

  c.   Finance charge $= \$1420 - \$1100 = \$320$

5. In the row for 12 monthly payments, find the value \$6.90. That value is in the column for 12.5%.

7. In the row for 24 monthly payments, find the value \$15.80. That value is in the column for 14.5%.

9. Finance charge per \$100 financed $= \dfrac{\text{Finance charge}}{\text{Amount financed}} \cdot \$100 = \dfrac{\$1279}{\$4450} \cdot \$100 \approx \$28.74$
Using Table 8.3 with 48 monthly payments the APR is 13.0%.

11. a.   Amount financed $= \$17,500 - \$500 = \$17,000$

  b.   Total installment price $= 60 \cdot \$360.55 + \$500 = \$22,133$

  c.   Finance charge $= \$22,133 - \$17,500 = \$4633$

  d.   Finance charge per \$100 financed $= \dfrac{\$4633}{\$17,000} \cdot \$100 \approx \$27.25$. The APR is 10.0%.

Chapter 8: Consumer Mathematics and Financial Management

13. **a.** $u = \dfrac{kRV}{100+V}$, where $k = 60 - 24 = 36$, $R = \$360.55$, and $V = \$16.16$

V is found by looking up the APR from Exercise 11d in the "36-payments" row of table 8.3

Interest saved $= u = \dfrac{36(\$360.55)(\$16.16)}{100+16.16} \approx \$1805.73$

b. Payoff amount = (Payment number 24) + (Total of remaining payments after payment 24) − (Interest saved)
$= \$360.55 + 36 \cdot \$360.55 - \$1805.73 = \$11,534.62$

c. $u = \dfrac{k(k+1)}{n(n+1)} \cdot F$, where $k = 36$, $n = 60$, and $F = \$4633$ (Computed in exercise 11c)

Interest saved $= u = \dfrac{36(36+1)}{60(60+1)} \cdot 4633 \approx \$1686.11$

d. Payoff amount = (Payment number 24) + (Total of remaining payments after payment 24) − (Interest saved)
$= \$360.55 + 36 \cdot \$360.55 - \$1686.11 = \$11,654.24$

15. **a.** Unpaid balance = \$950 − \$100 = \$850
$I = Prt = (\$850)(0.013)(1) = \$11.05$

b. Balance due June 9 = \$850 + \$11.05 + \$85 + \$67 = \$1013.05

c. Minimum monthly payment $= \dfrac{\$1013.05}{36} \approx \$28$

17. **a.** $I = Prt = (\$330.90)(0.015)(1) = \$4.96$

b. New Balance = \$445.59 + \$278.06 − \$110 + \$4.96 = \$618.61

c. Minimum monthly payment $= \dfrac{\$618.61}{10} = \$61.86$ because the new balance is over \$500.

19. **a.** Unpaid balance = \$3000 − \$2500 = \$500: $I = Prt = (\$500)(0.015)(1) = \$7.50$

b. Previous balance: $I = Prt = (\$3000)(0.015)(1) = \$45$

c. Average daily balance $= \dfrac{(\$3000)(5)+(\$500)(25)}{30} \approx \$916.67$: $I = Prt = (\$916.67)(0.015)(1) = \$13.75$

21-29. Answers will vary.

31. The payments will be greater than $\dfrac{\$1400-\$200}{30} = \$40$ because a finance charge must be paid.
Answer b is reasonable.

33. Answers will vary.

35. **a.** Number of days = 31
Average daily balance $= \dfrac{\$11,664.15}{30} = \$376.26$

b. Finance charge = $(0.013)(376.26) = \$4.89$

c. Balance due = \$466.15 + \$4.89 = \$471.04

Check Points 8.4

1. a. Down payment $= 0.10(\$240,000) = \$24,000$

    b. Amount of mortgage $= \$240,000 - \$24,000 = \$216,000$

    c. Three points $= 0.03(\$216,000) = \$6480$

    d. Thousands of dollars of mortgage: $\dfrac{\$216,000}{\$1000} = 216$

    Monthly payment $= \$6.32 \cdot 216 = \$1365.12$

    e. Total cost of interest $=$ (Total of all monthly payments) $-$ (Amount of the mortgage)
    $$= 360 \times \$1365.12 - \$216,000$$
    $$= 275,443.20$$

2. Interest for first month $= Prt = \$200,000 \times 0.07 \times \dfrac{1}{12} \approx \$1166.67$

    Principle payment $= \$1550.00 - \$1166.67 = \$383.33$
    Balance of loan $= \$200,000 - \$383.33 = \$199,616.67$

    Interest for second month $= Prt = \$199,616.67 \times 0.07 \times \dfrac{1}{12} \approx \$1164.43$

    Principle payment $= \$1550.00 - \$1164.43 = \$385.57$
    Balance of loan $= \$199,616.67 - \$385.57 = \$199,231.10$

Payment Number	Interest Payment	Principal Payment	Balance of Loan
1	$1166.67	$383.33	$199,616.67
2	$1164.43	$385.57	$199,231.10

3. A person with a yearly income of $35,000 qualifies for a $115,300 mortgage at 6.5%

Exercise Set 8.4

1. a. Down payment $= 0.20(\$125,000)$
    $$= \$25,000$$

    b. Amount of mortgage $= \$125,000 - \$25,000$
    $$= \$100,000$$

    c. Three points $= 0.03(\$100,000) = \$3000$

    d. $\dfrac{\$100,000}{\$1000} = 100$ thousands of dollars of mortgage

    Monthly payment $= \$6.65 \cdot 100 = \$665$

    e. Total cost of interest $=$ (Total of all monthly payments) $-$ (Amount of the mortgage)
    $$= 360 \cdot \$665 - \$100,000$$
    $$= \$139,400$$

3. Down payment = 0.05($40,000) = $2000
Mortgage amount = $40,000 − $2000 = $38,000
$\dfrac{\$38,000}{\$1000}$ = 38 thousands of dollars of mortgage

20-year option:
Monthly payment = $8.36 · 38 = $317.68
Total cost of interest = 20 · 12 · $317.68 − $38,000 = $38,243.20

30-year option:
Monthly payment = $7.34 · 38 = $278.92
Total cost of interest = 360 · $278.92 − $38,000 = $62,411.20

Interest savings with 20-year option = $62,411.20 − $38,243.20 = $24,168

5. $\dfrac{\$100,000}{\$1000}$ = 100 thousands of dollars of mortgage

30-year mortgage:
Monthly payment = $7.34 · 100 = $734
Total cost of interest = 360 · $734 − $100,000 = $164,240

20-year mortgage:
Monthly payment = $8.06 · 100 = $806
Total cost of interest = 20 · 12 · $806 − $100,000 = $93,440

The 20-year mortgage at 7.5% is more economical.

7. Amortization schedule:

Payment Number	Interest Payment	Principal Payment	Balance of Loan
1	$666.67	$67.33	$99,932.67
2	$666.22	$67.78	$99,864.89
3	$665.77	$68.23	$99,796.66

Calculations for Table:
Payment number 1:

$I = Prt = (\$100,000)(0.08)\left(\dfrac{1}{12}\right) \approx \$666.67$

Principal Payment = $734 − $666.67 = $67.33
Balance of Loan = $100,000 − $67.33 = $99,932.67

Payment number 2:

$I = Prt = (\$99,932.67)(0.08)\left(\dfrac{1}{12}\right) \approx \$666.22$

Principal Payment = $734 − $666.22 = $67.78
Balance of Loan = $99,932.67 − $67.78 = $99,864.89

Payment number 3

$I = Prt = (\$99,864.89)(0.08)\left(\dfrac{1}{12}\right) \approx \$665.77$

Principal Payment = $734 − $665.77 = $68.23
Balance of Loan = $99,864.89 − $68.23 = $99,796.66

9. Using Table 8.6, the person qualifies for a $172,300 mortgage at 7.0%.

11. Looking in the 6.5% row of Table 8.6, find the value $98,800. The column heading for that value is $30,000. So the buyer's annual income should be at least $30,000.

13. Down payment $= 0.30(\$80,000,000) = \$24,000,000$
    Amount of mortgage
    $= \$80,000,000 - \$24,000,000 = \$56,000,000$
    $$\frac{\$56,000,000}{\$1000} = 56,000 \text{ thousands of dollars of mortgage}$$
    Monthly payments $= \$10.29 \cdot 56,000 = \$576,240$
    Total cost of interest $= 360 \cdot \$576,240 - \$56,000,000 = \$151,446,400$

15-21. Answers will vary.

23. **a.** $85,100 mortgage for 30 years.

    b. $\$7.34 \times 85.1 = \$624.63$ monthly

    c. Answers will vary.

Check Points 8.5

1. **a.** High price = $63.38,
   Low price = $42.37

   b. Dividend $= \$0.72 \cdot 3000 = \$2160$

   c. Annual return for dividends alone = 1.5%
   1.5% is much lower than the 3.5% bank rate.

   d. Shares traded =
   $72,032 \cdot 100 = 7,203,200$ shares

   e. High price = $49.94,
   Low price = $48.33

   f. Price at close = $49.50

   g. The price went up $0.03 per share.

   h. Annual earnings per share $= \dfrac{\$49.50}{37} \approx \$1.34$

2. **a.** Earnings per share
   $\$19.16 - \$6.88 = \$12.28$
   Total earnings
   $\$12.28 \times 500 = \$6140$

   b. Total sale price $\$19.16 \times 500 = \$9580$
   Broker's commission
   $\$9580 \times 0.025 = \$239.50$

Exercise Set 8.5

1. **a.** High price = $73.25,
   Low price = $45.44

   b. Dividend $= \$1.20 \cdot 700 = \$840$

   c. Annual return for dividends alone = 2.2%
   2.2% is lower than a 3% bank rate.

   d. Shares traded $= 5915 \cdot 100 = 591,500$ shares

   e. High price = $56.38,
   Low price = $54.38

   f. Price at close = $55.50

   g. The price went up $1.25 per share.

   h. Annual earnings per share $= \dfrac{\$55.50}{17}$
   $\approx \$3.26$

3. **a.** Earnings per share $= \$65.00 - \$39.06$
   $= \$25.94$
   Total earnings $= 250(\$25.94) = \$6485$

   b. Total sale price $= 250(\$65.00)$
   $= \$16,250$
   Broker's commission $= 0.025(\$16,250)$
   $= \$406.25$

5. **a.** Cost of stock $= 400(\$37.50) = \$15,000$

   b. Broker's commission $= 0.02(\$15,000)$
   $= \$300$

7. a. Cost of stock = $240($17.75) = $4260

b. Broker's commission
= 0.025($4260) + 0.125(40)
= $111.50

9-19. Answers will vary.

Chapter 8 Review Exercises

1. $\frac{4}{5} = 4 \div 5 = 0.80 = 80\%$

2. $\frac{1}{8} = 1 \div 8 = 0.125 = 12.5\%$

3. $\frac{3}{4} = 3 \div 4 = 0.75 = 75\%$

4. $0.72 = 72\%$

5. $0.0035 = 0.35\%$

6. $4.756 = 475.6\%$

7. $65\% = 0.65$

8. $99.7\% = 0.997$

9. $150\% = 1.50$

10. $3\% = 0.03$

11. $0.65\% = 0.0065$

12. $\frac{1}{4}\% = 0.25\% = 0.0025$

13. $\frac{\$35,000}{\$50,000} = 0.70 = 70\%$ in stock

14. Tip = 0.15($78) = $11.70

15. a. Tax = 0.06($24) = $1.44

b. Total cost = $24 + $1.44 = $25.44

16. a. Amount of discount = 0.35($850)
= $297.50

b. Sale price = $850 − $297.50 = $552.50

17. $\frac{45-40}{40} = 0.125 = 12.5\%$ increase.

18. $\frac{\$56.00-\$36.40}{\$56.00} = 0.35 = 35\%$ decrease.

19. $I = Prt = (\$6000)(0.03)(1) = \$180$

20. $I = Prt = (\$8400)(0.05)(6) = \$2520$

21. $I = Prt = (\$20,000)(0.08)\left(\frac{9}{12}\right) = \$1200$

22. $I = Prt = (\$36,000)(0.15)\left(\frac{60}{360}\right) = \$900$

23. a. $I = Prt$
$= (\$3500)(0.105)\left(\frac{4}{12}\right)$
$= \$122.50$

b. Maturity value = $3500 + $122.50
= $3622.50

24. a. $A = \$7000(1+0.03)^5$
$= \$7000(1.03)^5$
$\approx \$8114.92$

b. Interest = $8114.92 − $7000
= $1114.92

25. a. $A = \$30,000\left(1+\frac{0.025}{4}\right)^{4 \cdot 10}$
$= \$30,000(1.00625)^{40}$
$\approx \$38,490.80$

b. Interest = $38,490.80 − $30,000
= $8490.80

26. a. $A = \$2500\left(1+\frac{0.04}{12}\right)^{12 \cdot 20}$
$= \$2500(1.003333)^{240}$
$\approx \$5556.46$

b. Interest = $5556.46 − $2500
= $3056.46

27. 6.25% rate compounded 12 times a year:
$A = \$5000\left(1+\frac{0.0625}{12}\right)^{12 \cdot 5} \approx \$6828.65$
6.3% compounded once a year:
$A = \$5000(1+0.063)^5 \approx \$6786.35$
A 6.25% rate compounded 12 times a year is a better choice.

28. $P = \dfrac{100,000}{\left(1+\dfrac{0.10}{12}\right)^{12\cdot18}} \approx \$16,653.64$

29. $P = \dfrac{75,000}{\left(1+\dfrac{0.05}{4}\right)^{4\cdot35}} \approx \$13,175.19$

30. $Y = \left(1+\dfrac{0.055}{4}\right)^{4} - 1 \approx 0.0561 = 5.61\%$

5.5% compounded quarterly is equivalent to 5.61% compounded annually.

31. $A = \$600\,\dfrac{\left(1+\dfrac{0.08}{4}\right)^{4\cdot18}-1}{\dfrac{0.08}{4}} \approx \$94,834$

32. a. Amount financed $= \$16,500 - \$500$
$$= \$16,000$$

b. Total installment price
$$= 60(\$350) + \$500$$
$$= \$21,500$$

c. Finance charge $= \$21,500 - \$16,500$
$$= \$5000$$

d. Finance charge per $100 financed
$$\dfrac{\$5000}{\$16,000} \times \$100 = \$31.25$$
The APR is approximately 11.5%.

33. a. Interest saved
$k = 12$, $R = \$350$, $v = \$6.34$
$$u = \dfrac{(12)(350)(6.34)}{100+6.34} = \$250.40$$

b. Payoff amount
$$= \$350 + (12)(\$350) - \$250.40$$
$$= \$4299.60$$

34. a. Interest saved
$k = 12$, $n = 60$, $F = \$5000$
$$u = \dfrac{12(12+1)}{60(60+1)} \cdot 5000 \approx \$213.11$$

b. Payoff amount
$$= \$350 + 12(\$350) - \$213.11 = \$4336.89$$

35. The actuarial method saves the borrower more money.

36. a. Unpaid balance $= \$1300 - \$200$
$$= \$1100$$
$I = Prt = (\$1100)(0.015)(1) = \$16.50$

b. Balance due
$$= \$1100 + \$380 + \$120 + \$140 + \$16.50$$
$$= \$1756.50$$

c. Minimum monthly payment
$$= \left(\dfrac{1}{36}\right)(\$1756.50)$$
$$= \$48.79$$
$$\approx \$49$$

37. a. Unpaid $= \$3600 - \$2000$
$$= \$1600$$
$I = Prt = (\$1600)(0.018)(1) = \$28.80$

b. Previous balance $= \$3600$
$I = Prt = (\$3600)(0.018)(1) = \$64.80$

c. Average daily balance
$$= \dfrac{(\$3600)(5) + (\$1600)(26)}{31} \approx \$1922.58$$
$I = Prt = (\$1922.58)(0.018)(1) = \$34.61$

38. a. Down payment $= 0.20(\$145,000) = \$29,000$

b. Amount of the mortgage
$$= \$145,000 - \$29,000 = \$116,000$$

c. Two points $= 0.02(\$116,000) = \$2320$

d. $\dfrac{\$116,000}{\$1000}$
$= 116$ thousands of dollars of mortgage
Monthly payment $\$6.65 \cdot 116 = \$771.40$

e. Total cost of interest
$360(\$771.40) - \$116,000$
$$= \$161,704$$

39. 30-year mortgage at 8.5%:

$$\frac{\$70,000}{\$1000} = 70 \text{ thousands of dollars of mortgage}$$

Monthly payment = $\$7.69 \cdot 70 = \$538.30$
Total cost of interest
$= 360(\$538.30) - \$70,000 = \$123,788$

20-year mortgage at 8%:
Monthly payment = $\$8.36 \cdot 70 = \$585.20$
Total cost of interest = $240(\$585.20) - \$70,000 = \$70,448$

A fixed-rate 20-year mortgage at 8% is more economical.

40. Amortization schedule:

Payment Number	Interest Payment	Principal Payment	Balance of Loan
1	$500	$59.20	$79,940.80
2	$499.63	$59.57	$79,881.23

Calculations for table:
Payment number 1:

$$I = Prt = (\$80,000)(0.075)\left(\frac{1}{12}\right) = \$500$$

Principal Payment = $\$559.20 - \$500 = \$59.20$
Balance of Loan = $\$80,000 - \$59.20 = \$79,940.80$

Payment number 2:

$$I = Prt = (\$79,940.80)(0.075)\left(\frac{1}{12}\right) = \$499.63$$

Principal Payment = $\$559.20 - \$499.63 = \$59.57$
Balance of Loan = $\$79,940.80 - \$59.57 = \$79,881.23$

41. $\dfrac{\$100,000}{\$1000} = 100$ thousands of dollars of mortgage
Option A: Monthly payment = $\$7.69 \cdot 100 = \$769$
Option B: Monthly Payment = $\$6.99 \cdot 100 = \$699$

42. High = $64.06, Low = $26.13

43. Dividend = $\$0.16(900) = \$144$

44. Annual return for dividends alone = 0.3%

45. Shares traded yesterday = $5458 \cdot 100$
$= 545,800$ shares

46. High = $61.25, Low = $59.25

47. Price at close = $61

48. Change in price = $1.75 increase

49. Annual earnings per share $\dfrac{\$61}{41} \approx \$1.49$

50. a. Earnings per share = $\$43.75 - \$27.50$
$= \$16.25$
Total earnings = $600(\$16.25) = \$9750$

b. Total sale price = $600(\$43.75)$
$= \$26,250$
Brokers Commission = $0.025(\$26,250)$
$= \$656.25$

51-52. Answers will vary.

Chapter 8 Test

1. a. Discount = $0.15(\$120) = \$18$

b. Sale price = $\$120 - \$18 = \$102$

2. Percent increase $\dfrac{12-10}{10} = 0.2 = 20\%$

3. $I = Prt = (\$2400)(0.12)\left(\dfrac{3}{12}\right) = \$72$

4. Maturity value = $\$2400 + \$72 = \$2472.$

5. $A = \$2000\left(1 + \dfrac{0.06}{12}\right)^{12 \cdot 5} = \$2697.70$

6. Interest earned = $\$2697.70 - \$2000$
$= \$697.70$

7. $P = \dfrac{100,000}{\left(1 + \frac{0.10}{2}\right)^{2 \cdot 20}} = \dfrac{100,000}{(1.05)^{40}} \approx \$14,204.57$

8. $Y = \left(1 + \dfrac{0.045}{4}\right)^4 - 1 \approx 0.0458 = 4.58\%$

4.5% compounded quarterly is equivalent to 4.58% compounded annually.

9. Amount financed = $\$16,000 - \$3000$
$= \$13,000$

10. Total installment price = $60(\$300) + \$3,000$
$= \$21,000$

11. Finance charge = $\$21,000 - \$16,000$
$= \$5000$

12. Finance charge per $100 financed
$$= \frac{\$5000}{\$13,000} \cdot \$100 = \$38.46$$
The APR is approximately 13.5%.

13. Interest saved
$k = 60 - 36 = 24$, $n = 60$, $F = \$5000$
$$u = \frac{24(24 + 1)}{60(60 + 1)} \cdot \$5000 = \$819.67$$

14. Payoff amount
$= \$300 + (24)(\$300) - \$819.67$
$= \$6680.33$

15. Unpaid balance $= \$880 - \$100 = \$780$
$I = Prt = \$780 \, (0.02)(1) = \$15.60$

16. Balance due
$= \$780 + \$350 + \$70 + \$120 + \$15.60$
$= \$1335.60$

17. Minimum monthly payment
$$= \frac{1}{36} (\$1335.60)$$
$= \$37$

18. Average daily balance
$$= \frac{(\$2400)(3) + (\$900)(27)}{30} = \$1050$$
$I = Prt = (\$1050)(0.016)(1) = \$16.80$

19. Down payment $= 0.10(\$120,000) = \$12,000$

20. Amount of mortgage $= \$120,000 - \$12,000$
$\qquad\qquad\qquad\qquad = \$108,000$

21. Two points $= 0.02(\$108,000) = \$2160$

22. $\dfrac{\$108,000}{\$1000} = 108$ thousands of dollars of mortgage

Monthly payment $= \$7.69 \cdot 108 = \$830.52$

23. Total cost of interest
$= 360(\$830.52) - \$108,000$
$= \$190,987.20$

24. $\dfrac{\$100,000}{\$1000} = 100$ thousands of dollars of mortgage

Option A 30-year mortgage:
Monthly payment $= \$6.99 \cdot 100 = \$699$

Option B 15-year mortgage:
Monthly payment $= \$8.99 \cdot 100 = \$899$

25. High $= \$25.75$, Low $= \$25.50$

26. Dividend $= \$2.03 \cdot 1000 = \$2030$

27. Total price paid $= 600(\$25.75) = \$15,450$
Broker's commission $= 0.025(\$15,450) = \$386.25$

Chapter 9
Measurement

Check Points 9.1

1. a. $78 \text{ in.} = \dfrac{78 \text{ in.}}{1} \cdot \dfrac{1 \text{ ft}}{12 \text{ in.}} = 6.5 \text{ ft}$

b. $17{,}160 \text{ ft} = \dfrac{17{,}160 \text{ ft}}{1} \cdot \dfrac{1 \text{ mi}}{5280 \text{ ft}} = 3.25 \text{ mi}$

c. $3 \text{ in.} = \dfrac{3 \text{ in.}}{1} \cdot \dfrac{1 \text{ yd}}{36 \text{ in.}} = 0.08\overline{3} \text{ yd} \approx 0.08 \text{ yd}$

2. a. $8000 \text{ m} = 8 \text{ km}$

b. $53 \text{ m} = 53{,}000 \text{ mm}$

c. $604 \text{ cm} = 0.0604 \text{ hm}$

d. $6.72 \text{ dam} = 6720 \text{ cm}$

3. a. $8 \text{ ft} = \dfrac{8 \text{ ft}}{1} \cdot \dfrac{30.48 \text{ cm}}{1 \text{ ft}} = 243.84 \text{ cm}$

b. $20 \text{ m} = \dfrac{20 \text{ m}}{1} \cdot \dfrac{1 \text{ yd}}{0.9 \text{ m}} \approx 22.22 \text{ yd}$

c. $30 \text{ m} = 3000 \text{ cm}$

$= \dfrac{3000 \text{ cm}}{1} \cdot \dfrac{1 \text{ in.}}{2.54 \text{ cm}}$

$\approx 1181.1 \text{ in.}$

4. $\dfrac{60 \text{ km}}{\text{hr}} = \dfrac{60 \text{ km}}{\text{hr}} \cdot \dfrac{1 \text{ mi}}{1.6 \text{ km}} = 37.5 \text{ mi/hr}$

Exercise Set 9.1

1. $30 \text{ in.} = \dfrac{30 \text{ in.}}{1} \cdot \dfrac{1 \text{ ft}}{12 \text{ in.}} = 2.5 \text{ ft}$

3. $30 \text{ ft} = \dfrac{30 \text{ ft}}{1} \cdot \dfrac{12 \text{ in.}}{1 \text{ ft}} = 360 \text{ in.}$

5. $6 \text{ in.} = \dfrac{6 \text{ in.}}{1} \cdot \dfrac{1 \text{ yd}}{36 \text{ in.}} \approx 0.17 \text{ yd}$

7. $6 \text{ yd} = \dfrac{6 \text{ yd}}{1} \cdot \dfrac{36 \text{ in.}}{1 \text{ yd}} = 216 \text{ in.}$

9. $6 \text{ yd} = \dfrac{6 \text{ yd}}{1} \cdot \dfrac{3 \text{ ft}}{1 \text{ yd}} = 18 \text{ ft}$

11. $6 \text{ ft} = \dfrac{6 \text{ ft}}{1} \cdot \dfrac{1 \text{ yd}}{3 \text{ ft}} = 2 \text{ yd}$

13. $23{,}760 \text{ ft} = \dfrac{23{,}760 \text{ ft}}{1} \cdot \dfrac{1 \text{ mi}}{5280 \text{ ft}} = 4.5 \text{ mi}$

15. $0.75 \text{ mi} = \dfrac{0.75 \text{ mi}}{1} \cdot \dfrac{5280 \text{ ft}}{1 \text{ mi}} = 3960 \text{ ft}$

17. $5 \text{ m} = 500 \text{ cm}$

19. $16.3 \text{ hm} = 1630 \text{ m}$

21. $317.8 \text{ cm} = 0.03178 \text{ hm}$

23. $0.023 \text{ mm} = 0.000023 \text{ m}$

25. $2196 \text{ mm} = 21.96 \text{ dm}$

27. $14 \text{ in.} = \dfrac{14 \text{ in.}}{1} \cdot \dfrac{2.54 \text{ cm}}{1 \text{ in.}} \approx 35.56 \text{ cm}$

29. $14 \text{ cm} = \dfrac{14 \text{ cm}}{1} \cdot \dfrac{1 \text{ in.}}{2.54 \text{ cm}} \approx 5.51 \text{ in.}$

31. $265 \text{ mi} = \dfrac{265 \text{ mi}}{1} \cdot \dfrac{1.6 \text{ km}}{1 \text{ mi}} \approx 424 \text{ km}$

33. $265 \text{ km} = \dfrac{265 \text{ km}}{1} \cdot \dfrac{1 \text{ mi}}{1.6 \text{ km}} \approx 165.625 \text{ mi}$

35. $12 \text{ m} = \dfrac{12 \text{ m}}{1} \cdot \dfrac{1 \text{ yd}}{0.9 \text{ m}} \approx 13.33 \text{ yd}$

37. $14 \text{ dm} = 140 \text{ cm} = \dfrac{140 \text{ cm}}{1} \cdot \dfrac{1 \text{ in.}}{2.54 \text{ cm}} \approx 55.12 \text{ in.}$

39. $160 \text{ in.} = \dfrac{160 \text{ in.}}{1} \cdot \dfrac{2.54 \text{ cm}}{1 \text{ in.}}$

$\approx 406.4 \text{ cm} = 0.4064 \text{ dam}$

41. $5 \text{ ft} = \dfrac{5 \text{ ft}}{1} \cdot \dfrac{30.48 \text{ cm}}{1 \text{ ft}} \approx 152.4 \text{ cm} \approx 1.5 \text{ m}$

43. $5 \text{ m} = 500 \text{ cm} = \dfrac{500 \text{ cm}}{1} \cdot \dfrac{1 \text{ ft}}{30.48 \text{ cm}} \approx 16.67 \text{ ft}$

45. $\dfrac{96 \text{ km}}{\text{hr}} = \dfrac{96 \text{ km}}{\text{hr}} \cdot \dfrac{1 \text{ mi}}{1.6 \text{ km}} \approx 60 \text{ mi/hr}$

47. $\dfrac{45 \text{ mi}}{\text{hr}} = \dfrac{45 \text{ mi}}{\text{hr}} \cdot \dfrac{1.6 \text{ km}}{1 \text{ mi}} \approx 72 \text{ km/hr}$

49. meter

51. millimeter

53. meter

55. millimeter

57. millimeter

59. b.

61. a.

63. c.

65. a.

67. $2 \cdot 4 \cdot 27 \text{ m} = 216 \text{ m} = 0.216 \text{ km}$

69. $93 \text{ million miles} = \dfrac{93,000,000 \text{ mi}}{1} \cdot \dfrac{1.6 \text{ km}}{1 \text{ mi}}$
$= 148.8 \text{ million kilometers}$

71-77. Answers will vary.

79. $900 \text{ m} = 9 \text{ hm}$

81. $11,000 \text{ mm} = 11 \text{ m}$

83. $5 \text{ yd} = \dfrac{5 \text{ yd}}{1} \cdot \dfrac{36 \text{ in.}}{1 \text{ yd}} \cdot \dfrac{2.54 \text{ cm}}{1 \text{ in.}} = 457.2 \text{ cm}$

Check Points 9.2

1. The area is 8 square units.

2. $\dfrac{103,000 \text{ people}}{10.5 \text{ square miles}} \approx 9809.5 \text{ people per square mile}$

3. a. $1.8 \text{ acres} = \dfrac{1.8 \text{ acres}}{1} \cdot \dfrac{0.4 \text{ ha}}{1 \text{ acre}} = 0.72 \text{ ha}$

    b. $\dfrac{\$415,000}{0.72 \text{ ha}} = \$576,389 \text{ per hectare}$

4. The volume is 9 cubic units.

5. $10,000 \text{ ft}^3 = \dfrac{10,000 \text{ ft}^3}{1} \cdot \dfrac{7.48 \text{ gal}}{1 \text{ ft}^3} = 74,800 \text{ gal}$

6. $220,000 \text{ cm}^3 = \dfrac{220,000 \text{ cm}^3}{1} \cdot \dfrac{1 \text{ L}}{1000 \text{ cm}^3} = 220 \text{ L}$

Exercise Set 9.2

1. $4 \cdot 4 = 16 \text{ square units}$

3. $6 \cdot 4 = 24 \text{ square units}$

5. 8 square units

7. $14 \text{ cm}^2 = \dfrac{14 \text{ cm}^2}{1} \cdot \dfrac{1 \text{ in.}^2}{6.5 \text{ cm}^2} \approx 2.15 \text{ in.}^2$

9. $30 \text{ m}^2 = \dfrac{30 \text{ m}^2}{1} \cdot \dfrac{1 \text{ yd}^2}{0.8 \text{ m}^2} = 37.5 \text{ yd}^2$

11. $10.2 \text{ ha} = \dfrac{10.2 \text{ ha}}{1} \cdot \dfrac{1 \text{ acre}}{0.4 \text{ ha}} = 25.5 \text{ acres}$

13. $14 \text{ in.}^2 = \dfrac{14 \text{ in.}^2}{1} \cdot \dfrac{6.5 \text{ cm}^2}{1 \text{ in.}^2} = 91 \text{ cm}^2$

15. $2 \cdot 4 \cdot 3 = 24 \text{ cubic units}$

17. $10,000 \text{ ft}^3 = \dfrac{10,000 \text{ ft}^3}{1} \cdot \dfrac{7.48 \text{ gal}}{1 \text{ ft}^3}$
$= 74,800 \text{ gal}$

19. $8 \text{ yd}^3 = \dfrac{8 \text{ yd}^3}{1} \cdot \dfrac{200 \text{ gal}}{1 \text{ yd}^3} = 1600 \text{ gal}$

21. $2079 \text{ in.}^3 = \dfrac{2079 \text{ in.}^3}{1} \cdot \dfrac{1 \text{ gal}}{231 \text{ in.}^3} = 9 \text{ gal}$

23. $2700 \text{ gal} = \dfrac{2700 \text{ gal}}{1} \cdot \dfrac{1 \text{ yd}^3}{200 \text{ gal}} = 13.5 \text{ yd}^3$

25. $45{,}000 \text{ cm}^3 = \dfrac{45{,}000 \text{ cm}^3}{1} \cdot \dfrac{1 \text{ L}}{1000 \text{ cm}^3} = 45 \text{ L}$

27. $17 \text{ cm}^3 = \dfrac{17 \text{ cm}^3}{1} \cdot \dfrac{1 \text{ L}}{1000 \text{ cm}^3} \cdot \dfrac{1 \text{ mL}}{0.001 \text{ L}} = 17 \text{ mL}$

29. $1.5 \text{ L} = \dfrac{1.5 \text{ L}}{1} \cdot \dfrac{1000 \text{ cm}^3}{1 \text{ L}} = 1500 \text{ cm}^3$

31. $150 \text{ mL} = \dfrac{150 \text{ mL}}{1} \cdot \dfrac{0.001 \text{ L}}{\text{mL}} \cdot \dfrac{1000 \text{ cm}^3}{1 \text{ L}}$
    $= 150 \text{ cm}^3$

33. $\dfrac{799{,}065 \text{ people}}{145{,}556 \text{ square miles}}$
    ≈ 5.5 people per square mile

35. **a.** $8 \text{ ha} = \dfrac{8 \text{ ha}}{1} \cdot \dfrac{1 \text{ acre}}{0.4 \text{ ha}} = 20 \text{ acres}$

    b. $\dfrac{\$250{,}000}{20 \text{ acres}} = \$12{,}500$ per acre

37. square centimeters or square meters

39. square kilometers

41. b.

43. b.

45. $45{,}000 \text{ ft}^3 = \dfrac{45{,}000 \text{ ft}^3}{1} \cdot \dfrac{7.48 \text{ gal}}{1 \text{ ft}^3}$
    $= 336{,}600 \text{ gal}$

47. $4000 \text{ cm}^3 = \dfrac{4000 \text{ cm}^3}{1} \cdot \dfrac{1 \text{ L}}{1000 \text{ cm}^3} = 4 \text{ L}$

49-53. Answers will vary.

55. $\dfrac{46{,}690 \text{ people}}{1000 \text{ ha}}$
    $= \dfrac{46{,}690 \text{ people}}{1000 \text{ ha}} \cdot \dfrac{0.4 \text{ ha}}{1 \text{ acre}} \cdot \dfrac{640 \text{ acres}}{1 \text{ square mile}}$
    $= 11{,}952.64$ people per square mile

57. Answers will vary.

59. $1 \text{ cm}^3 = \dfrac{1 \text{ cm}^3}{1} \cdot \dfrac{1 \text{ L}}{1000 \text{ cm}^3} \cdot \dfrac{1000 \text{ mL}}{1 \text{ L}}$
    $= 1$ milliliter
    $1 \text{ dm}^3 = \dfrac{1 \text{ dm}^3}{1} \cdot \dfrac{1000 \text{ cm}^3}{1 \text{ dm}^3} \cdot \dfrac{1 \text{ L}}{1000 \text{ cm}^3}$
    $= 1$ liter
    $1 \text{ m}^3 = \dfrac{1 \text{ m}^3}{1} \cdot \dfrac{1{,}000{,}000 \text{ cm}^3}{1 \text{ m}^3} \cdot \dfrac{1 \text{ L}}{1000 \text{ cm}^3}$
    $= 1000 \text{ L}$
    $= 1$ kiloliter

61. Approximately 6.5 liters. 6.5 mL is only a little more than a teaspoon and 6.5 kL is thousands of gallons.

Check Points 9.3

1. **a.** $4.2 \text{ dg} = 420 \text{ mg}$

   b. $620 \text{ cg} = 6.2 \text{ g}$

2. **a.** $186 \text{ lb} = \dfrac{186 \text{ lb}}{1} \cdot \dfrac{0.45 \text{ kg}}{1 \text{ lb}} = 83.7 \text{ kg}$

   b. $83.7 \times 1.2 \text{ mg} = 100.44 \text{ mg}$ dose

3. $F = \dfrac{9}{5} \cdot 50 + 32 = 122$
   $50°C = 122°F$

4. $C = \dfrac{5}{9}(59 - 32) = 15$
   $59°F = 15°C$

Exercise Set 9.3

1. 7.4 dg = 740 mg

3. 870 mg = 0.87 g

5. 8 g = 800 cg

7. 18.6 kg = 18,600 g

9. 0.018 mg = 0.000018 g

11. $36 \text{ oz} = \dfrac{36 \cancel{oz}}{1} \cdot \dfrac{1 \text{ lb}}{16 \cancel{oz}} = 2.25 \text{ lb}$

13. $36 \text{ oz} = \dfrac{36 \cancel{oz}}{1} \cdot \dfrac{28 \text{ g}}{1 \cancel{oz}} = 1008 \text{ g}$

15. $540 \text{ lb} = \dfrac{540 \cancel{lb}}{1} \cdot \dfrac{0.45 \text{ kg}}{1 \cancel{lb}} = 243 \text{ kg}$

17. $80 \text{ lb} = \dfrac{80 \cancel{lb}}{1} \cdot \dfrac{0.45 \text{ kg}}{1 \cancel{lb}} \cdot \dfrac{1000 \text{ g}}{1 \cancel{kg}} = 36,000 \text{ g}$

    or $80 \text{ lb} = \dfrac{80 \cancel{lb}}{1} \cdot \dfrac{16 \cancel{oz}}{1 \cancel{lb}} \cdot \dfrac{28 \text{ g}}{1 \cancel{oz}} = 35,840 \text{ g}$

19. $540 \text{ kg} = \dfrac{540 \cancel{kg}}{1} \cdot \dfrac{1 \text{ lb}}{0.45 \cancel{kg}} = 1200 \text{ lb}$

21. $200 \text{ t} = \dfrac{200 \cancel{t}}{1} \cdot \dfrac{1 \text{ T}}{0.9 \cancel{t}} \approx 222.22 \text{ T}$

23. 10° C
    $F = \dfrac{9}{5} \cdot 10 + 32$
    10° C = 50° F

25. 35° C
    $F = \dfrac{9}{5} \cdot 35 + 32$
    35° C = 95° F

27. 57° C
    $F = \dfrac{9}{5} \cdot 57 + 32$
    57° C = 134.6° F

29. –5° C
    $F = \dfrac{9}{5}(-5) + 32$
    –5° C = 23° F

31. 68° F
    $C = \dfrac{5}{9}(68 - 32)$
    68° F = 20° C

33. 41° F
    $C = \dfrac{5}{9}(41 - 32)$
    41° F = 5° C

35. 72° F
    $C = \dfrac{5}{9}(72 - 32)$
    72° F ≈ 22.2° C

37. 23° F
    $C = \dfrac{5}{9}(23 - 32)$
    23° F = –5° C

39. 350° F
    $C = \dfrac{5}{9}(350 - 32)$
    350° F ≈ 176.7° C

41. –22° F
    $C = \dfrac{5}{9}(-22 - 32)$
    –22° F = –30° C

43. milligram or gram

45. milligram or gram

47. kilogram

49. kilogram

51. b.

53. a.

55. c.

57. 720 g = 0.720 kg
    14 − 0.720 = 13.28 kg

193

59. $85 \text{ g} = \dfrac{85 \cancel{g}}{1} \cdot \dfrac{1 \text{ oz}}{28 \cancel{g}} \approx 3.04 \text{ oz}$

Cost $= 33 + 3 \cdot 22 = 99$ cents

61. $\dfrac{\$3.15}{3 \text{ kg}} = \$1.05$ per kg for economy size

$720 \text{ g} = 0.72 \text{ kg}$

$\dfrac{\$.60}{0.72 \text{ kg}} = \$.83$ per kg for regular size

It is more economical to purchase the regular size.

63. $80 \text{ lb} = \dfrac{80 \cancel{lb}}{1} \cdot \dfrac{0.45 \text{ kg}}{1 \cancel{lb}} = 36 \text{ kg}$

$36 \times 2.5 \text{ mg} = 90 \text{ mg}$ dose

65. a. $\dfrac{21.5 \text{ mg}}{\text{tsp}} = \dfrac{21.5 \text{ mg}}{\cancel{\text{tsp}}} \cdot \dfrac{2 \cancel{\text{tsp}}}{1 \text{ dose}} = 43 \text{ mg/dose}$

b. $\dfrac{21.5 \text{ mg}}{\text{tsp}} = \dfrac{21.5 \text{ mg}}{\cancel{\text{tsp}}} \cdot \dfrac{\cancel{\text{tsp}}}{5 \cancel{\text{ml}}} \cdot \dfrac{30 \cancel{\text{ml}}}{1 \cancel{\text{oz}}} \cdot \dfrac{4 \cancel{\text{oz}}}{1 \text{ bottle}}$

$= 516 \text{ mg/bottle}$

67. a.

69. c.

71. $-7° \text{ C}$

$F = \dfrac{9}{5}(-7) + 32$

$-7°\text{C} = 19.4° \text{ F}$

73-77. Answers will vary.

79. a. $\dfrac{3\cancel{\text{¢}}}{1\cancel{g}} \cdot \dfrac{1000 \cancel{g}}{1 \cancel{kg}} \cdot \dfrac{0.45 \cancel{kg}}{1 \text{ lb}} = 1350\text{¢ per pound}$

False

b. <u>True</u>

c. False

d. False

Chapter 9 Review Exercises

1. $69 \text{ in.} = \dfrac{69 \cancel{\text{in.}}}{1} \cdot \dfrac{1 \text{ ft}}{12 \cancel{\text{in.}}} = 5.75 \text{ ft}$

2. $9 \text{ in.} = \dfrac{9 \cancel{\text{in.}}}{1} \cdot \dfrac{1 \text{ yd}}{36 \cancel{\text{in.}}} = 0.25 \text{ yd}$

3. $21 \text{ ft} = \dfrac{21 \cancel{\text{ft}}}{1} \cdot \dfrac{1 \text{ yd}}{3 \cancel{\text{ft}}} = 7 \text{ yd}$

4. $13{,}200 \text{ ft} = \dfrac{13{,}200 \cancel{\text{ft}}}{1} \cdot \dfrac{1 \text{ mi}}{5280 \cancel{\text{ft}}} = 2.5 \text{ mi}$

5. $22.8 \text{ m} = 2280 \text{ cm}$

6. $7 \text{ dam} = 70 \text{ m}$

7. $19.2 \text{ hm} = 1920 \text{ m}$

8. $144 \text{ cm} = 0.0144 \text{ hm}$

9. $0.5 \text{ mm} = 0.0005 \text{ m}$

10. $18 \text{ cm} = 180 \text{ mm}$

11. $23 \text{ in.} = \dfrac{23 \cancel{\text{in.}}}{1} \cdot \dfrac{2.54 \text{ cm}}{1 \cancel{\text{in.}}} = 58.42 \text{ cm}$

12. $19 \text{ cm} = \dfrac{19 \cancel{\text{cm}}}{1} \cdot \dfrac{1 \text{ in.}}{2.54 \cancel{\text{cm}}} \approx 7.48 \text{ in.}$

13. $330 \text{ mi} = \dfrac{330 \cancel{\text{mi}}}{1} \cdot \dfrac{1.6 \text{ km}}{1 \cancel{\text{mi}}} = 528 \text{ km}$

14. $600 \text{ km} = \dfrac{600 \cancel{\text{km}}}{1} \cdot \dfrac{1 \text{ mi}}{1.6 \cancel{\text{km}}} = 375 \text{ mi}$

15. $14 \text{ m} = \dfrac{14 \cancel{\text{m}}}{1} \cdot \dfrac{1 \text{ yd}}{0.9 \cancel{\text{m}}} \approx 15.56 \text{ yd}$

16. $12 \text{ m} = \dfrac{12 \cancel{\text{m}}}{1} \cdot \dfrac{1 \cancel{\text{yd}}}{0.9 \cancel{\text{m}}} \cdot \dfrac{3 \text{ ft}}{1 \cancel{\text{yd}}} = 40 \text{ ft}$

17. $45 \text{ km per hour} = \dfrac{45 \cancel{\text{km}}}{1 \text{ hr}} \cdot \dfrac{1 \text{ mi}}{1.6 \cancel{\text{km}}}$

$\approx 28.13 \text{ miles/hour}$

18. $60 \text{ mi per hour} = \dfrac{60 \cancel{\text{mi}}}{1 \text{ hr}} \cdot \dfrac{1.6 \text{ km}}{1 \cancel{\text{mi}}}$

$= 96 \text{ km/hr}$

19. $0.024 \text{ km; } 24{,}000 \text{ cm; } 2400 \text{ m}$

20. $6 \cdot 800 \text{ m} = 4800 \text{ m} = 4.8 \text{ km}$

21. $3 \cdot 8 = 24$ square units

22. $\dfrac{29,760,000 \text{ people}}{155,973 \text{ square miles}}$

    ≈ 190.8 people per square mile;
    Answers will vary

23. $7.2 \text{ ha} = \dfrac{7.2 \text{ ha}}{1} \cdot \dfrac{1 \text{ acre}}{0.4 \text{ ha}} = 18$ acres

24. $30 \text{ m}^2 = \dfrac{30 \text{ m}^2}{1} \cdot \dfrac{1 \text{ ft}^2}{0.09 \text{ m}^2} \approx 333.33 \text{ ft}^2$

25. $12 \text{ mi}^2 = \dfrac{12 \text{ mi}^2}{1} \cdot \dfrac{2.6 \text{ km}^2}{1 \text{ mi}^2} = 31.2 \text{ km}^2$

26. a

27. $2 \cdot 4 \cdot 3 = 24$ cubic units

28. $33,600$ cubic feet $= \dfrac{33,600 \text{ ft}^3}{1} \cdot \dfrac{7.48 \text{ gal}}{1 \text{ ft}^3}$

    $= 251,328$ gal

29. $76,000 \text{ cm}^3 = \dfrac{76,000 \text{ cm}^3}{1} \cdot \dfrac{1 \text{ L}}{1000 \text{ cm}^3} = 76 \text{ L}$

30. c

31-32. Answers will vary.

33. $12.4 \text{ dg} = 1240 \text{ mg}$

34. $12 \text{ g} = 1200 \text{ cg}$

35. $0.012 \text{ mg} = 0.000012 \text{ g}$

36. $450 \text{ mg} = 0.00045 \text{ kg}$

37. $210 \text{ lb} = \dfrac{210 \text{ lb}}{1} \cdot \dfrac{0.45 \text{ kg}}{1 \text{ lb}} = 94.5 \text{ kg}$

38. $392 \text{ g} = \dfrac{392 \text{ g}}{1} \cdot \dfrac{1 \text{ oz}}{28 \text{ g}} = 14 \text{ oz}$

39. Kilograms; Answers will vary.

40. $36 \text{ oz} = \dfrac{36 \text{ oz}}{1} \cdot \dfrac{1 \text{ lb}}{16 \text{ oz}} = 2.25 \text{ lb}$

41. a

42. c

43. 15° C
    $$F = \dfrac{9}{5} \cdot 15 + 32 = 59° \text{ F}$$

44. 100° C
    $$F = \dfrac{9}{5} \cdot 100 + 32$$
    $$100° \text{C} = 212° \text{ F}$$

45. 5° C
    $$F = \dfrac{9}{5} \cdot 5 + 32$$
    $$5° \text{ C} = 41° \text{F}$$

46. 0° C
    $$F = \dfrac{9}{5} \cdot 0 + 32$$
    $$0° \text{C} = 32° \text{ F}$$

47. −25° C
    $$F = \dfrac{9}{5}(-25) + 32$$
    $$-25° \text{ C} = -13° \text{ F}$$

48. 59° F
    $$C = \dfrac{5}{9}(59 - 32)$$
    $$59° \text{F} = 15° \text{ C}$$

49. 41° F
    $$C = \dfrac{5}{9}(41 - 32)$$
    $$41° \text{ F} = 5° \text{C}$$

50. 212° F
    $$C = \dfrac{5}{9}(212 - 32)$$
    $$212° \text{F} = 100° \text{ C}$$

51. 98.6° F
    $$C = \dfrac{5}{9}(98.6 - 32)$$
    $$98.6° \text{ F} = 37° \text{ C}$$

52. 0° F
    $$C = \dfrac{5}{9}(0 - 32)$$
    $$0° \text{F} \approx -17.8° \text{C}$$

53. $14°$ F

$$C = \frac{5}{9}(14 - 32)$$

$14°$ F $= -10°$ C

54. A decrease of $15°C$ is more than a decrease of $15°F$; Explanations will vary.

Chapter 9 Test

1. 807 mm = 0.00807 hm

2. 635 cm $= \dfrac{635 \,\cancel{cm}}{1} \cdot \dfrac{1 \text{ in.}}{2.54 \,\cancel{cm}} = 250$ in.

3. $8 \cdot 600$ m = 4800 m = 4.8 km

4. mm

5. cm

6. km

7. 80 miles per hour $= \dfrac{80 \,\cancel{mi}}{1 \text{ hr}} \cdot \dfrac{1.6 \text{ km}}{1 \,\cancel{mi}}$

$= 128$ km/hr

8. $1 \text{ yd}^2 = (3 \text{ ft})(3 \text{ ft}) = 9 \text{ ft}^2$

A square yard is 9 times greater than a square foot.

9. $\dfrac{39,133,966 \text{ people}}{195,365 \text{ square miles}}$

≈ 200.3 people per square mile

10. 18 ha $= \dfrac{18 \,\cancel{ha}}{1} \cdot \dfrac{1 \text{ acre}}{0.4 \,\cancel{ha}} = 45$ acres

11. b

12. Answers will vary.

$1 \text{ m}^3 = (10 \text{ dm})(10 \text{ dm})(10 \text{ dm}) = 1000 \text{ dm}^3$

A cubic meter is 1000 times greater than a cubic decimeter.

13. $10,000 \text{ ft}^3 = \dfrac{10,000 \,\cancel{ft^3}}{1} \cdot \dfrac{7.48 \text{ gal}}{1 \,\cancel{ft^3}}$

$= 74,800$ gal

14. b

15. 137 g = 0.137 kg

16. 90 lb $= \dfrac{90 \,\cancel{lb}}{1} \cdot \dfrac{0.45 \text{ kg}}{1 \,\cancel{lb}} = 40.5$ kg

17. kg

18. mg

19. $30°$ C

$$F = \frac{9}{5} \cdot 30 + 32$$

$30°$C $= 86°$ F

20. $176°$ F

$$C = \frac{5}{9}(176 - 32)$$

$176°$F $= 80°$C

21. d

Chapter 10
Geometry

Check Points 10.1

1. a. F

   b. ray FB and ray FR

   c. $\angle 3$, $\angle F$, $\angle BFR$, $\angle RFB$

2. Hand moves $\dfrac{1}{12}$ of a rotation

   $\dfrac{1}{12} \cdot 360° = 30°$

3. $90° - 19° = 71°$

4. $180° - 46° = 134°$

5. $m\angle 1 = 57°$
   $m\angle 2 = 180° - 57° - 123°$
   $m\angle 3 = m\angle 2 = 123°$

6. $m\angle 1 = m\angle 8 = 29°$
   $m\angle 5 = m\angle 8 = 29°$
   $m\angle 2 = m\angle 8 = 29°$
   $m\angle 6 = 180° - m\angle 8 = 180° - 29° = 151°$
   $m\angle 7 = m\angle 6 = 151°$
   $m\angle 3 = m\angle 7 = 151°$
   $m\angle 4 = m\angle 3 = 151°$

Exercise Set 10.1

1. a. C

   b. ray CD and ray CB

   c. $\angle C$, $\angle DCB$, $\angle BCD$

3. Hand moves $\dfrac{5}{12}$ of a rotation

   $\dfrac{5}{12} \cdot 360° = 150°$

5. Hand moves $\dfrac{4-1}{12} = \dfrac{3}{12} = \dfrac{1}{4}$
   of a rotation
   $\dfrac{1}{4} \cdot 360° = 90°$

7. The measure of the angle is 65°.
   The angle is acute.

9. Obtuse

11. Straight

13. $90° - 25° = 65°$

15. $108° - 34° = 146°$

17. $90° - 48° = 42°$

19. $90° - 89° = 1°$

21. $90° - 37.4° = 52.6°$

23. $180° - 111° = 69°$

25. $180° - 16° = 164°$

27. $180° - 90° = 90°$

29. $180° - 93\dfrac{1}{4}° = 86\dfrac{3}{4}°$

31. $m\angle 1 = 180° - 72° = 108°$
    $m\angle 2 = 72°$
    $m\angle 3 = m\angle 1 - 108°$

33. $m\angle 1 = 90° - 40° = 50°$
    $m\angle 2 = 90°$
    $m\angle 3 = m\angle 1 = 50°$

35. $m\angle 1 = 180° - 112° = 68°$
    $m\angle 2 = m\angle 1 = 68°$
    $m\angle 3 = 112°$
    $m\angle 4 = 112°$
    $m\angle 5 = m\angle 1 = 68°$
    $m\angle 6 = m\angle 2 = 68°$
    $m\angle 7 = m\angle 3 = 112°$

37-51. Answers will vary.

53. $85° = 40° + m\angle 1$
    $m\angle 1 = 45°$

55. $m\angle BGD = m\angle BGC + m\angle CGD$
    We know
    $m\angle AGB + m\angle BGC + m\angle CGD + m\angle DGE = 180°$
    Since $m\angle AGB = m\angle BGC$ and
    $m\angle CGD = m\angle DGE$, this becomes
    $m\angle BGC + m\angle BGC + m\angle CGD + m\angle CGD = 180°$
    or $2(m\angle BGC) + 2(m\angle CGD) = 180°$
    or $m\angle BGC + m\angle CGD = 90°$
    Therefore, $m\angle BGD = 90°$.

Check Points 10.2

1. $m\angle A + 116° + 15° = 180°$

   $m\angle A + 131° = 180°$

   $m\angle A = 180° - 131°$

   $m\angle A = 49°$

2. $m\angle 1 = 180° - 90° = 90°$

   $m\angle 2 = 180° - 36° - m\angle 1$

   $= 180° - 36° - 90°$

   $= 54°$

   $m\angle 3 = m\angle 2 = 54°$

   $m\angle 4 = 180° - 58° - m\angle 3$

   $= 180° - 58° - 54°$

   $= 68°$

   $m\angle 5 = 180° - m\angle 4$

   $= 180° - 68°$

   $= 112°$

3. $\dfrac{3}{8} = \dfrac{12}{x}$

   $3 \cdot x = 8 \cdot 12$

   $3x = 96$

   $\dfrac{3x}{3} = \dfrac{96}{3}$

   $x = 32$ in.

4. $\dfrac{h}{2} = \dfrac{56}{3.5}$

   $3.5 \cdot h = 2 \cdot 56$

   $3.5h = 112$

   $\dfrac{3.5h}{3.5} = \dfrac{112}{3.5}$

   $h = 32$ yd

5. $c^2 = a^2 + b^2$

   $c^2 = 7^2 + 24^2$

   $c^2 = 49 + 576$

   $c^2 = 625$

   $c = \sqrt{625}$

   $c = 25$ ft

6. $w^2 + 9^2 = 15^2$

   $w^2 + 81 = 225$

   $w^2 = 144$

   $w = \sqrt{144}$

   $w = 12$ in.

Exercise Set 10.2

1. $m\angle A = 180° - 46° - 67° = 67°$

3. $m\angle A = 180° - 58° - 90° = 32°$

5. $m\angle 1 = 180° - 40° - 90° = 50°$

   $m\angle 2 = 180° - m\angle 1 = 180° - 50° = 130°$

   $m\angle 3 = m\angle 1 = 50°$

   $m\angle 4 = m\angle 2 = 130°$

   $m\angle 5 = 180° - 80° - m\angle 3$

   $= 180° - 80° - 50°$

   $= 50°$

7. Corresponding angles:

   Angles A and D

   Angles B and E

   Angles C and F

   Corresponding sides:

   $\overline{AB}$ and $\overline{DE}$

   $\overline{AC}$ and $\overline{DF}$

   $\overline{BC}$ and $\overline{EF}$

9. Corresponding angles:

   Angles N and R

   Angles P and S

   Angles M and T

   Corresponding sides:

   $\overline{NM}$ and $\overline{RT}$

   $\overline{NP}$ and $\overline{RS}$

   $\overline{MP}$ and $\overline{TS}$

11. $\dfrac{18}{9} = \dfrac{10}{x}$

    $18 \cdot x = 9 \cdot 10$

    $18x = 90$

    $\dfrac{18x}{18} = \dfrac{90}{18}$

    $x = 5$ in.

13. $\dfrac{30}{10} = \dfrac{18}{x}$

    $30 \cdot x = 10 \cdot 18$

    $30x = 180$

    $\dfrac{30x}{30} = \dfrac{180}{30}$

    $x = 6$ m

15. $\dfrac{20}{15} = \dfrac{x}{12}$

$15x = 20 \cdot 12$

$15x = 240$

$\dfrac{15x}{15} = \dfrac{240}{15}$

$x = 16$ in.

17. $c^2 = 8^2 + 15^2$

$c^2 = 64 + 225$

$c^2 = 289$

$c = 17$ m

19. $c^2 = 15^2 + 36^2$

$c^2 = 225 + 1296$

$c^2 = 1521$

$c = 39$ m

21. $a^2 + 16^2 = 20^2$

$a^2 + 256 = 400$

$a^2 = 144$

$a = 12$ cm

23. $b^2 + 9^2 = 16^2$

$b^2 + 81 = 256$

$b^2 = 175$

$b \approx 13.2$ m

25. No; $46.1° + 58.3° + 75.9° = 180.3°$
There is an error of $180.3° - 180° = 0.3°$

27. Let x = height of tree.

$\dfrac{x}{5} = \dfrac{86}{6}$

$6 \cdot x = 5 \cdot 86$

$6x = 430$

$x \approx 71.7$ ft

29. Let x = height of the plane.

$x^2 + 8^2 = 10^2$

$x^2 + 64 = 100$

$x^2 = 36$

$x = 6$ km

31. Let x = the length of the ladder.

$x^2 + 15^2 = 20^2$

$x^2 + 225 = 400$

$x^2 = 175$

$x \approx 13.2$ ft

33. Let x = measure of the side of the screen.

$x^2 + x^2 = 25^2$

$2x^2 = 729$

$\dfrac{2x^2}{2} = \dfrac{729}{2}$

$x^2 = 364.5$

$x \approx 19.1$ in.

35. Let c = the length of one support wire.

$c^2 = 8^2 + 15^2$

$c^2 = 64 + 225$

$c^2 = 289$

$c = 17$ ft

$2c = 34$ ft

37-43. Answers will vary.

45. $m\angle PQT = 180° - 70° - 60° = 50°$

$m\angle SQR = 180° - m\angle PQT - 50°$

$\qquad = 180° - 50° - 50°$

$\qquad = 80°$

$m\angle R = 180° - m\angle SQR - 30°$

$\qquad = 180° - 80° - 30°$

$\qquad = 70°$

47. The lengths should satisfy the Pythagorean theorem.

$c^2 = 16^2 + 20^2$

$c^2 = 256 + 400$

$c^2 = 656$

$c \approx 25.6$ ft, not $24\frac{1}{4}$ ft

No, the floor is not squared off properly.

Check Points 10.3

1. $P = 2l + 2w$

$P = 2 \cdot 160$ ft $+ 2 \cdot 90$ ft $= 500$ ft

$\text{Cost} = \dfrac{500 \text{ feet}}{1} \cdot \dfrac{\$6.50}{\text{foot}} = \$3250$

2. a. Sum $= (n-2)180°$
$= (12-2)\,180°$
$= 10 \cdot 180°$
$= 1800°$

b. $m\angle A = \dfrac{1800°}{12} = 150°$

Exercise Set 10.3

1. Quadrilateral (4 sides)

3. Pentagon (5 sides)

5. a (square), b (rhombus), d (rectangle), and e (parallelogram) all have two pairs of parallel sides.

7. a (square), d (rectangle)

9. c (trapezoid)

11. $P = 2 \cdot 3 \text{ cm} + 2 \cdot 12 \text{ cm}$
$= 6 \text{ cm} + 24 \text{ cm}$
$= 30 \text{ cm}$

13. $P = 2 \cdot 6 \text{ yd} + 2 \cdot 8 \text{ yd}$
$= 12 \text{ yd} + 16 \text{ yd}$
$= 28 \text{ yd}$

15. $P = 4 \cdot 250 \text{ in.} = 1000 \text{ in.}$

17. $P = 9 \text{ ft} + 7 \text{ ft} + 11 \text{ ft} = 27 \text{ ft}$

19. $P = 3 \cdot 6 \text{ yd} = 18 \text{ yd}$

21. $P = 12 \text{ yd} + 12 \text{ yd} + 9 \text{ yd} + 9 \text{ yd} + 21 \text{ yd} + 21 \text{ yd}$
$= 84 \text{ yd}$

23. First determine lengths of unknown sides.

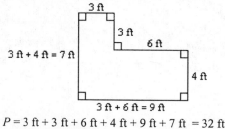

$P = 3 \text{ ft} + 3 \text{ ft} + 6 \text{ ft} + 4 \text{ ft} + 9 \text{ ft} + 7 \text{ ft} = 32 \text{ ft}$

25. Sum $= (n-2)180°$
$= (5-2)\,180°$
$= 3 \cdot 180°$
$= 540°$

27. Sum $= (n-2)180°$
$= (4-2)180°$
$= 2 \cdot 180°$
$= 360°$

29. From Exercise 25, we know the sum of the measures of the angles of a pentagon is 540°. Since all 5 angles have the same degree measure, $m\angle 1 = \dfrac{540°}{5} = 108°$.

31. a. From Exercise 25, we know the sum of the measures of the angles of a pentagon is 540°.

b. $m\angle 1 = 540° - 70° - 150° - 90° - 90°$
$= 140°$

33. No entry

35. Stop, yield, deer crossing

37. Equilateral triangle

39. Yield

41. $P = 2 \cdot 400 \text{ ft} + 2 \cdot 200 \text{ ft}$
$= 800 \text{ ft} + 400 \text{ ft}$
$= 1200 \text{ ft}$

$\text{Cost} = \dfrac{1200 \text{ ft}}{1} \cdot \dfrac{1 \text{ yd}}{3 \text{ ft}} \cdot \dfrac{\$14}{1 \text{ yd}} = \$5600$

43. First determine lengths of unknown sides.

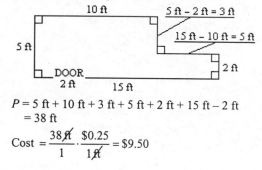

$P = 5 \text{ ft} + 10 \text{ ft} + 3 \text{ ft} + 5 \text{ ft} + 2 \text{ ft} + 15 \text{ ft} - 2 \text{ ft}$
$= 38 \text{ ft}$

$\text{Cost} = \dfrac{38 \text{ ft}}{1} \cdot \dfrac{\$0.25}{1 \text{ ft}} = \$9.50$

45-49. Answers will vary.

51. All sides have length a, therefore $P = 6a$.

Check Points 10.4

1. Area of large rectangle:
$$A_{large} = lw$$
$$= (13 \text{ ft} + 3 \text{ ft}) \times (3 \text{ ft} + 6 \text{ ft})$$
$$= 16 \text{ ft} \cdot 9 \text{ ft}$$
$$= 144 \text{ ft}^2$$
Area of small rectangle:
$$A_{small} = lw$$
$$= 13 \text{ ft} \cdot 6 \text{ ft}$$
$$= 78 \text{ ft}^2$$
Area of path $= 144 \text{ ft}^2 - 78 \text{ ft}^2 = 66 \text{ ft}^2$

2. First convert the linear measures in feet to linear yards.
$$18 \text{ ft} = \frac{18 \text{ ft}}{1} \cdot \frac{1 \text{ yd}}{3 \text{ ft}} = 6 \text{ yd}$$
$$21 \text{ ft} = \frac{21 \text{ ft}}{1} \cdot \frac{1 \text{ yd}}{3 \text{ ft}} = 7 \text{ yd}$$
Area of floor $= 6 \text{ yd} \cdot 7 \text{ yd} = 42 \text{ yd}^2$
$$\text{Cost of carpet} = \frac{42 \text{ yd}^2}{1} \cdot \frac{\$16}{1 \text{ yd}^2} = \$672$$

3. $A = bh$
$A = 10 \text{ in.} \cdot 6 \text{ in.} = 60 \text{ in.}^2$

4. $A = \frac{1}{2}bh$
$A = \frac{1}{2} \cdot 12 \text{ ft} \cdot 5 \text{ ft} = 30 \text{ ft}^2$

5. $A = \frac{1}{2}h(a+b)$
$= \frac{1}{2} \cdot 7 \text{ ft} \cdot (20 \text{ ft} + 10 \text{ ft})$
$= \frac{1}{2} \cdot 7 \text{ ft} \cdot 30 \text{ ft}$
$= 105 \text{ ft}^2$

6. $C = \pi d$
$\approx \pi (10 \text{ in.})$
$\approx 31.4 \text{ in.}$

7. Find the circumference of the semicircle:
$$C_{semicircle} = \frac{1}{2}\pi d$$
$$\approx \frac{1}{2}\pi (10 \text{ ft})$$
$$\approx 15.7 \text{ ft}$$
Length of trim $= 10 \text{ ft} + 12 \text{ ft} + 12 \text{ ft} + 15.7 \text{ ft}$
$= 49.7 \text{ ft}$

8. First, find the area of pizzas.

Large:	Medium:
$A = \pi r^2$	$A = \pi r^2$
$= \pi (9 \text{ in.})^2$	$= \pi (7 \text{ in.})^2$
$= 81\pi \text{ in.}^2$	$= 49\pi \text{ in.}^2$
$\approx 254 \text{ in.}^2$	$\approx 154 \text{ in.}^2$

Next, find the price per square inch.

Large:	Medium:
$\frac{\$20.00}{81\pi \text{ in.}^2}$	$\frac{\$14.00}{49\pi \text{ in.}^2}$
$\approx \frac{\$20.00}{254 \text{ in.}^2}$	$\approx \frac{\$14.00}{154 \text{ in.}^2}$
$\approx \frac{\$0.08}{\text{in.}^2}$	$\approx \frac{\$0.09}{\text{in.}^2}$

The large pizza is a better buy.

Exercise Set 10.4

1. $A = 6 \text{ m} \cdot 3 \text{ m} = 18 \text{ m}^2$

3. $A = (4 \text{ in.})^2 = 16 \text{ in.}^2$

5. $A = 50 \text{ cm} \cdot 42 \text{ cm} = 2100 \text{ cm}^2$

7. $A = \frac{1}{2} \cdot 14 \text{ in.} \cdot 8 \text{ in.} = 56 \text{ in.}^2$

9. $A = \frac{1}{2} \cdot 9.8 \text{ yd} \cdot 4.2 \text{ yd} = 20.58 \text{ yd}^2$

11. $A = \frac{1}{2} \cdot 7 \text{ m} \cdot (16 \text{ m} + 10 \text{ m})$
$= \frac{7}{2} \text{ m} \cdot (26 \text{ m})$
$= 91 \text{ m}^2$

13. $C = 2\pi \cdot 4 \text{ cm} \approx 25 \text{ cm}$
$A = \pi (4 \text{ cm})^2 \approx 50 \text{ cm}^2$

15. $C = \pi \cdot 12 \text{ yd} \approx 38 \text{ yd}$

$r = \dfrac{d}{2} = \dfrac{12 \text{ yd}}{2} = 6 \text{ yd}$

$A = \pi(6 \text{ yd})^2 \approx 113 \text{ yd}^2$

17. The figure breaks into a lower rectangle and an upper rectangle.

Area of lower rectangle: Area of upper rectangle:

$A = lw$ $A = lw$

$A = (12 \text{ m})(3 \text{ m})$ $A = (9 \text{ m})(4 \text{ m})$

$A = 36 \text{ m}^2$ $A = 36 \text{ m}^2$

Total area $= 36 \text{ m}^2 + 36 \text{ m}^2 = 72 \text{ m}^2$

19. The figure breaks into a lower rectangle and an upper triangle.

Area of rectangle: Area of triangle:

$A = lw$ $A = \frac{1}{2}bh$

$A = (24 \text{ m})(10 \text{ m})$ $A = \frac{1}{2}(24 \text{ m})(5 \text{ m})$

$A = 240 \text{ m}^2$ $A = 60 \text{ m}^2$

Total area $= 240 \text{ m}^2 + 60 \text{ m}^2 = 300 \text{ m}^2$

21. First convert the linear measures in feet to linear yards.

$9 \text{ ft} = \dfrac{9 \text{ ft}}{1} \cdot \dfrac{1 \text{ yd}}{3 \text{ ft}} = 3 \text{ yd}$

$21 \text{ ft} = \dfrac{21 \text{ ft}}{1} \cdot \dfrac{1 \text{ yd}}{3 \text{ ft}} = 7 \text{ yd}$

Area of floor $= 3 \text{ yd} \cdot 7 \text{ yd} = 21 \text{ yd}^2$

Cost of carpet $= \dfrac{21 \text{ yd}^2}{1} \cdot \dfrac{\$26.50}{1 \text{ yd}^2} = \$556.50$

23. Area of tile $=$ (Area of floor) $-$ (Area of store) $-$ (Area of refrigerator)

$= (12 \text{ ft} \cdot 15 \text{ ft}) - (3 \text{ ft} \cdot 4 \text{ ft}) - (4 \text{ ft} \cdot 5 \text{ ft})$

$= 180 \text{ ft}^2 - 12 \text{ ft}^2 - 20 \text{ ft}^2 = 148 \text{ ft}^2$

25. Area of pool $= 14 \text{ m} \cdot 10 \text{ m} = 140 \text{ m}^2$

Cost of cover $= \dfrac{\$6.50}{1 \text{ m}^2} \cdot \dfrac{140 \text{ m}^2}{1} = \$910$

27. a. Area of lawn $=$ (Area of lot) $-$ (Area of house) $-$ (Area of shed) $-$ (Area of driveway)

$= 200 \text{ ft} \cdot 500 \text{ ft} - 60 \text{ ft} \cdot 100 \text{ ft} - (20 \text{ ft})^2 - 100 \text{ ft} \cdot 20 \text{ ft}$

$= 100,000 \text{ ft}^2 - 6000 \text{ ft}^2 - 400 \text{ ft}^2 - 2000 \text{ ft}^2$

$= 91,600 \text{ ft}^2$

Maximum number of bags of fertilizer $= \dfrac{1 \text{ bag}}{4000 \text{ ft}^2} \cdot \dfrac{91,600 \text{ ft}^2}{1} = 22.9 \text{ bags} \rightarrow 23 \text{ bags}$

b. Total cost of fertilizer $= \dfrac{\$25.00}{1 \text{ bag}} \cdot \dfrac{23 \text{ bags}}{2} = \$575$

29. Amount of fencing $= C = 2\pi \cdot 20 \text{ m} \approx 126 \text{ m}$

31. $C = 2\pi \cdot 30 \text{ ft} \approx 188.5 \text{ ft}$

$188.5 \text{ ft} = \dfrac{188.5 \text{ ft}}{1} \cdot \dfrac{12 \text{ in.}}{1 \text{ ft}} = 2262 \text{ in.}$

Number of plants $= \dfrac{1 \text{ plant}}{6 \text{ in.}} \cdot \dfrac{2262 \text{ in.}}{1} = 377 \text{ plants}$

33. First, find the area of pizzas.

    Large: Medium:

$A = \pi r^2$ $A = \pi r^2$

$\quad = \pi (7 \text{ in.})^2$ $= \pi (3.5 \text{ in.})^2$

$\quad = 49\pi \text{ in.}^2$ $= 12.25\pi \text{ in.}^2$

$\quad \approx 154 \text{ in.}^2$ $\approx 38 \text{ in.}^2$

Next, find the price per square inch.

    Large: Medium:

$\dfrac{\$12.00}{49\pi \text{ in.}^2}$ $\dfrac{\$5.00}{12.25\pi \text{ in.}^2}$

$\approx \dfrac{\$12.00}{154 \text{ in.}^2}$ $\approx \dfrac{\$5.00}{38 \text{ in.}^2}$

$\approx \dfrac{\$0.08}{\text{in.}^2}$ $\approx \dfrac{\$0.13}{\text{in.}^2}$

The large pizza is a better buy.

35. Area of road $=$ (Area of large circle) $-$ (Area of small circle)

$\quad = \pi \cdot (50 \text{ ft})^2 - \pi (40 \text{ ft})^2 = 900\pi \text{ ft}^2 \approx 2827.433 \text{ ft}^2$

Cost of asphalt $= \dfrac{2827.433 \text{ ft}^2}{1} \cdot \dfrac{\$0.80}{\text{ft}^2} = \$2261.95$

37-41. Answers will vary.

43. First, find the area of the large rectangle that includes the path and the pool.
The length of the large rectangle is 30 ft + 3 ft + 3 ft = 36 ft.
The width of the large rectangle is 14 ft + 3 ft + 3 ft = 20 ft.

Area of large rectangular $= 36 \text{ ft} \cdot 20 \text{ ft} = 720 \text{ ft}^2$
Next, find the area of the pool.

Area of pool $= 30 \text{ ft} \cdot 14 \text{ ft} = 420 \text{ ft}^2$
Area of path $=$ (Area of large Rectangle) $-$ (Area of Pool)

$\quad = 720 \text{ ft}^2 - 420 \text{ ft}^2 = 300 \text{ ft}^2$

Cost of resurfacing $= \dfrac{300 \text{ ft}^2}{1} \cdot \dfrac{\$2}{\text{ft}^2} = \$600$

45. Area of large square $= (4 \text{ cm})^2 = 16 \text{ cm}^2$

Area of one corner triangle $= \dfrac{1}{2} \cdot 2 \text{ cm} \cdot 2 \text{ cm} = 2 \text{ cm}^2$

Area of shaded region $=$ (Area of large square) $- (4 \times$ Area of corner triangle)

$\quad = 16 \text{ cm}^2 - 4 \cdot 2 \text{ cm}^2$

$\quad = 8 \text{ cm}^2$

Check Points 10.5

1. $V = 5 \text{ ft} \cdot 3 \text{ ft} \cdot 7 \text{ ft} = 105 \text{ ft}^3$

2. $= \dfrac{6 \cancel{\text{ft}}}{1} \cdot \dfrac{1 \text{ yd}}{3 \cancel{\text{ft}}} = 2 \text{ yd}$

   $V = (2 \text{ yd})^3 = 8 \text{ yd}^3$

3. $B = (6 \text{ ft})^2 = 36 \text{ ft}^2$

   $V = \dfrac{1}{3} \cdot 36 \text{ ft}^2 \cdot 4 \text{ ft}$

   $= 48 \text{ ft}^3$

4. $r = \dfrac{1}{2}(8 \text{ cm}) = 4 \text{ cm}$

   $V = \pi(4 \text{ in.})^2 \cdot 6 \text{ in.} \approx 302 \text{ in.}^3$

5. $V = \dfrac{1}{3}\pi(4 \text{ in.})^2 \cdot 6 \text{ in.} \approx 101 \text{ in.}^3$

6. No, it is not enough air.

   $V = \dfrac{4}{3}\pi(4.5 \text{ in.})^3 \approx 382 \text{ in.}^3$

7. New dimensions: $l = 16 \text{ yd}$, $w = 10 \text{ yd}$, $h = 6 \text{ yd}$

   $SA = 2lw + 2lh + 2wh$

   $= 2 \cdot 16 \text{ yd} \cdot 10 \text{ yd} + 2 \cdot 16 \text{ yd} \cdot 6 \text{ yd} + 2 \cdot 10 \text{ yd} \cdot 6 \text{ yd}$

   $= 320 \text{ yd}^2 + 192 \text{ yd}^2 + 120 \text{ yd}^2$

   $= 632 \text{ yd}^2$

Exercise Set 10.5

1. $V = 3 \text{ in.} \cdot 3 \text{ in.} \cdot 4 \text{ in.} = 36 \text{ in.}^3$

3. $V = (4 \text{ cm})^3 = 64 \text{ cm}^3$

5. $B = 7 \text{ yd} \cdot 5 \text{ yd} = 35 \text{ yd}^2$

   $V = \dfrac{1}{3} \cdot 35 \text{ yd}^2 \cdot 15 \text{ yd}$

   $= 175 \text{ yd}^3$

7. $B = 4 \text{ in.} \cdot 7 \text{ in.} = 28 \text{ in.}^2$

   $V = \dfrac{1}{3} \cdot 28 \text{ in.}^2 \cdot 6 \text{ in.}$

   $= 56 \text{ in.}^3$

9. $V = \pi(5 \text{ cm})^2 \cdot 6 \text{ cm} \approx 471 \text{ cm}^3$

11. $r = \dfrac{1}{2}(24 \text{ in.}) = 12 \text{ in.}$

    $V = \pi(12 \text{ in.})^2 \cdot 21 \text{ in.} \approx 9500 \text{ in.}^3$

13. $V = \dfrac{1}{3}\pi(4 \text{ m})^2 \cdot 9 \text{ m} \approx 151 \text{ m}^3$

15. $r = \dfrac{1}{2} \cdot 6 \text{ yd} = 3 \text{ yd}$

    $V = \dfrac{1}{3}\pi(3 \text{ yd})^2 \cdot 5 \text{ yd}$

    $\approx 47 \text{ yd}^3$

17. $V = \dfrac{4}{3}\pi(6 \text{ m})^3 \approx 905 \text{ m}^3$

19. $r = \dfrac{1}{2} \cdot 18 \text{ cm} = 9 \text{ cm}$

    $V = \dfrac{4}{3}\pi(9 \text{ cm})^3 \approx 3054 \text{ cm}^3$

21. Surface Area

    $= 2(5 \text{ m} \cdot 3 \text{ m}) + 2(2 \text{ m} \cdot 3 \text{ m}) + 2(5 \text{ m} \cdot 2 \text{ m})$

    $= 2 \cdot 15 \text{ m}^2 + 2 \cdot 6 \text{ m}^2 + 2 \cdot 10 \text{ m}^2$

    $= 30 \text{ m}^2 + 12 \text{ m}^2 + 20 \text{ m}^2$

    $= 62 \text{ m}^2$

23. Surface Area $= 6(4 \text{ ft})^2 = 96 \text{ ft}^2$

25. Volume = (volume of cone) + (volume of hemisphere)

    $V = \dfrac{1}{3}\pi(6 \text{ cm})^2 \cdot 15 \text{ cm} + \dfrac{1}{2}\left[\dfrac{4}{3}\pi(6 \text{ cm})^3 \right]$

    $\approx 1018 \text{ cm}^3$

27. First convert all linear measures in feet to linear yards.

    $12 \text{ ft} = \dfrac{12 \cancel{\text{ft}}}{1} \cdot \dfrac{1 \text{ yd}}{3 \cancel{\text{ft}}} = 4 \text{ yd}$

    $9 \text{ ft} = \dfrac{9 \cancel{\text{ft}}}{1} \cdot \dfrac{1 \text{ yd}}{3 \cancel{\text{ft}}} = 3 \text{ yd}$

    $6 \text{ ft} = \dfrac{6 \cancel{\text{ft}}}{1} \cdot \dfrac{1 \text{ yd}}{3 \cancel{\text{ft}}} = 2 \text{ yd}$

    Total dirt $= 4 \text{ yd} \cdot 3 \text{ yd} \cdot 2 \text{ yd} = 24 \text{ yd}^3$

    Total cost $= \dfrac{24 \cancel{\text{yd}^3}}{1} \cdot \dfrac{1 \cancel{\text{truck}}}{6 \cancel{\text{yd}^3}} \cdot \dfrac{\$10}{1 \cancel{\text{truck}}} = \$40$

29. Volume of house $= 1400 \text{ ft}^2 \cdot 9 \text{ ft} = 12,600 \text{ ft}^3$
No. This furnace will not be adequate.

31. a. First convert linear measures in feet to linear yards.

$$756 \text{ ft} = \frac{756 \text{ ft}}{1} \cdot \frac{1 \text{ yd}}{3 \text{ ft}} = 252 \text{ yd}$$

$$480 \text{ ft} = \frac{480 \text{ ft}}{1} \cdot \frac{1 \text{ yd}}{3 \text{ ft}} = 160 \text{ yd}$$

$$B = (252 \text{ yd})^2 = 63,504 \text{ yd}^2$$

$$V = \frac{1}{3} \cdot 63,504 \text{ yd}^2 \cdot 160 \text{ yd}$$
$$= 3,386,880 \text{ yd}^3$$

b. $\frac{1 \text{ block}}{1.5 \text{ yd}^3} \cdot \frac{3,386,880 \text{ yd}^3}{1} = 2,257,920 \text{ blocks}$

33. Volume of tank $= \pi(3 \text{ ft})^2 \cdot \frac{7}{3} \text{ ft} \approx 66 \text{ ft}^3$

Yes. The volume of the tank is less than 67 cubic feet.

35. $V = \frac{4}{3}\pi(3960 \text{ mi})^3 \approx 260,120,252,600 \text{ mi}^3$

37. Answers will vary.

39. New volume $= \frac{4}{3}\pi(2r)^3 = \frac{4}{3}\pi \cdot 8r^3 = 8\left(\frac{4}{3}\pi r^3\right)$

The volume is multiplied by 8.

41. Volume of darkly shaded region
$= $ (Volume of rectangular solid)
$\quad - $ (Volume of pyramid)

$= 6 \text{ cm} \cdot 6 \text{ cm} \cdot 7 \text{ cm} - \frac{1}{3}(6 \text{ cm})^2 \cdot 7 \text{ cm}$

$= 168 \text{ cm}^3$

43. Surface area
$= $ (Areas of 3 rectangles) + (Area of 2 triangles)
$= (5 \text{ cm} \cdot 6 \text{ cm} + 4 \text{ cm} \cdot 6 \text{ cm} + 3 \text{ cm} \cdot 6 \text{ cm})$
$\quad + 2\left(\frac{1}{2} \cdot 3 \text{ cm} \cdot 4 \text{ cm}\right)$
$= 72 \text{ cm}^2 + 12 \text{ cm}^2$
$= 84 \text{ cm}^2$

Check Points 10.6

1. $c^2 = a^2 + b^2 = 3^2 + 4^2 = 25$
$c = \sqrt{25} = 5$
$\sin A = \frac{3}{5}$
$\cos A = \frac{4}{5}$
$\tan A = \frac{3}{4}$

2. $\tan A = \frac{a}{b}$
$\tan 62° = \frac{a}{140}$
$a = 140 \tan 62° \approx 263 \text{ cm}$

3. $\cos A = \frac{b}{c}$
$\cos 62° = \frac{140}{c}$
$c \cos 62° = 140$
$c = \frac{140}{\cos 62°}$
$c \approx 298 \text{ cm}$

4. Let $a = $ the height of the tower.
$\tan 85.4° = \frac{a}{80}$
$a = 80 \tan 85.4° \approx 994 \text{ ft}$

5. $\tan A = \frac{14}{10}$
$A = \tan^{-1}\left(\frac{14}{10}\right) \approx 54°$

Chapter 10: Geometry

Exercise Set 10.6

1. $\sin A = \dfrac{3}{5}$

   $\cos A = \dfrac{4}{5}$

   $\tan A = \dfrac{3}{4}$

3. First find the length of missing side.

   $a^2 = 29^2 - 21^2 = 400$

   $a = 20$

   $\sin A = \dfrac{20}{29}$

   $\cos A = \dfrac{21}{29}$

   $\tan A = \dfrac{20}{21}$

5. First find the length of missing side.

   $b^2 = 26^2 - 10^2 = 576$

   $b = 24$

   $\sin A = \dfrac{10}{26} = \dfrac{5}{13}$

   $\cos A = \dfrac{24}{26} = \dfrac{12}{13}$

   $\tan A = \dfrac{10}{24} = \dfrac{5}{12}$

7. First find the length of missing side.

   $a^2 = 35^2 - 21^2 = 784$

   $a = 28$

   $\sin A = \dfrac{28}{35} = \dfrac{4}{5}$

   $\cos A = \dfrac{21}{35} = \dfrac{3}{5}$

   $\tan A = \dfrac{28}{21} = \dfrac{4}{3}$

9. $\tan A = \dfrac{a}{b}$

   $\tan 37° = \dfrac{a}{250}$

   $a = 250 \tan 37° \approx 188$ cm

11. $\cos 34° = \dfrac{b}{220}$

    $b = 220 \cos 34° \approx 182$ in.

13. $\sin 34° = \dfrac{a}{13}$

    $a = 13 \sin 34° \approx 7$ m

15. $\tan 33° = \dfrac{14}{b}$

    $b = \dfrac{14}{\tan 33°} \approx 22$ yd

17. $\sin 30° = \dfrac{20}{c}$

    $c = \dfrac{20}{\sin 30°} = 40$ m

19. $m\angle B = 90° - 40° = 50°$

    Side a: $\tan 40° = \dfrac{a}{22}$

    $a = 22 \tan 40° \approx 18$ yd

    Side c: $\cos 40° = \dfrac{22}{c}$

    $c = \dfrac{22}{\cos 40°} \approx 29$ yd

    $m\angle B = 50°,\ a \approx 18$ yd, $c \approx 28$ yd

21. $m\angle B = 90° - 52° = 38°$

    Side a: $\sin 52° = \dfrac{a}{54}$

    $a = 54 \sin 52° \approx 43$ cm

    Side b: $\cos 52° = \dfrac{b}{54}$

    $b = 54 \cos 52° \approx 33$ cm

    $m\angle B = 38°,\ a \approx 43$ cm, $b \approx 33$ cm

23. $\sin A = \dfrac{30}{50}$

    $A = \sin^{-1}\left(\dfrac{30}{50}\right) \approx 37°$

25. $\cos A = \dfrac{15}{17}$

    $A = \cos^{-1}\left(\dfrac{15}{17}\right) \approx 28°$

27. $\tan 40° = \dfrac{a}{630}$

    $a = 630 \tan 40° \approx 529$ yd

29. $\sin 10° = \dfrac{500}{c}$

    $c = \dfrac{500}{\sin 10°} \approx 2879$ ft

31. Let h = the height of the tower.

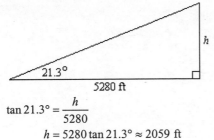

$$\tan 21.3° = \frac{h}{5280}$$
$$h = 5280 \tan 21.3° \approx 2059 \text{ ft}$$

33. Let x = the distance.

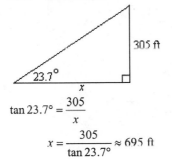

$$\tan 23.7° = \frac{305}{x}$$
$$x = \frac{305}{\tan 23.7°} \approx 695 \text{ ft}$$

35. $\tan x = \frac{125}{172}$

$$x = \tan^{-1}\left(\frac{125}{172}\right) \approx 36°$$

37. $m\angle P = 36°$

$$\tan 36° = \frac{1000}{d}$$
$$d = \frac{1000}{\tan 36°} \approx 1376 \text{ ft}$$

39. Let A = the angle of elevation.

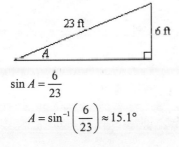

$$\sin A = \frac{6}{23}$$
$$A = \sin^{-1}\left(\frac{6}{23}\right) \approx 15.1°$$

41–47. Answers will vary.

49. The sine and cosine of an acute angle cannot be greater than or equal to 1 because they are each the ratio of a leg of a right triangle to the hypotenuse. The hypotenuse of a right triangle is always the longest side; this results in a value less than 1.

51.

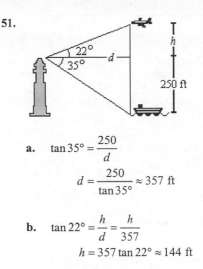

a. $\tan 35° = \frac{250}{d}$

$$d = \frac{250}{\tan 35°} \approx 357 \text{ ft}$$

b. $\tan 22° = \frac{h}{d} = \frac{h}{357}$

$$h = 357 \tan 22° \approx 144 \text{ ft}$$

Height of plane = 250 ft + 144 ft = 394 ft.

Check Points 10.7

1. Answers will vary. Possible answer:

The upper left and lower right vertices are odd.
The lower left and upper right vertices are even.
One possible tracing:
Start at the upper left, trace around the square, then trace down the diagonal.

Exercise Set 10.7

1. a. A and C are even vertices.
B and D are odd vertices.
Because this graph has two odd vertices, by Euler's second rule, it is traversable.

b. Sample path: $DABDCB$

3. a. C, D, E are even vertices.
A and B are odd vertices.
Because this graph has two odd vertices, by Euler's second rule, it is traversable.

b. Sample path: $ADCBDEAB$

5. A, B, D, E are odd vertices.
Because this graph has more than two odd vertices, by Euler's third rule, it is not traversable.

7. 3

9. 2

11. Pitcher and wrench

13. The sum of the angles in such a triangle is greater than $180°$.

15. Yes

17. No

19. Each vertex for a carbon atom is even, with degree 4.
    Each vertex for a hydrogen atom is odd with degree 1.

21. 2 doors connect room A to the outside. This is shown in the graph by connecting 2 edges from A to E.

23. Yes. B and D are the only two odd vertices. It is traversable by Euler's second rule.

25-41. Answers will vary.

Chapter 10 Review Exercises

1. Acute

2. Obtuse

3. Right

4. $180° - 115° = 65°$

5. $90° - 41° = 49°$

6. Measure of complement $= 90° - 73° = 17°$

7. Measure of supplement $= 180° - 46° = 134°$

8. $m\angle 1 = 180° - 70° = 110°$
   $m\angle 2 = 70°$
   $m\angle 3 = m\angle 1 = 110°$

9. $m\angle 1 = 180° - 42° = 138°$
   $m\angle 2 = 42°$
   $m\angle 3 = m\angle 1 = 138°$
   $m\angle 4 = m\angle 1 = 138°$
   $m\angle 5 = m\angle 2 = 42°$
   $m\angle 6 = 42°$
   $m\angle 7 = m\angle 3 = 138°$

10. $m\angle A = 180° - 60° - 48° = 72°$

11. $m\angle A = 90° - 39° = 51°$

12. $m\angle 1 = 180° - 50° - 40° = 90°$
    $m\angle 2 = 180° - 90° = 90°$
    $m\angle 3 = 180° - 40° = 140°$
    $m\angle 4 = 40°$
    $m\angle 5 = m\angle 3 = 140°$

13. $\dfrac{8}{4} = \dfrac{10}{x}$
    $8x = 40$
    $x = 5$ ft

14. $\dfrac{9}{x} = \dfrac{7+5}{5}$
    $\dfrac{9}{x} = \dfrac{12}{5}$
    $12x = 45$
    $x = \dfrac{45}{12} = 3.75$ ft

15. $c^2 = 8^2 + 6^2$
    $c^2 = 64 + 36$
    $c^2 = 100$
    $c = 10$ ft

16. $c^2 = 6^2 + 4^2$
    $c^2 = 36 + 16$
    $c^2 = 52$
    $c \approx 7.2$ in.

17. $b^2 = 15^2 - 11^2$
    $b^2 = 225 - 121$
    $b^2 = 104$
    $b \approx 10.2$ cm

18. $\dfrac{x}{5} = \dfrac{9+6}{6}$
    $\dfrac{x}{5} = \dfrac{15}{6}$
    $6x = 75$
    $x = 12.5$ ft

19. $a^2 = 25^2 - 20^2$
    $a^2 = 625 - 400$
    $a^2 = 225$
    $a = 15$ ft

20. $b^2 = 13^2 + 5^2$

$b^2 = 169 - 25$

$b^2 = 144$

$b = 12$ yd

21. Rectangle, square

22. Rhombus, square

23. A trapezoid does not four angles with the same measure.

24. $P = 2 \ (6 \text{ cm}) + 2(9 \text{ cm})$

$= 12 \text{ cm} + 18 \text{ cm}$

$= 30 \text{ cm}$

25. $P = 2 \cdot 1000 \text{ yd} + 2 \cdot 1240 \text{ yd}$

$= 2000 \text{ yd} + 2480 \text{ yd}$

$= 4480 \text{ yd}$

26. First find the lengths of missing sides.

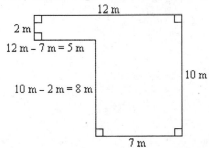

$P = 12 \text{ m} + 10 \text{ m} + 7 \text{ m} + 8 \text{ m} + 5 \text{ m} + 2 \text{ m}$

$= 44 \text{ m}$

27. Sum $= (n - 2) \ 180°$

$= (12 - 2) \ 180°$

$= 10 \cdot 180°$

$= 1800°$

28. Sum $= (n - 2) \ 180°$

$= (8 - 2) \ 180°$

$= 6 \cdot 180°$

$= 1080°$

29. Sum of measures of angles $= (n - 2)180°$

$= (8 - 2)180°$

$= 6 \cdot 180°$

$= 1080°$

$m\angle 1 = \dfrac{1080°}{8} = 135°$

30. Amount of baseboard

=Perimeter of room–Lengths of doorways

$= 2 \cdot 35 \text{ ft} + 2 \cdot 15 \text{ ft} - 4 \cdot 3 \text{ ft}$

$= 70 \text{ ft} + 30 \text{ ft} - 12 \text{ ft}$

$= 88 \text{ ft}$

Cost $= \dfrac{\$1.50}{1 \text{ ft}} \cdot \dfrac{88 \text{ ft}}{1} = \$132$

31. $A = 5 \text{ ft} \cdot 6.5 \text{ ft} = 32.5 \text{ ft}^2$

32. $A = 5 \text{ m} \cdot 4 \text{ m} = 20 \text{ m}^2$

33. $A = \dfrac{1}{2} \cdot 20 \text{ cm} \cdot 5 \text{ cm} = 50 \text{ cm}^2$

34. $A = \dfrac{1}{2} \cdot 10 \text{ yd} \cdot (22 \text{ yd} + 5 \text{ yd})$

$= \dfrac{1}{2} \cdot 10 \text{ yd} \cdot (27 \text{ yd})$

$= 135 \text{ yd}^2$

35. $C = \pi \cdot 20 \text{ m} \approx 63 \text{ m}$

$r = \dfrac{1}{2} d = \dfrac{1}{2} \cdot 20 \text{ m} = 10 \text{ m}$

$A = \pi r^2 = \pi (10 \text{ m})^2 \approx 314 \text{ m}^2$

36. Area = (Area of square) + (Area of triangle)

$= (12 \text{ in.})^2 + \dfrac{1}{2} \cdot 12 \text{ in.} \cdot 8 \text{ in.}$

$= 144 \text{ in.}^2 + 48 \text{ in.}^2$

$= 192 \text{ in.}^2$

37. Area = (Area of top rectangle) + (Area of bottom rectangle)

$= 8 \text{ m} \cdot 2 \text{ m} + 6 \text{ m} \cdot 2 \text{ m}$

$= 16 \text{ m}^2 + 12 \text{ m}^2$

$= 28 \text{ m}^2$

38. First convert linear measurements in feet to linear yards.

$15 \text{ ft} = \dfrac{15 \text{ ft}}{1} \cdot \dfrac{1 \text{ yd}}{3 \text{ ft}} = 5 \text{ yd}$

$21 \text{ ft} = \dfrac{21 \text{ ft}}{1} \cdot \dfrac{1 \text{ yd}}{3 \text{ ft}} = 7 \text{ yd}$

Area $= 5 \text{ yd} \cdot 7 \text{ yd} = 35 \text{ yd}^2$

Cost $= \dfrac{\$22.50}{1 \text{ yd}^2} \cdot \dfrac{35 \text{ yd}^2}{1} = \$787.50$

39. Area of floor $= 40 \text{ ft} \cdot 50 \text{ ft} = 2000 \text{ ft}^2$

Area of each tile $= (2 \text{ ft})^2 = 4 \text{ ft}^2$

Number of tiles $= \dfrac{2000 \text{ ft}^2}{4 \text{ ft}^2} = 500 \text{ tiles}$

Cost $= \dfrac{\$13}{10 \text{ tiles}} \cdot \dfrac{500 \text{ tiles}}{1} = \$650$

40. $C = \pi d = \pi \cdot 10 \text{ yd} \approx 31 \text{ yd}$

41. $V = 5 \text{ cm} \cdot 3 \text{ cm} \cdot 4 \text{ cm} = 60 \text{ cm}^3$

42. $V =$ (Volume of rectangular solid)
    $+$ (Volume of Pyramid)

$= 8 \text{ m} \cdot 9 \text{ m} \cdot 10 \text{ m} + \dfrac{1}{3}(8 \text{ m} \cdot 9 \text{ m}) \, 10 \text{ m}$

$\approx 720 \text{ m}^3 + 240 \text{ m}^3 = 960 \text{ m}^3$

43. $V = \pi(4 \text{ yd})^2 \cdot 8 \text{ yd} \approx 402 \text{ yd}^3$

44. $V = \pi(40 \text{ in.})^2 \cdot 28 \text{ in.} \approx 140,743 \text{ in.}^3$

45. $V = \dfrac{4}{3}\pi(6 \text{ m})^3 \approx 905 \text{ m}^3$

46. Surface area $= 2 \cdot 5 \text{ m} \cdot 3 \text{ m} + 2 \cdot 3 \text{ m} \cdot 6 \text{ m} + 2 \cdot 5 \text{ m} \cdot 6 \text{ m}$

$= 30 \text{ m}^2 + 36 \text{ m}^2 + 60 \text{ m}^2$

$= 126 \text{ m}^2$

47. Volume of one box $= 8 \text{ m} \cdot 4 \text{ m} \cdot 3 \text{ m} = 96 \text{ m}^3$

Volume of 50 boxes $= 50 \cdot 96 \text{ m}^3 = 4800 \text{ m}^3$

48. $V = \dfrac{1}{3}(145 \text{ m})^2 \cdot 93 \text{ m} = 651,775 \text{ m}^3$

49. First convert linear measures in feet to linear yards.

$27 \text{ ft} = \dfrac{27 \text{ ft}}{1} \cdot \dfrac{1 \text{ yd}}{3 \text{ ft}} = 9 \text{ yd}$

$4 \text{ ft} = \dfrac{4 \text{ ft}}{1} \cdot \dfrac{1 \text{ yd}}{3 \text{ ft}} = \dfrac{4}{3} \text{ yd}$

$6 \text{ in.} = \dfrac{6 \text{ in.}}{1} \cdot \dfrac{1 \text{ yd}}{36 \text{ in.}} = \dfrac{1}{6} \text{ yd}$

Volume $= 9 \text{ yd} \cdot \dfrac{4}{3} \text{ yd} \cdot \dfrac{1}{6} \text{ yd} = 2 \text{ yd}^3$

Cost $= \dfrac{\$40}{1 \text{ yd}} \cdot \dfrac{2 \text{ yd}}{1} = \$80$

50. First compute length of hypotenuse

$c^2 = 12^2 + 9^2 = 144 + 81 = 225$

$c = 15$

$\sin A = \dfrac{9}{15} = \dfrac{3}{5}$

$\cos A = \dfrac{12}{15} = \dfrac{4}{5}$

$\tan A = \dfrac{9}{12} = \dfrac{3}{4}$

51. $\tan 23° = \dfrac{a}{100}$

$a = 100 \tan 23° \approx 42 \text{ mm}$

52. $\sin 61° = \dfrac{20}{c}$

$c = \dfrac{20}{\sin 61°} \approx 23 \text{ cm}$

53. $\sin 48° = \dfrac{a}{50}$

$a = 50 \sin 48° \approx 37 \text{ in.}$

54. $\sin A = \dfrac{17}{20}$

$A = \sin^{-1}\left(\dfrac{17}{20}\right) \approx 58°$

55. $\dfrac{1}{2} \text{ mi} = \dfrac{0.5 \text{ mi}}{1} \cdot \dfrac{5280 \text{ ft}}{1 \text{ mi}} = 2640 \text{ ft}$

$\sin 17° = \dfrac{h}{2640}$

$h = 2640 \sin 17° \approx 772 \text{ ft}$

56. $\tan 32° = \dfrac{d}{50}$

$d = 50 \tan 32° \approx 31 \text{ m}$

57.

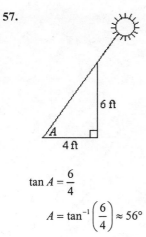

6 ft

4 ft

$\tan A = \dfrac{6}{4}$

$A = \tan^{-1}\left(\dfrac{6}{4}\right) \approx 56°$

58. The graph is not traversable because there are more than two odd vertices.

59. All vertices have even degrees, so the graph is traversable. Possible path: $ABCDABCDA$

60. 0

61. 2

62. 1

63. 2

Chapter 10 Test

1. Measure of supplement $= 180° - 54° = 126°$

2. $m\angle 1 = 133°$

because alternate exterior angles are equal.

3. $m\angle 1 = 180° - 40° - 70° = 70°$

4. First find measures of other angles of triangle.

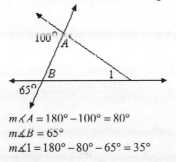

$m\angle A = 180° - 100° = 80°$
$m\angle B = 65°$
$m\angle 1 = 180° - 80° - 65° = 35°$

5. $\dfrac{x}{8} = \dfrac{4}{10}$

$10x = 4 \cdot 8$

$10x = 32$

$x = \dfrac{32}{10} = 3.2$ in.

6. $b^2 = 26^2 - 24^2$

$b^2 = 676 - 576$

$b^2 = 100$

$b = 10$ ft

7. Sum $= (n - 2)\,180°$
$= (10 - 2)\,180°$
$= 8 \cdot 180°$
$= 1440°$

8. First find lengths of missing sides.

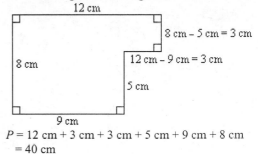

$P = 12$ cm $+ 3$ cm $+ 3$ cm $+ 5$ cm $+ 9$ cm $+ 8$ cm
$= 40$ cm

9. d

10. $A = \dfrac{1}{2}bh$

$A = \dfrac{1}{2} \cdot 47\text{ m} \cdot 22\text{ m} = 517\text{m}^2$

11. $A = \dfrac{1}{2} \cdot 15\text{ in.}(40\text{ in.} + 30\text{ in.})$

$= \dfrac{1}{2} \cdot 15\text{ in.}(70\text{ in.})$

$= 525$ in.^{2}

12. $C = \pi d = \pi \cdot 40\text{ m} \approx 126\text{ m}$

$A = \pi r^2 = \pi(20\text{ m})^2 \approx 1257\text{ m}^2$

13. Area of floor $8\,\text{ft}\cdot 6\,\text{ft} = 48\ \text{ft}^2$
Convert inches to feet:

$$8\ \text{in.}=\frac{8\ \text{in.}}{1}\cdot\frac{1\ \text{ft}}{12\ \text{in.}}=\frac{2}{3}\ \text{ft}$$

Area of one tile

$$=\left(\frac{2}{3}\ \text{ft}\right)^2=\frac{4}{9}\ \text{ft}^2$$

Number of tiles

$$=\frac{48\ \text{ft}^2}{\frac{4}{9}\text{ft}^2}=108\ \text{tiles}$$

14. $V = 3\ \text{ft}\cdot 2\ \text{ft}\cdot 3\ \text{ft} = 18\ \text{ft}^3$

15. $V =\dfrac{1}{3}(4\ \text{m}\cdot 3\ \text{m})\,4\ \text{m} = 16\ \text{m}^3$

16. $V = \pi(5\,\text{cm})^2\cdot 7\ \text{cm} \approx 550\ \text{cm}^3$

17. $\sin 28° = \dfrac{40}{c}$

$$c = \frac{40}{\sin 28°} \approx 85\ \text{cm}$$

18.

$$\tan 34° = \frac{h}{104}$$
$$h = 104\tan 34° \approx 70\ \text{ft}$$

19. The graph is traversable because there are two odd vertices (B and E). Sample path: $BCAECDE$

Chapter 11
Counting Methods and Probability Theory

Check Points 11.1

1. Multiply the number of choices for each of the two courses of the meal:

   Appetizers :   Main Courses:

   $\underline{10}$ × $\underline{15}$ = 150

2. Multiply the number of choices for each of the two courses:

   Psychology :   Social Science:

   $\underline{10}$ × $\underline{4}$ = 40

3. Multiply the number of choices for each of the three decisions:

   Size :   Crust :   Topping:

   $\underline{2}$ × $\underline{3}$ × $\underline{5}$ = 30

4. Multiply the number of choices for each of the five options:

   Color: A/C: Electric/Gas: Onboard Computer: Global Positioning System:

   $\underline{10}$ × $\underline{2}$ × $\underline{2}$ × $\underline{2}$ × $\underline{2}$ = 160

5. Multiply the number of choices for each of the six questions:

   Question #1:   Question #2:   Question #3:   Question #4:   Question #5:   Question #6:

   $\underline{3}$ × $\underline{3}$ × $\underline{3}$ × $\underline{3}$ × $\underline{3}$ × $\underline{3}$ $= 3^6 = 729$

6. Multiply the number of choices for each of the five digits:

   $\overbrace{\underline{\text{Digit 1:}}}^{1-9}$ $\overbrace{\underline{\text{Digit 2:}}}^{0-9}$ $\overbrace{\underline{\text{Digit 3:}}}^{0-9}$ $\overbrace{\underline{\text{Digit 4:}}}^{0-9}$ $\overbrace{\underline{\text{Digit 5:}}}^{0-9}$

   $\underline{9}$ × $\underline{10}$ × $\underline{10}$ × $\underline{10}$ × $\underline{10}$ = 90,000

Exercise Set 11.1

1. $8 \cdot 10 = 80$

3. $3 \cdot 4 = 12$

5. $3 \cdot 2 = 6$

7. Multiply the number of choices for each of the three decisions:

   Drink: Size: Flavor:

   $\underline{2}$ × $\underline{4}$ × $\underline{5}$ = 40

9. Multiply the number of choices for each of the four menu categories:

   Main Course: Vegetables: Beverages: Desserts:

   $\underline{4}$ × $\underline{3}$ × $\underline{4}$ × $\underline{3}$ = 144

   This includes, for example, an order of ham and peas with tea and cake.
   This also includes an order of beef and peas with milk and pie.

11. Multiply the number of choices for each of the three categories:

   Gender: Age: Payment method:

   $\underline{2}$ × $\underline{2}$ × $\underline{2}$ = 8

13. Multiply the number of choices for each of the five options:

Color:	A/C:	Transmission:	Windows:	CD Player:
6	× 2	× 2	× 2	× 2	= 96

15. Multiply the number of choices for each of the five questions:

Question 1:	Question 2:	Question 3:	Question 4:	Question 5:
3	× 3	× 3	× 3	× 3	= 243

17. Multiply the number of choices for each of the three digits:

Digit 1:	Digit 2:	Digit 3:
8	× 2	× 9	= 144

19. Multiply the number of choices for each of the letters and digits:

Letter 1:	Letter 2:	Digit 1:	Digit 2:	Digit 3:
26	× 26	× 10	× 10	× 10	= 676,000

21. This situation involves making choices with seven groups of items. Each stock is a group, and each group has three choices. Multiply choices: $3 \times 3 \times 3 \times 3 \times 3 \times 3 \times 3 = 3^7 = 2187$

23-25. Answers will vary.

27. Multiply the number of choices for each of the four groups of items:

Bun:	Sauce:	Lettuce:	Tomatoes:
12	× 30	× 4	× 3	= 4320

Total time $= 10 \times 4320 = 43,200$ minutes, which is $43,200 \div 60 = 720$ hours.

Check Points 11.2

1. There is only one choice each for the first and last performers. This leaves two choices for the second performer and then one for the third.

Offspring	Pink Floyd, Sublime		Rolling Stones
1st Performer:	2nd Performer:	3rd Performer:	4th Performer:
1	× 2	× 1	× 1	= 2

2. The number of choices decreases by 1 each time a book is selected.

1st Book:	2nd Book:	3rd Book:	4th Book:	5th Book:
5	× 4	× 3	× 2	× 1	= 120

3. a. $\dfrac{9!}{6!} = \dfrac{9 \cdot 8 \cdot 7 \cdot 6!}{6!} = \dfrac{9 \cdot 8 \cdot 7 \cdot \cancel{6!}}{\cancel{6!}} = 9 \cdot 8 \cdot 7 = 504$

   b. $\dfrac{16!}{11!} = \dfrac{16 \cdot 15 \cdot 14 \cdot 13 \cdot 12 \cdot 11!}{11!} = \dfrac{16 \cdot 15 \cdot 14 \cdot 13 \cdot 12 \cdot \cancel{11!}}{\cancel{11!}} = 16 \cdot 15 \cdot 14 \cdot 13 \cdot 12 = 524,160$

   c. $\dfrac{100!}{99!} = \dfrac{100 \cdot 99!}{99!} = \dfrac{100 \cdot \cancel{99!}}{\cancel{99!}} = 100$

4. $_7P_4 = \dfrac{7!}{(7-4)!} = \dfrac{7!}{3!} = \dfrac{7 \cdot 6 \cdot 5 \cdot 4 \cdot 3!}{3!} = \dfrac{7 \cdot 6 \cdot 5 \cdot 4 \cdot \cancel{3!}}{\cancel{3!}} = 7 \cdot 6 \cdot 5 \cdot 4 = 840$

5. $_9P_5 = \dfrac{9!}{(9-5)!} = \dfrac{9!}{4!} = \dfrac{9 \cdot 8 \cdot 7 \cdot 6 \cdot 5 \cdot 4!}{4!} = \dfrac{9 \cdot 8 \cdot 7 \cdot 6 \cdot 5 \cdot \cancel{4!}}{\cancel{4!}} = 9 \cdot 8 \cdot 7 \cdot 6 \cdot 5 = 15,120$

Exercise Set 11.2

1. The number of choices decreases by 1 each time a performer is selected.

1st Performer:	2nd Performer:	3rd Performer:	4th Performer:	5th Performer:	6th Performer:	
6	× 5	× 4	× 3	× 2	× 1	= 720

3. The number of choices decreases by 1 each time a sentence is selected.

1st Sentence:	2nd Sentence:	3rd Sentence:	4th Sentence:	5th Sentence:	
5	× 4	× 3	× 2	× 1	= 120

5. There is only one choice for the 6th performer. The number of choices decreases by 1 each time a performer is selected.

1st Performer:	2nd Performer:	3rd Performer:	4th Performer:	5th Performer:	6th Performer:	
5	× 4	× 3	× 2	× 1	× 1	= 120

7. The number of choices decreases by 1 each time a book is selected.

1st Book:	2nd:	3rd Book:	4th:	5th Book:	6th:	7th Book:	8th:	9th Book:	
9	× 8	× 7	× 6	× 5	× 4	× 3	× 2	× 1	= 362,880

9. There is only one choice each for the first and last sentences. For the other values, the number of choices decreases by 1 each time a sentence is selected.

1st Sentence:	2nd Sentence:	3rd Sentence:	4th Sentence:	5th Sentence:	
1	× 3	× 2	× 1	× 1	= 6

11. There are two choices for the first movie and one for the second. There is only one choice for the last movie. This leaves two choices for the third movie and one for the fourth.

G rated		Other two movies		NC-17 Rated	
1st Movie:	2nd Movie:	3rd Movie:	4th Movie:	5th Movie:	
2	× 1	× 2	× 1	× 1	= 4

13. $\dfrac{9!}{6!} = \dfrac{9 \cdot 8 \cdot 7 \cdot 6!}{6!} = 9 \cdot 8 \cdot 7 = 504$

15. $\dfrac{29!}{25!} = \dfrac{29 \cdot 28 \cdot 27 \cdot 26 \cdot 25!}{25!}$

$= 29 \cdot 28 \cdot 27 \cdot 26$

$= 570,024$

17. $\dfrac{19!}{11!} = \dfrac{19 \cdot 18 \cdot 17 \cdot 16 \cdot 15 \cdot 14 \cdot 13 \cdot 12 \cdot 11!}{11!}$

$= 19 \cdot 18 \cdot 17 \cdot 16 \cdot 15 \cdot 14 \cdot 13 \cdot 12$

$- 3,047,466,240$

19. $\dfrac{600!}{599!} = \dfrac{600 \cdot 599!}{599!} = 600$

21. $\dfrac{104!}{102!} = \dfrac{104 \cdot 103 \cdot 102!}{102!} = 104 \cdot 103 = 10,712$

23. $7! - 3! = 5040 - 6 = 5034$

25. $(7-3)! = 4! = 4 \cdot 3 \cdot 2 \cdot 1 = 24$

27. $\left(\dfrac{12}{4}\right)! = 3! = 3 \cdot 2 \cdot 1 = 6$

29. $\dfrac{7!}{(7-2)!} = \dfrac{7!}{5!} = \dfrac{7 \cdot 6 \cdot 5!}{5!} = 7 \cdot 6 = 42$

31. $\dfrac{13!}{(13-3)!} = \dfrac{13!}{10!}$

$= \dfrac{13 \cdot 12 \cdot 11 \cdot 10!}{10!}$

$= 13 \cdot 12 \cdot 11$

$= 1716$

33. $_9 P_4 = \dfrac{9!}{(9-4)!}$

$= \dfrac{9!}{5!}$

$= \dfrac{9 \cdot 8 \cdot 7 \cdot 6 \cdot 5!}{5!}$

$= 9 \cdot 8 \cdot 7 \cdot 6$

$= 3024$

35. $_8 P_5 = \dfrac{8!}{(8-5)!}$

$= \dfrac{8!}{3!}$

$= \dfrac{8 \cdot 7 \cdot 6 \cdot 5 \cdot 4 \cdot 3!}{3!}$

$= 8 \cdot 7 \cdot 6 \cdot 5 \cdot 4$

$= 6720$

37. $_6 P_6 = \dfrac{6!}{(6-6)!} = \dfrac{6!}{0!} = \dfrac{6 \cdot 5 \cdot 4 \cdot 3 \cdot 2 \cdot 1}{1} = 720$

39. $_8 P_0 = \dfrac{8!}{(8-0)!} = \dfrac{8!}{8!} = 1$

41. $_{10} P_3 = \dfrac{10!}{(10-3)!}$

$= \dfrac{10!}{7!}$

$= \dfrac{10 \cdot 9 \cdot 8 \cdot 7!}{7!}$

$= 10 \cdot 9 \cdot 8$

$= 720$

43. $_{13} P_7 = \dfrac{13!}{(13-7)!}$

$= \dfrac{13!}{6!}$

$= \dfrac{13 \cdot 12 \cdot 11 \cdot 10 \cdot 9 \cdot 8 \cdot 7 \cdot 6!}{6!}$

$= 13 \cdot 12 \cdot 11 \cdot 10 \cdot 9 \cdot 8 \cdot 7$

$= 8,648,640$

45. $_6 P_3 = \dfrac{6!}{(6-3)!}$

$= \dfrac{6!}{3!}$

$= \dfrac{6 \cdot 5 \cdot 4 \cdot 3!}{3!}$

$= 6 \cdot 5 \cdot 4$

$= 120$

47. $_9 P_5 = \dfrac{9!}{(9-5)!}$

$= \dfrac{9!}{4!}$

$= \dfrac{9 \cdot 8 \cdot 7 \cdot 6 \cdot 5 \cdot 4!}{4!}$

$= 9 \cdot 8 \cdot 7 \cdot 6 \cdot 5$

$= 15,120$

49-53. Answers will vary.

55. $_{12} P_{10} = \dfrac{12!}{(12-10)!}$

$= \dfrac{12!}{2!}$

$= \dfrac{12 \cdot 11 \cdot 10 \cdot 9 \cdot 8 \cdot 7 \cdot 6 \cdot 5 \cdot 4 \cdot 3 \cdot 2!}{2!}$

$= 12 \cdot 11 \cdot 10 \cdot 9 \cdot 8 \cdot 7 \cdot 6 \cdot 5 \cdot 4 \cdot 3$

$= 239,500,800$

57. First select 3 out of the 8 jazz groups.

There are $_8P_3 = \dfrac{8!}{(8-3)!} = \dfrac{8!}{5!} = 336$ ways to arrange the 1st, 3rd, and 8th performers.

This leaves 13 groups (5 remaining jazz groups and 8 rock groups) to be arranged.

There are $_{13}P_{13} = \dfrac{13!}{(13-13)!} = \dfrac{13!}{0!} = 13! = 6,227,020,800$ ways to arrange the remaining performers.

The total number of arrangements is found by multiplying these values: $336 \times 6,227,020,800 = 2.0923 \times 10^{12}$

59. Multiply the number of choices for each of the four digits:

2, 4	2, 4, 6, 7, 8, 9	2, 4, 6, 7, 8, 9	7, 9
Digit 1:	Digit 2:	Digit 3:	Digit 4:
2 $\times$	6 $\times$	6 $\times$	2 = 144

61. $8! = 40,320$

63. $13! = 6,227,020,800$

65. $_{13}P_6 = 1,235,520$

67. $_5P_5 = 120$

Check Points 11.3

1. a. Order does not matter. This problem involves combinations.

b. Order matters. This problem involves permutations.

2. $_7C_3 = \dfrac{7!}{(7-3)!3!} = \dfrac{7!}{4!3!} = \dfrac{7 \cdot 6 \cdot 5 \cdot 4!}{4! \cdot 3 \cdot 2 \cdot 1} = \dfrac{7 \cdot 6 \cdot 5 \cdot \cancel{4!}}{\cancel{4!} \cdot 3 \cdot 2 \cdot 1} = \dfrac{7 \cdot 6 \cdot 5}{3 \cdot 2 \cdot 1} = 35$

3. $_{16}C_4 = \dfrac{16!}{(16-4)!4!} = \dfrac{16!}{12!4!} = \dfrac{16 \cdot 15 \cdot 14 \cdot 13 \cdot 12!}{12! \cdot 4 \cdot 3 \cdot 2 \cdot 1} = \dfrac{16 \cdot 15 \cdot 14 \cdot 13 \cdot \cancel{12!}}{\cancel{12!} \cdot 4 \cdot 3 \cdot 2 \cdot 1} = \dfrac{16 \cdot 15 \cdot 14 \cdot 13}{4 \cdot 3 \cdot 2 \cdot 1} = 1820$

4. Choose the Democrats: $_{50}C_3 = \dfrac{50!}{(50-3)!3!} = \dfrac{50!}{47!3!} = \dfrac{50 \cdot 49 \cdot 48 \cdot 47!}{47! \cdot 3 \cdot 2 \cdot 1} = \dfrac{50 \cdot 49 \cdot 48 \cdot \cancel{47!}}{\cancel{47!} \cdot 3 \cdot 2 \cdot 1} = \dfrac{50 \cdot 49 \cdot 48}{3 \cdot 2 \cdot 1} = 19,600$

Choose the Republicans: $_{49}C_2 = \dfrac{49!}{(49-2)!2!} = \dfrac{49!}{47!2!} = \dfrac{49 \cdot 48 \cdot 47!}{47! \cdot 2 \cdot 1} = \dfrac{49 \cdot 48 \cdot \cancel{47!}}{\cancel{47!} \cdot 2 \cdot 1} = \dfrac{49 \cdot 48}{2 \cdot 1} = 1176$

Multiply the choices: $19,600 \times 1176 = 23,049,600$

Exercise Set 11.3

1. Order matters. This problem involves permutations.

3. Order does not matter. This problem involves combinations.

5. Order does not matter. This problem involves combinations.

7. Order matters. This problem involves permutations.

9. Order matters. This problem involves permutations.

11. $_6C_5 = \dfrac{6!}{(6-5)!5!} = \dfrac{6!}{1!5!} = \dfrac{6\cdot5!}{1\cdot5!} = 6$

13. $_9C_5 = \dfrac{9!}{(9-5)!5!} = \dfrac{9!}{4!5!} = \dfrac{9\cdot8\cdot7\cdot6\cdot5!}{4\cdot3\cdot2\cdot1\cdot5!} = 126$

15. $_{11}C_4 = \dfrac{11!}{(11-4)!4!} = \dfrac{11!}{7!4!} = \dfrac{11\cdot10\cdot9\cdot8\cdot7!}{7!\cdot4\cdot3\cdot2\cdot1} = 330$

17. $_8C_1 = \dfrac{8!}{(8-1)!1!} = \dfrac{8!}{7!1!} = \dfrac{8\cdot7!}{7!1} = 8$

19. $_7C_7 = \dfrac{7!}{(7-7)!7!} = \dfrac{7!}{0!7!} = 1$

21. $_{30}C_3 = \dfrac{30!}{(30-3)!3!} = \dfrac{30!}{27!3!} = \dfrac{30\cdot29\cdot28\cdot27!}{27!\cdot3\cdot2\cdot1} = 4060$

23. $_5C_0 = \dfrac{5!}{(5-0)!0!} = \dfrac{5!}{5!0!} = 1$

25. $\dfrac{_7C_3}{_5C_4} = \dfrac{\frac{7!}{(7-3)!3!}}{\frac{5!}{(5-4)!4!}} = \dfrac{\frac{7!}{4!3!}}{\frac{5!}{1!4!}} = \dfrac{\frac{7\cdot6\cdot5\cdot4!}{4!\cdot3\cdot2\cdot1}}{\frac{5\cdot4!}{1\cdot4!}} = \dfrac{35}{5} = 7$

27. $_6C_3 = \dfrac{6!}{(6-3)!3!} = \dfrac{6!}{3!3!} = \dfrac{6\cdot5\cdot4\cdot3!}{3!3\cdot2\cdot1} = 20$

29. $_{12}C_4 = \dfrac{12!}{(12-4)!4!} = \dfrac{12!}{8!4!} = \dfrac{12\cdot11\cdot10\cdot9\cdot8!}{8!4\cdot3\cdot2\cdot1} = 495$

31. $_{17}C_8 = \dfrac{17!}{(17-8)!8!} = \dfrac{17!}{9!8!} = \dfrac{17\cdot16\cdot15\cdot14\cdot13\cdot12\cdot11\cdot10\cdot9!}{9!8\cdot7\cdot6\cdot5\cdot4\cdot3\cdot2\cdot1} = 24,310$

33. $_{31}C_3 = \dfrac{31!}{(31-3)!3!} = \dfrac{31!}{28!3!} = \dfrac{31\cdot30\cdot29\cdot28!}{28!3\cdot2\cdot1} = 4495$

35. $_{53}C_6 = \dfrac{53!}{(53-6)!6!} = \dfrac{53!}{47!6!} = \dfrac{53\cdot52\cdot51\cdot50\cdot49\cdot48\cdot47!}{47!6\cdot5\cdot4\cdot3\cdot2\cdot1} = 22,957,480$

37. Choose the men: $_7C_4 = \dfrac{7!}{(7-4)!4!} = \dfrac{7!}{3!4!} = \dfrac{7\cdot6\cdot5\cdot4!}{3\cdot2\cdot1\cdot4!} = 35$

Choose the women: $_7C_5 = \dfrac{7!}{(7-5)!5!} = \dfrac{7!}{2!5!} = \dfrac{7\cdot6\cdot5!}{2\cdot1\cdot5!} = 21$

Multiply the choices: $35\cdot21 = 735$

39. Choose the Republicans: $_{55}C_4 = \dfrac{55!}{(55-4)!4!} = \dfrac{55!}{51!4!} = \dfrac{55 \cdot 54 \cdot 53 \cdot 52 \cdot 51!}{51!4 \cdot 3 \cdot 2 \cdot 1} = 341,055$

Choose the Democrats: $_{45}C_3 = \dfrac{45!}{(45-3)!3!} = \dfrac{45!}{42!3!} = \dfrac{45 \cdot 44 \cdot 43 \cdot 42!}{42!3 \cdot 2 \cdot 1} - 14,190$

Multiply the choices: $341,055 \times 14,190 = 4,839,570,450$

41-43. Answers will vary.

45. Selections for 6/53 lottery: $_{53}C_6 = \dfrac{53!}{(53-6)!6!} = \dfrac{53!}{47!6!} = \dfrac{53 \cdot 52 \cdot 51 \cdot 50 \cdot 49 \cdot 48 \cdot 47!}{47!6 \cdot 5 \cdot 4 \cdot 3 \cdot 2 \cdot 1} = 22,957,480$

Selections for 5/36 lottery: $_{36}C_5 = \dfrac{36!}{(36-5)!5!} = \dfrac{36!}{31!5!} = \dfrac{36 \cdot 35 \cdot 34 \cdot 33 \cdot 32 \cdot 31!}{31!5 \cdot 4 \cdot 3 \cdot 2 \cdot 1} = 376,992$

The 5/36 lottery is easier to win because there are fewer possible selections.

47. For a group of 20 people:

$_{20}C_2 = \dfrac{20!}{(20-2)!2!} = \dfrac{20!}{18!2!} = \dfrac{20 \cdot 19 \cdot 18!}{18!2 \cdot 1} = 190$ handshakes

Time $= 3 \times 190 = 570$ seconds, which gives $570 \div 60 = 9.5$ minutes.

For a group of 40 people:

$_{40}C_2 = \dfrac{40!}{(40-2)!2!} = \dfrac{40!}{38!2!} = \dfrac{40 \cdot 39 \cdot 38!}{38!2 \cdot 1} = 780$ handshakes

Time $= 3 \times 780 = 2340$ seconds, which gives $2340 \div 60 = 39$ minutes.

49. $_{10}C_6 = 210$

51. $_{40}C_6 = 3,838,380$

Check Points 11.4

1. a. The event of getting a 2 can occur in one way.

$P(2) = \dfrac{\text{number of ways a 2 can occur}}{\text{total number of possible outcomes}} = \dfrac{1}{6}$

b. The event of getting a number less than 4 can occur in three ways: 1, 2, 3.

$P(\text{less than 4}) = \dfrac{\text{number of ways a number less than 4 can occur}}{\text{total number of possible outcomes}} = \dfrac{3}{6} = \dfrac{1}{2}$

c. The event of getting a number greater than 7 cannot occur.

$P(\text{greater than 7}) = \dfrac{\text{number of ways a number greater than 7 can occur}}{\text{total number of possible outcomes}} = \dfrac{0}{6} = 0$

The probability of an event that cannot occur is 0.

d. The event of getting a number less than 7 can occur in six ways: 1, 2, 3, 4, 5, 6.

$P(\text{less than 7}) = \dfrac{\text{number of ways a number less than 7 can occur}}{\text{total number of possible outcomes}} = \dfrac{6}{6} = 1$

The probability of any certain event is 1.

2. a. $P(\text{ace}) = \dfrac{\text{number of ways a ace can occur}}{\text{total number of possibilities}} = \dfrac{4}{52} = \dfrac{1}{13}$

    b. $P(\text{red card}) = \dfrac{\text{number of ways a red card can occur}}{\text{total number of possible outcomes}} = \dfrac{26}{52} = \dfrac{1}{2}$

    c. $P(\text{red king}) = \dfrac{\text{number of ways a red king can occur}}{\text{total number of possible outcomes}} = \dfrac{2}{52} = \dfrac{1}{26}$

3. The table shows the four equally likely outcomes. The Cc child will be a carrier who is not actually sick.

    $P(\text{carrier, not sick}) = P(Cc) = \dfrac{\text{number of ways } Cc \text{ can occur}}{\text{total number of possible outcomes}} = \dfrac{2}{4} = \dfrac{1}{2}$

4. $P(\text{selecting a Muslim from the Arab American population})$

    $= \dfrac{\text{number of Arab Americans who are Muslims}}{\text{total number of Arab Americans}} = \dfrac{0.69}{3.00} = \dfrac{69}{300} = \dfrac{23}{100} = 0.23$

Exercise Set 11.4

1. $P(4) = \dfrac{\text{number of ways a 4 can occur}}{\text{total number of possible outcomes}} = \dfrac{1}{6}$

3. $P(\text{odd number}) = \dfrac{\text{number of ways an odd number can occur}}{\text{total number of possible outcomes}} = \dfrac{3}{6} = \dfrac{1}{2}$

5. $P(\text{less than 3}) = \dfrac{\text{number of ways a number less than 3 can occur}}{\text{total number of possible outcomes}} = \dfrac{2}{6} = \dfrac{1}{3}$

7. $P(\text{less than 7}) = \dfrac{\text{number of ways a number less than 7 can occur}}{\text{total number of possible outcomes}} = \dfrac{6}{6} = 1$

9. $P(\text{greater than 7}) = \dfrac{\text{number of ways a number greater than 7 can occur}}{\text{total number of possible outcomes}} = \dfrac{0}{6} = 0$

11. $P(\text{queen}) = \dfrac{\text{number of ways a queen can occur}}{\text{total number of possibilities}} = \dfrac{4}{52} = \dfrac{1}{13}$

13. $P(\text{club}) = \dfrac{\text{number of ways a club can occur}}{\text{total number of possibilities}} = \dfrac{13}{52} = \dfrac{1}{4}$

15. $P(\text{picture card}) = \dfrac{\text{number of ways a picture card can occur}}{\text{total number of possibilities}} = \dfrac{12}{52} = \dfrac{3}{13}$

17. $P(\text{queen of spades}) = \dfrac{\text{number of ways a queen of spades can occur}}{\text{total number of possibilities}} = \dfrac{1}{52}$

19. $P(\text{diamond and spade}) = \dfrac{\text{number of ways a diamond and a spade can occur}}{\text{total number of possibilities}} = \dfrac{0}{52} = 0$

21. $P(\text{two heads}) = \dfrac{\text{number of ways two heads can occur}}{\text{total number of possibilities}} = \dfrac{1}{4}$

23. $P(\text{same on each toss}) = \dfrac{\text{number of ways the same outcome on each toss can occur}}{\text{total number of possibilities}} = \dfrac{2}{4} = \dfrac{1}{2}$

25. $P(\text{head on second toss}) = \dfrac{\text{number of ways a head on the second toss can occur}}{\text{total number of possibilities}} = \dfrac{2}{4} = \dfrac{1}{2}$

27. $P(\text{exactly one female child}) = \dfrac{\text{number of ways exactly one female child can occur}}{\text{total number of possibilities}} = \dfrac{3}{8}$

29. $P(\text{exactly two male children}) = \dfrac{\text{number of ways exactly two male children can occur}}{\text{total number of possibilities}} = \dfrac{3}{8}$

31. $P(\text{at least one male child}) = \dfrac{\text{number of ways at least one male child can occur}}{\text{total number of possiblities}} = \dfrac{7}{8}$

33. $P(\text{four male children}) = \dfrac{\text{number of ways four male children can occur}}{\text{total number of possibilities}} = \dfrac{0}{8} = 0$

35. $P(\text{two even numbers}) = \dfrac{\text{number of ways two even numbers can occur}}{\text{total number of possibilities}} = \dfrac{9}{36} = \dfrac{1}{4}$

37. $P(\text{two numbers whose sum is 5}) = \dfrac{\text{number of ways two numbers whose sum is 5 can occur}}{\text{total number of possibilities}} = \dfrac{4}{36} = \dfrac{1}{9}$

39. $P(\text{two numbers whose sum exceeds 12})$

$= \dfrac{\text{number of ways two numbers whose sum exceeds 12 can occur}}{\text{total number of possibilities}} = \dfrac{0}{36} = 0$

41. $P(\text{red region}) = \dfrac{\text{number of ways a red region can occur}}{\text{total number of possibilities}} = \dfrac{3}{10}$

43. $P(\text{blue region}) = \dfrac{\text{number of ways a blue region can occur}}{\text{total number of possibilities}} = \dfrac{2}{10} = \dfrac{1}{5}$

45. $P(\text{region that is red or blue}) = \dfrac{\text{number of ways a region that is red or blue can occur}}{\text{total number of possibilities}} = \dfrac{5}{10} = \dfrac{1}{2}$

47. $P(\text{region that is red and blue}) = \dfrac{\text{number of ways a region that is red and blue can occur}}{\text{total number of possibilities}} = \dfrac{0}{10} = 0$

49. $P(\text{sickle cell anemia}) = \dfrac{\text{number of ways sickle cell anemia can occur}}{\text{total number of possibilities}} = \dfrac{1}{4}$

51. $P(\text{healthy}) = \dfrac{\text{number of ways a healthy child can occur}}{\text{total number of possibilities}} = \dfrac{1}{4}$

53. $P(\text{sickle cell trait}) = \dfrac{\text{number of ways sickle cell trait can occur}}{\text{total number of possibilities}} = \dfrac{2}{4} = \dfrac{1}{2}$

55. $P(\text{weight training}) = \dfrac{\text{number surveyed who weight train}}{\text{total number surveyed}} = \dfrac{320}{2000} = \dfrac{4}{25} = 0.16$

57. $P(\text{biking}) = \dfrac{\text{number surveyed who bike}}{\text{total number surveyed}} = \dfrac{240}{2000} = \dfrac{3}{25} = 0.12$

59. $P(\text{African}) = \dfrac{\text{number of Africans}}{\text{total world population}} = \dfrac{784,400,000}{6,054,900,000} \approx 0.130$

61. $P(\text{North American}) = \dfrac{\text{number of North Americans}}{\text{total world population}} = \dfrac{309,600,000}{6,054,900,000} \approx 0.051$

63. Saturn had the greatest percentage satisfied at approximately 81% or $\dfrac{81}{100}$.

65. Ford

67-71. Answers will vary.

73. The area of the target is $(12 \text{ in.})^2 = 144 \text{ in.}^2$

The area of the yellow region is $(9 \text{ in.})^2 - (6 \text{ in.})^2 + (3 \text{ in.})^2 = 54 \text{ in.}^2$

The probability that the dart hits a yellow region is $\dfrac{54 \text{ in.}^2}{144 \text{ in.}^2} = 0.375$

Check Points 11.5

1. total number of permutations $= 5! = 5 \cdot 4 \cdot 3 \cdot 2 \cdot 1 = 120$
number of arrangements with Offspring first, the Rolling Stones fourth, and the Beatles last.

Offspring	Pink Floyd, Sublime		Rolling Stones	Beatles
1st:	2nd:	3rd:	4th:	5th:
1	$\times$ 2	$\times$ 1	$\times$ 1	$\times$ 1 = 2

$P(\text{Offspring 1st, Rolling Stones 4th, and the Beatles last}) = \dfrac{2}{120} = \dfrac{1}{60}$

2. Number of LOTTO selections: $_{49}C_6 = \dfrac{49!}{(49-6)!6!} = \dfrac{49!}{43!6!} = \dfrac{49 \cdot 48 \cdot 47 \cdot 46 \cdot 45 \cdot 44 \cdot 43!}{43!6 \cdot 5 \cdot 4 \cdot 3 \cdot 2 \cdot 1} = 13,983,816$

$P(\text{winning}) = \dfrac{\text{one LOTTO ticket}}{\text{total number of LOTTO combinations}} = \dfrac{1}{13,983,816} \approx 0.0000000715$

3. total number of combinations: $_{10}C_3 = \dfrac{10!}{(10-3)!3!} = \dfrac{10!}{7!3!} = \dfrac{10 \cdot 9 \cdot 8 \cdot 7!}{7!3 \cdot 2 \cdot 1} = 120$

a. total number of combinations of 3 men: $_6C_3 = \dfrac{6!}{(6-3)!3!} = \dfrac{6!}{3!3!} = \dfrac{6 \cdot 5 \cdot 4 \cdot 3!}{3 \cdot 2 \cdot 1 \cdot 3!} = 20$

$P(\text{3 men}) = \dfrac{\text{number of combinations with 3 men}}{\text{total number of combinations}} = \dfrac{20}{120} = \dfrac{1}{6}$

b. Select 2 out of 6 men: $_6C_2 = \dfrac{6!}{(6-2)!2!} = \dfrac{6!}{4!2!} = \dfrac{6 \cdot 5 \cdot 4!}{4!2 \cdot 1} = 15$

Select 1 out of 4 women: $_4C_1 = \dfrac{4!}{(4-1)!1!} = \dfrac{4!}{3!1!} = \dfrac{4 \cdot 3!}{3!} = \dfrac{4 \cdot 3!}{3!} = 4$

total number of combinations of 2 men and 1 woman: $15 \times 4 = 60$

$P(\text{2 men, 1 woman}) = \dfrac{\text{number of combinations with 2 men, 1 woman}}{\text{total number of combinations}} = \dfrac{60}{120} = \dfrac{1}{2}$

Exercise Set 11.5

1. a. $5! = 5 \cdot 4 \cdot 3 \cdot 2 \cdot 1 = 120$

b.

$$\underbrace{\text{Martha}}_{\underset{1}{\underline{\text{1st:}}}} \quad \underbrace{\text{Lee, Nancy, Paul}}_{\underset{\times\ 3\ \times\ 2\ \times\ 1}{\underline{\text{2nd:}}\ \underline{\text{3rd:}}\ \underline{\text{4th:}}}} \quad \underbrace{\text{Armando}}_{\underset{\times\ \ 1}{\underline{\text{5th:}}}} \quad = 6$$

c. $P(\text{Martha first and Armando last}) = \dfrac{6}{120} = \dfrac{1}{20}$

3. a. total number of permutations $= 6! = 6 \cdot 5 \cdot 4 \cdot 3 \cdot 2 \cdot 1 = 720$

number of permutations with E first $= 1 \cdot 5 \cdot 4 \cdot 3 \cdot 2 \cdot 1 = 120$

$P(\text{E first}) = \dfrac{\text{number of permutations with E first}}{\text{total number of permutations}} = \dfrac{120}{720} = \dfrac{1}{6}$

b. number of permutations with C fifth and B last $= 4 \cdot 3 \cdot 2 \cdot 1 \cdot 1 \cdot 1 = 24$

$P(\text{C fifth and B last}) = \dfrac{\text{number of permutations with C fifth and B last}}{\text{total number of permutations}} = \dfrac{24}{720} = \dfrac{1}{30}$

c. $P(\text{D, E, C, A, B, F}) = \dfrac{\text{number of permutations with order D, E, C, A, B, F}}{\text{total number of permutations}} = \dfrac{1}{720}$

d. number of permutations with A or B first $= 2 \cdot 5 \cdot 4 \cdot 3 \cdot 2 \cdot 1 = 240$

$P(\text{A or B first}) = \dfrac{\text{number of permutations with A or B first}}{\text{total number of permutations}} = \dfrac{240}{720} = \dfrac{1}{3}$

5. a. $_9C_3 = \dfrac{9!}{(9-3)!3!} = \dfrac{9!}{6!3!} = \dfrac{9 \cdot 8 \cdot 7 \cdot 6!}{6!3 \cdot 2 \cdot 1} = 84$

b. $_5C_3 = \dfrac{5!}{(5-3)!3!} = \dfrac{5!}{2!3!} = \dfrac{5 \cdot 4 \cdot 3!}{2 \cdot 1 \cdot 3!} = 10$

c. $P(\text{all women}) = \dfrac{\text{number of ways to select 3 women}}{\text{total number of possible combinations}} = \dfrac{10}{84} = \dfrac{5}{42}$

7. $_{51}C_6 = \dfrac{51!}{(51-6)!6!} = \dfrac{51!}{45!6!} = \dfrac{51\cdot50\cdot49\cdot48\cdot47\cdot46\cdot45!}{45!6\cdot5\cdot4\cdot3\cdot2\cdot1} = 18,009,460$

$P(\text{winning}) = \dfrac{\text{number of ways of winning}}{\text{total number of possible combinations}} = \dfrac{1}{18,009,460} \approx 0.0000000555$

If 100 different tickets are purchased, $P(\text{winning}) = \dfrac{100}{18,009,460} \approx 0.00000555$

9. a. $_{25}C_6 = \dfrac{25!}{(25-6)!6!} = \dfrac{25!}{19!6!} = \dfrac{25\cdot24\cdot23\cdot22\cdot21\cdot20\cdot19!}{19!6\cdot5\cdot4\cdot3\cdot2\cdot1} = 177,100$

$P(\text{all are defective}) = \dfrac{\text{number of ways to choose 6 defective transistors}}{\text{total number of possible combinations}} = \dfrac{1}{177,100} \approx 0.00000565$

b. $_{19}C_6 = \dfrac{19!}{(19-6)!6!} = \dfrac{19!}{13!6!} = \dfrac{19\cdot18\cdot17\cdot16\cdot15\cdot14\cdot13!}{13!6\cdot5\cdot4\cdot3\cdot2\cdot1} = 27,132$

$P(\text{none are defective}) = \dfrac{\text{number of ways to choose 6 good transistors}}{\text{total number of possible permutations}} = \dfrac{27,132}{177,100} = \dfrac{969}{6325} \approx 0.153$

11. total number of possible combinations: $_{10}C_3 = \dfrac{10!}{(10-3)!3!} = \dfrac{10!}{7!3!} = \dfrac{10\cdot9\cdot8\cdot7!}{7!3\cdot2\cdot1} = 120$

number of ways to select one Democrat: $_6C_1 = \dfrac{6!}{(6-1)!1!} = \dfrac{6!}{5!1!} = \dfrac{6\cdot5!}{5!1} = 6$

number of ways to select two Republicans: $_4C_2 = \dfrac{4!}{(4-2)!2!} = \dfrac{4!}{2!2!} = \dfrac{4\cdot3\cdot2!}{2!2\cdot1} = 6$

number of ways to select one Democrat and two Republicans: $_6C_1 \cdot _4C_2 = 6\cdot6 = 36$

$P(\text{one Democrat and two Republicans}) = \dfrac{36}{120} = \dfrac{3}{10} = 0.3$

13. a. $_{52}C_5 = \dfrac{52!}{(52-5)!5!} = \dfrac{52!}{47!5!} = \dfrac{52\cdot51\cdot50\cdot49\cdot48\cdot47!}{47!5\cdot4\cdot3\cdot2\cdot1} = 2,598,960$

b. $_{13}C_5 = \dfrac{13!}{(13-5)!5!} = \dfrac{13!}{8!5!} = \dfrac{13\cdot12\cdot11\cdot10\cdot9\cdot8!}{8!5\cdot4\cdot3\cdot2\cdot1} = 1287$

c. $P(\text{diamond flush}) = \dfrac{\text{number of possible 5-card diamond flushes}}{\text{total number of possible combinations}} = \dfrac{1287}{2,598,960} \approx 0.000495$

15. total number of possible combinations: $_{52}C_3 = \dfrac{52!}{(52-3)!3!} = \dfrac{52!}{49!3!} = \dfrac{52\cdot51\cdot50\cdot49!}{49!3\cdot2\cdot1} = 22,100$

number of ways to select 3 picture cards: $_{12}C_3 = \dfrac{12!}{(12-3)!3!} = \dfrac{12!}{9!3!} = \dfrac{12\cdot11\cdot10\cdot9!}{9!3\cdot2\cdot1} = 220$

$P(\text{3 picture cards}) = \dfrac{220}{22,100} = \dfrac{11}{1105} \approx 0.00995$

17. total number of possible combinations: $_{52}C_4 = \dfrac{52!}{(52-4)!4!} = \dfrac{52!}{48!4!} = \dfrac{52 \cdot 51 \cdot 50 \cdot 49 \cdot 48!}{48!4 \cdot 3 \cdot 2 \cdot 1} = 270,725$

number of ways to select 2 queens: $_4C_2 = \dfrac{4!}{(4-2)!2!} = \dfrac{4!}{2!2!} = \dfrac{4 \cdot 3 \cdot 2!}{2!2 \cdot 1} = 6$

number of ways to select 2 kings: $_4C_2 = 6$

number of ways to select 3 jacks and 1 queen: $_4C_2 \cdot _4C_2 = 6 \cdot 6 = 36$

$P(\text{2 queens and 2 kings}) = \dfrac{36}{270,725} \approx 0.000133$

19-21. Answers will vary.

23. Refer to solution 7: $_{51}C_6 = 18,009,460$

$P(\text{winning}) = \dfrac{\text{number of ways of winning } (x)}{\text{total number of possible combinations}} = \dfrac{x}{18,009,460} = \dfrac{1}{2}$, therefore $x = 9,004,730$.

At \$1 per ticket, a person must spend \$9,004,730 to have a probability of winning of $\dfrac{1}{2}$.

25. total number of possible combinations:

Digit 1:		Digit 2:		Digit 3:		
5	×	4	×	3	=	60

number of even numbers greater than 500:

	5		1, 3, and 2 or 4		2 or 4		
Digit 1:			Digit 2:		Digit 3:		
1	×		3	×	2	=	6

$P(\text{even and greater than 500}) = \dfrac{\text{number of even numbers greater than 500}}{\text{total number of possible combinations}} = \dfrac{6}{60} = \dfrac{1}{10}$

Check Points 11.6

1. $P(\text{not a diamond}) = 1 - P(\text{diamond}) = 1 - \dfrac{13}{52} = \dfrac{39}{52} = \dfrac{3}{4}$

2. $P(\text{30-year-old not dying}) = 1 - P(\text{30-year-old dying}) = 1 - \dfrac{1}{1000} = \dfrac{999}{1000}$

3. $P(\text{4 or 5}) = P(4) + P(5) = \dfrac{1}{6} + \dfrac{1}{6} = \dfrac{2}{6} = \dfrac{1}{3}$

4. $P(\text{math or psychology}) = P(\text{math}) + P(\text{psychology}) - P(\text{math and psychology}) = \dfrac{23}{50} + \dfrac{11}{50} - \dfrac{7}{50} = \dfrac{27}{50}$

5. $P(\text{odd or less than 5}) = P(\text{odd}) + P(\text{less than 5}) - P(\text{odd and less than 5}) = \dfrac{4}{8} + \dfrac{4}{8} - \dfrac{2}{8} = \dfrac{6}{8} = \dfrac{3}{4}$

6. Note: total in group $= 14 + 12 + 6 + 8 = 40$, Muslims $= 14 + 6 = 20$, African-Americans $= 14 + 12 = 26$

$P(\text{Muslim or African-American})$

$= P(\text{Muslim}) + P(\text{African-American}) - P(\text{Muslim and African-American}) = \dfrac{20}{40} + \dfrac{26}{40} - \dfrac{14}{40} = \dfrac{32}{40} = \dfrac{4}{5}$

7. There are 2 red queens. Number of favorable outcomes = 2, Number of unfavorable outcomes = 50

    a.   Odds in favor of getting a red queen are 2 to 50 or 2:50 which reduces to 1:25.

    b.   Odds against getting a red queen are 50 to 2 or 50:2 which reduces to 25:1.

8. number of unfavorable outcomes = 995, number of favorable outcomes = 5
Odds against winning the scholarship are 995 to 5 or 995:5 which reduces to 199:1.

9. number of favorable outcomes = 17, number of unfavorable outcomes = 8

$$P(\text{domestic flight arriving on time}) = \frac{17}{17+8} = \frac{17}{25} = 0.68 \text{ or } 68\%.$$

Exercise Set 11.6

1. $P(\text{not an ace}) = 1 - P(\text{ace}) = 1 - \dfrac{4}{52} = \dfrac{48}{52} = \dfrac{12}{13}$

3. $P(\text{not a heart}) = 1 - P(\text{heart}) = 1 - \dfrac{13}{52} = \dfrac{39}{52} = \dfrac{3}{4}$

5. $P(\text{not a picture card}) = 1 - P(\text{picture card}) = 1 - \dfrac{12}{52} = \dfrac{40}{52} = \dfrac{10}{13}$

7. $P(\text{not a straight flush}) = 1 - P(\text{straight flush}) = 1 - \dfrac{36}{2,598,960} = \dfrac{2,598,924}{2,598,960} \approx 0.999986$

9. $P(\text{not a full house}) = 1 - P(\text{full house}) = 1 - \dfrac{3744}{2,598,960} = \dfrac{2,595,216}{2,598,960} \approx 0.998559$

11. a. 0.10 (read from graph)

     b. $1.00 - 0.10 = 0.90$

13. a. $100\% - 29\% = 71\% = 0.71$

     b. $100\% - 22\% = 78\% = 0.78$

15. $30.3\% + 23.0\% = 53.3\% = 0.533$

17. $P(2 \text{ or } 3) = P(2) + P(3) = \dfrac{4}{52} + \dfrac{4}{52} = \dfrac{8}{52} = \dfrac{2}{13}$

19. $P(\text{red 2 or black 3}) = P(\text{red 2}) + P(\text{black 3}) = \dfrac{2}{52} + \dfrac{2}{52} = \dfrac{4}{52} = \dfrac{1}{13}$

21. $P(\text{2 of hearts or 3 of spades}) = P(\text{2 of hearts}) + P(\text{3 of spades}) = \dfrac{1}{52} + \dfrac{1}{52} = \dfrac{2}{52} = \dfrac{1}{26}$

23. $P(\text{sum of 2 or 3}) = P(\text{sum of 2}) + P(\text{sum of 3}) = \dfrac{1}{36} + \dfrac{2}{36} = \dfrac{3}{36} = \dfrac{1}{12}$

25. $P(\text{two evens or two odds}) = P(\text{two evens}) + P(\text{two odds}) = \dfrac{9}{36} + \dfrac{9}{36} = \dfrac{18}{36} = \dfrac{1}{2}$

27. $P(\text{professor or instructor}) = P(\text{professor}) + P(\text{instructor}) = \dfrac{8}{44} + \dfrac{10}{44} = \dfrac{18}{44} = \dfrac{9}{22}$

29. $P(\text{even or less than 5}) = P(\text{even}) + P(\text{less than 5}) - P(\text{even and less than 5}) = \dfrac{3}{6} + \dfrac{4}{6} - \dfrac{2}{6} = \dfrac{5}{6}$

31. $P(7 \text{ or red}) = P(7) + P(\text{red}) - P(\text{red } 7) = \dfrac{4}{52} + \dfrac{26}{52} - \dfrac{2}{52} = \dfrac{28}{52} = \dfrac{7}{13}$

33. $P(\text{heart or picture card}) = P(\text{heart}) + P(\text{picture card}) - P(\text{heart and picture card}) = \dfrac{13}{52} + \dfrac{12}{52} - \dfrac{3}{52} = \dfrac{22}{52} = \dfrac{11}{26}$

35. $P(\text{odd or less than 6}) = P(\text{odd}) + P(\text{less than 6}) - P(\text{odd and less than 6}) = \dfrac{4}{8} + \dfrac{5}{8} - \dfrac{3}{8} = \dfrac{6}{8} = \dfrac{3}{4}$

37. $P(\text{even or greater than 5}) = P(\text{even}) + P(\text{greater than 5}) - P(\text{even and greater than 5}) = \dfrac{4}{8} + \dfrac{3}{8} - \dfrac{2}{8} = \dfrac{5}{8}$

39. $P(\text{professor or male}) = P(\text{professor}) + P(\text{male}) - P(\text{male professor}) = \dfrac{19}{40} + \dfrac{22}{40} - \dfrac{8}{40} = \dfrac{33}{40}$

41. $P(\text{teach. assist. or female}) = P(\text{teach. assist.}) + P(\text{female}) - P(\text{female teach. assist.}) = \dfrac{21}{40} + \dfrac{18}{40} - \dfrac{7}{40} = \dfrac{32}{40} = \dfrac{4}{5}$

43. $P(\text{Democrat or business major}) = P(\text{Democrat}) + P(\text{business major}) - P(\text{Democrat and business major})$
$$= \dfrac{29}{50} + \dfrac{11}{50} - \dfrac{5}{50} = \dfrac{35}{50} = \dfrac{7}{10}$$

45. number of favorable outcomes = 4, number of unfavorable outcomes = 2
Odds in favor of getting a number greater than 2 are 4:2, or 2:1.

47. number of unfavorable outcomes = 2, number of favorable outcomes = 4
Odds against getting a number greater than 2 or 2:4, or 1:2.

49. number of favorable outcomes = 9, number of unfavorable outcomes $= 100 - 9 = 91$

    a. Odds in favor of a child in a one-parent household having a parent who is a college graduate are 9:91.

    b. Odds against a child in a one-parent household having a parent who is a college graduate are 91:9.

51. number of favorable outcomes = 13, number of unfavorable outcomes = 39
Odds in favor of a heart are 13:39, or 1:3.

53. number of favorable outcomes = 26, number of unfavorable outcomes = 26
Odds in favor of a red card are 26:26, or 1:1.

55. number of unfavorable outcomes = 48, number of favorable outcomes = 4
Odds against a 9 are 48:4, or 12:1.

57. number of unfavorable outcomes = 50, number of favorable outcomes = 2
Odds against a black king are 50:2, or 25:1.

59. number of unfavorable outcomes = 47, number of favorable outcomes = 5
    Odds against a spade greater than 3 and less than 9 are 47:5.

61. number of unfavorable outcomes = 980, number of favorable outcomes = 20
    Odds against winning are 980:20, or 49:1.

63. number of favorable outcomes = 1, number of unfavorable outcomes = 19
    Odds in favor of being a victim are 1:19.

65. $P(\text{winning}) = \dfrac{3}{3+4} = \dfrac{3}{7}$

67. $P(\text{miss free throw}) = \dfrac{4}{21+4} = \dfrac{4}{25} = 0.16 = 16\%$

    In 100 free throws, on average he missed 16, so he made $100 - 16 = 84$.

69-75. Answers will vary.

77. $P(\text{driving intoxicated or driving accident})$

    $= P(\text{driving intoxicated}) + P(\text{driving accident}) - P(\text{driving accident while intoxicated})$

    Substitute the three given probabilities and solve for the unknown probability:
    $$0.35 = 0.32 + 0.09 - P(\text{driving accident while intoxicated})$$
    $P(\text{driving accident while intoxicated}) = 0.32 + 0.09 - 0.35$
    $P(\text{driving accident while intoxicated}) = 0.06$

Check Points 11.7

1. $P(\text{green and green}) = P(\text{green}) \cdot P(\text{green}) = \dfrac{2}{38} \cdot \dfrac{2}{38} = \dfrac{1}{19} \cdot \dfrac{1}{19} = \dfrac{1}{361} \approx 0.00277$

2. $P(\text{4 boys in a row}) = P(\text{boy and boy and boy and boy}) = P(\text{boy}) \cdot P(\text{boy}) \cdot P(\text{boy}) \cdot P(\text{boy}) = \dfrac{1}{2} \cdot \dfrac{1}{2} \cdot \dfrac{1}{2} \cdot \dfrac{1}{2} = \dfrac{1}{16}$

3. **a.** $P(\text{hit four years in a row}) = P(\text{hit}) \cdot P(\text{hit}) \cdot P(\text{hit}) \cdot P(\text{hit}) = \dfrac{5}{19} \cdot \dfrac{5}{19} \cdot \dfrac{5}{19} \cdot \dfrac{5}{19} = \dfrac{625}{130,321} \approx 0.00480 \approx 0.48\%$

   b. Note: $P(\text{not hit in any single year}) = 1 - P(\text{hit in any single year}) = 1 - \dfrac{5}{19} = \dfrac{14}{19}$,. Therefore,

   $P(\text{not hit in next four years})$

   $= P(\text{not hit}) \cdot P(\text{not hit}) \cdot P(\text{not hit}) \cdot P(\text{not hit}) = \dfrac{14}{19} \cdot \dfrac{14}{19} \cdot \dfrac{14}{19} \cdot \dfrac{14}{19} = \dfrac{38,416}{130,321} \approx 0.295 \approx 29.5\%$

   c. $P(\text{hit at least once in next four years}) = 1 - P(\text{not hit in next four years}) \approx 1 - 0.295 \approx 0.705 \approx 70.5\%$

4. $P(\text{2 kings}) = P(\text{king}) \cdot P(\text{king given the first card was a king}) = \dfrac{4}{52} \cdot \dfrac{3}{51} = \dfrac{1}{13} \cdot \dfrac{1}{17} = \dfrac{1}{221} \approx 0.00452 \approx 0.452\%$

5. $P(\text{3 hearts}) = P(\text{heart}) \cdot P(\text{heart given the first card was a heart}) \cdot P(\text{heart given the first two cards were hearts})$

   $= \dfrac{13}{52} \cdot \dfrac{12}{51} \cdot \dfrac{11}{50} = \dfrac{1}{4} \cdot \dfrac{4}{17} \cdot \dfrac{11}{50} = \dfrac{1}{1} \cdot \dfrac{1}{17} \cdot \dfrac{11}{50} = \dfrac{11}{850} \approx 0.0129 \approx 1.29\%$

6. $P(\text{heart}|\text{red}) = \dfrac{13}{26} = \dfrac{1}{2}$

7. a. $P(\text{Air Force}|\text{male}) = \dfrac{\text{number of males in Air Force}}{\text{number of males}}$

$= \dfrac{288,271}{402,602 + 316,858 + 161,571 + 288,271} = \dfrac{288,271}{1,169,302} \approx 0.247 \approx 24.7\%$

b. $P(\text{female}|\text{Navy}) = \dfrac{\text{number of females in the Navy}}{\text{number of people in the Navy}}$

$= \dfrac{51,582}{316,858 + 51,582} = \dfrac{51,582}{368,440} \approx 0.140 \approx 14.0\%$

Exercise Set 11.7

1. $P(\text{green and red}) = P(\text{green}) \cdot P(\text{red}) = \dfrac{2}{6} \cdot \dfrac{3}{6} = \dfrac{1}{3} \cdot \dfrac{1}{2} = \dfrac{1}{6}$

3. $P(\text{yellow and yellow}) = P(\text{yellow}) \cdot P(\text{yellow}) = \dfrac{1}{6} \cdot \dfrac{1}{6} = \dfrac{1}{36}$

5. $P(\text{color other than red each time}) = P(\text{not red}) \cdot P(\text{not red}) = \dfrac{3}{6} \cdot \dfrac{3}{6} = \dfrac{1}{2} \cdot \dfrac{1}{2} = \dfrac{1}{4}$

7. $P(\text{green and red and yellow}) = P(\text{green}) \cdot P(\text{red}) \cdot P(\text{yellow}) = \dfrac{2}{6} \cdot \dfrac{3}{6} \cdot \dfrac{1}{6} = \dfrac{1}{3} \cdot \dfrac{1}{2} \cdot \dfrac{1}{6} = \dfrac{1}{36}$

9. $P(\text{red every time}) = P(\text{red}) \cdot P(\text{red}) \cdot P(\text{red}) = \dfrac{3}{6} \cdot \dfrac{3}{6} \cdot \dfrac{3}{6} = \dfrac{1}{2} \cdot \dfrac{1}{2} \cdot \dfrac{1}{2} = \dfrac{1}{8}$

11. $P(2 \text{ and } 3) = P(2) \cdot P(3) = \dfrac{1}{6} \cdot \dfrac{1}{6} = \dfrac{1}{36}$

13. $P(\text{even and greater than 2}) = P(\text{even}) \cdot P(\text{greater than 2}) = \dfrac{3}{6} \cdot \dfrac{4}{6} = \dfrac{1}{2} \cdot \dfrac{2}{3} = \dfrac{1}{3}$

15. $P(\text{picture card and heart}) = P(\text{picture card}) \cdot P(\text{heart}) = \dfrac{12}{52} \cdot \dfrac{13}{52} = \dfrac{3}{13} \cdot \dfrac{1}{4} = \dfrac{3}{52}$

17. $P(2 \text{ kings}) = P(\text{king}) \cdot P(\text{king}) = \dfrac{4}{52} \cdot \dfrac{4}{52} = \dfrac{1}{13} \cdot \dfrac{1}{13} = \dfrac{1}{169}$

19. $P(\text{red each time}) = P(\text{red}) \cdot P(\text{red}) = \dfrac{26}{52} \cdot \dfrac{26}{52} = \dfrac{1}{2} \cdot \dfrac{1}{2} = \dfrac{1}{4}$

21. $P(\text{all heads}) = P(\text{heads}) \cdot P(\text{heads}) \cdot P(\text{heads}) \cdot P(\text{heads}) \cdot P(\text{heads}) \cdot P(\text{heads}) = \dfrac{1}{2} \cdot \dfrac{1}{2} \cdot \dfrac{1}{2} \cdot \dfrac{1}{2} \cdot \dfrac{1}{2} \cdot \dfrac{1}{2} = \dfrac{1}{64}$

23. $P(\text{head and number greater than 4}) = P(\text{head}) \cdot P(\text{number greater than 4}) = \dfrac{1}{2} \cdot \dfrac{2}{6} = \dfrac{1}{6}$

25. a. $P(\text{hit two years in a row}) = P(\text{hit}) \cdot P(\text{hit}) = \dfrac{1}{16} \cdot \dfrac{1}{16} = \dfrac{1}{256}$

b. $P(\text{Hit three consecutive years}) = P(\text{hit}) \cdot P(\text{hit}) \cdot P(\text{hit}) = \dfrac{1}{16} \cdot \dfrac{1}{16} \cdot \dfrac{1}{16} = \dfrac{1}{4096}$

c. $P(\text{not hit in next ten years}) = [P(\text{not hit})]^{10} = \left(1 - \dfrac{1}{16}\right)^{10} = \left(\dfrac{15}{16}\right)^{10} \approx 0.524$

d. $P(\text{hit at least once in next ten years}) = 1 - P(\text{not hit in next ten years}) \approx 1 - 0.524 \approx 0.476$

27. $P(\text{all four rate their health as excellent}) = (0.22)(0.22)(0.22)(0.22) = (0.22)^4 \approx 0.00234 \approx 0.234\%$

29. $P(\text{first three are 65 or older, fourth is } 25 - 44, \text{ fifth is 24 or younger})$
$= (0.24)(0.24)(0.24)(0.22)(0.33) \approx 0.001 \approx 0.1\%$

31. $P(\text{solid and solid}) = P(\text{solid}) \cdot P(\text{solid given first was solid}) = \dfrac{15}{30} \cdot \dfrac{14}{29} = \dfrac{1}{2} \cdot \dfrac{14}{29} = \dfrac{7}{29}$

33. $P(\text{coconut then caramel}) = P(\text{coconut}) \cdot P(\text{caramel given first was coconut}) = \dfrac{5}{30} \cdot \dfrac{10}{29} = \dfrac{1}{6} \cdot \dfrac{10}{29} = \dfrac{5}{87}$

35. $P(\text{two Democrats}) = P(\text{Democrat}) \cdot P(\text{Democrat given first was Democrat}) = \dfrac{5}{15} \cdot \dfrac{4}{14} = \dfrac{1}{3} \cdot \dfrac{2}{7} = \dfrac{2}{21}$

37. $P(\text{Independent then Republican}) = P(\text{Independent}) \cdot P(\text{Republican given first was Independent})$
$= \dfrac{4}{15} \cdot \dfrac{6}{14} = \dfrac{4}{15} \cdot \dfrac{3}{7} = \dfrac{4}{35}$

39. $P(\text{no Independents}) = P(\text{not Independent}) \cdot P(\text{not Independent given first was not Independent})$
$= \dfrac{11}{15} \cdot \dfrac{10}{14} = \dfrac{11}{15} \cdot \dfrac{5}{7} = \dfrac{11}{21}$

41. $P(\text{three cans of apple juice})$
$= P(\text{apple juice}) \cdot P\left(\begin{matrix}\text{apple juice given} \\ \text{first was apple juice}\end{matrix}\right) \cdot P\left(\begin{matrix}\text{apple juice given first} \\ \text{two were apple juice}\end{matrix}\right) = \dfrac{6}{20} \cdot \dfrac{5}{19} \cdot \dfrac{4}{18} = \dfrac{1}{57}$

43. $P(\text{grape juice then orange juice then mango juice})$
$= P(\text{grape juice}) \cdot P\left(\begin{matrix}\text{orange juice given} \\ \text{first was grape juice}\end{matrix}\right) \cdot P\left(\begin{matrix}\text{mango juice given first was grape juice} \\ \text{and second war orange juice}\end{matrix}\right) = \dfrac{8}{20} \cdot \dfrac{4}{19} \cdot \dfrac{2}{18} = \dfrac{8}{855}$

45. $P(\text{no grape juice})$
$= P(\text{not grape juice}) \cdot P\left(\begin{matrix}\text{not grape juice given} \\ \text{first was not grape juice}\end{matrix}\right) \cdot P\left(\begin{matrix}\text{not grape juice given first} \\ \text{two were not grape juice}\end{matrix}\right) = \dfrac{12}{20} \cdot \dfrac{11}{19} \cdot \dfrac{10}{18} = \dfrac{11}{57}$

47. $P(3 \mid \text{red}) = \dfrac{1}{5}$

49. $P\left(\text{even}|\text{yellow}\right) = \dfrac{2}{3}$

51. $P\left(\text{red}|\text{odd}\right) = \dfrac{3}{4}$

53. $P\left(\text{red}|\text{at least 5}\right) = \dfrac{3}{4}$

55. $P\left(\text{red}|\text{even}\right) = \dfrac{2}{3}$

57. $P\left(\text{odd}|\text{yellow}\right) = \dfrac{1}{2}$

59. $P\left(\text{unemployed}|\text{male}\right) = \dfrac{\text{number of unemployed males}}{\text{number of males}} = \dfrac{2433}{67,761+2433} = \dfrac{2433}{70,194} \approx 0.0347 \approx 3.47\%$

61. $P\left(\text{female}|\text{employed}\right) = \dfrac{\text{number of employed females}}{\text{number of employed persons}} = \dfrac{58,655}{58,655+67,761} = \dfrac{58,655}{126,416} \approx 0.464 \approx 46.4\%$

63. $P\left(\text{Army officer}|\text{African American}\right)$

$= \dfrac{\text{number of African American Army officers}}{\text{number of African Americans}} = \dfrac{9162}{9162+3524+1341+4282} = \dfrac{9162}{18,309} \approx 0.500 \approx 50.0\%$

65. $P\left(\text{Hispanic American}|\text{Marine Corps officer}\right)$

$= \dfrac{\text{number of Hispanic American Marine Corps officers}}{\text{number of Marine Corps officers}} = \dfrac{914}{1341+914+599+15,043} = \dfrac{914}{17,897} \approx 0.0511 \approx 5.11\%$

67-73. Answers will vary.

75. $P(\text{no one hospitalized}) = [P(\text{not hospitalized})]^5 = (0.9)(0.9)(0.9)(0.9)(0.9) = (0.9)^5 \approx 0.59049 \approx 59.0\%$

77. a. Answers will vary.

b. $P(\text{three different birthdays}) = \dfrac{365}{365} \cdot \dfrac{364}{365} \cdot \dfrac{363}{365} \approx 0.992$

c. $P(\text{at least two have same birthday}) = 1 - P(\text{three different birthdays}) = 1 - 0.992 = 0.008$

d. $P(20 \text{ different birthdays})$

$= \dfrac{365 \cdot 364 \cdot 363 \cdot 362 \cdot 361 \cdot 360 \cdot 359 \cdot 358 \cdot 357 \cdot 356 \cdot 355 \cdot 354 \cdot 353 \cdot 352 \cdot 351 \cdot 350 \cdot 349 \cdot 348 \cdot 347 \cdot 346}{365 \cdot 365 \cdot 365 \cdot 365 \cdot 365 \cdot 365 \cdot 365 \cdot 365 \cdot 365 \cdot 365 \cdot 365 \cdot 365 \cdot 365 \cdot 365 \cdot 365 \cdot 365 \cdot 365 \cdot 365 \cdot 365 \cdot 365} \approx 0.589$

$P(\text{at least two have same birthday}) = 1 - P(20 \text{ different birthdays}) = 1 - 0.589 = 0.411$

e. 23 people (determine by trial-and-error using method shown in part d)

Check Points 11.8

1. $E = 1 \cdot \dfrac{1}{4} + 2 \cdot \dfrac{1}{4} + 3 \cdot \dfrac{1}{4} + 4 \cdot \dfrac{1}{4} = \dfrac{1+2+3+4}{4} = \dfrac{10}{4} = 2.5$

2. $E = 0 \cdot \dfrac{1}{16} + 1 \cdot \dfrac{4}{16} + 2 \cdot \dfrac{6}{16} + 3 \cdot \dfrac{4}{16} + 4 \cdot \dfrac{1}{16} = \dfrac{0+4+12+12+4}{16} = \dfrac{32}{16} = 2$

3. a. $E = \$0(0.01) + \$2000(0.15) + \$4000(0.08) + \$6000(0.05) + \$8000(0.01) + \$10,000(0.70) = \$8000$
    This means that in the long run, the average cost of a claim is expected to be \$8000

    b. An average premium charge of \$8000 would cause the company to neither lose nor gain money.

4. $E = (1)\left(\dfrac{1}{5}\right) + \left(-\dfrac{1}{4}\right)\left(\dfrac{4}{5}\right) = \dfrac{1}{5} + \left(-\dfrac{1}{5}\right) = 0$
    Since the expected value is 0, there is nothing to gain or lose on average by guessing.

5. Values of gain or loss:
    Grand Prize: $\$1000 - \$2 = \$998$, Consolation Prize: $\$50 - \$2 = \$48$, Nothing: $\$0 - \$2 = -\$2$
    $E = (-\$2)\left(\dfrac{997}{1000}\right) + (\$48)\left(\dfrac{2}{1000}\right) + (\$998)\left(\dfrac{1}{1000}\right) = \dfrac{-\$1994 + \$96 + \$998}{1000} = -\dfrac{\$900}{1000} = -\$0.90$
    The expected value for one ticket is $-\$0.90$. This means that in the long run a player can expect to lose \$0.90 for each ticket bought. Buying five tickets will make your likelihood of winning five times greater, however there is no advantage to this strategy because the *cost* of five tickets is also five times greater than one ticket.

6. $E = (\$2.20)\left(\dfrac{20}{80}\right) + (-\$1.00)\left(\dfrac{60}{80}\right) = \dfrac{\$44 - \$60}{80} = \dfrac{-\$16}{80} = -\$0.20$
    This means that in the long run a player can expect to lose an average of \$0.20 for each \$1 bet.

Exercise Set 11.8

1. $E = 1 \cdot \dfrac{1}{2} + 2 \cdot \dfrac{1}{4} + 3 \cdot \dfrac{1}{4} = 1.75$

3. a. $E = \$0(0.65) + \$50,000(0.20) + \$100,000(0.10) + \$150,000(0.03) + \$200,000(0.01) + \$250,000(0.01) = \$29,000$

    b. \$29,000

    c. \$29,050

5. $E = -\$10,000(0.9) + \$90,000(0.1) = \$0$. This means on the average there will be no gain or loss.

7. $E = -\$99,999\left(\dfrac{27}{10,000,000}\right) + \$1\left(\dfrac{9,999,973}{10,000,000}\right) = \$0.73$

9. Probabilities after eliminating one possible answer: Guess Correctly: $\dfrac{1}{4}$, Guess Incorrectly: $\dfrac{3}{4}$
    $E = (1)\left(\dfrac{1}{4}\right) + \left(-\dfrac{1}{4}\right)\left(\dfrac{3}{4}\right) = \dfrac{1}{4} + \left(-\dfrac{3}{16}\right) = \dfrac{1}{16}$ expected points on a guess if one answer is eliminated.
    Yes, it is advantageous to guess after eliminating one possible answer.

11. First mall: $E = \$300,000\left(\dfrac{1}{2}\right) - \$100,000\left(\dfrac{1}{2}\right) = \$100,000$

    Second mall: $E = \$200,000\left(\dfrac{3}{4}\right) - \$60,000\left(\dfrac{1}{4}\right) = \$135,000$

    Choose the second mall.

13. a. $E = \$700,000(0.2) + \$0(0.8) = \$140,000$

    b. No

15. $E = \$4\left(\dfrac{1}{6}\right) - \$1\left(\dfrac{5}{6}\right) = -\$\dfrac{1}{6} \approx -\$0.17$. This means an expected loss of approximately \$0.17 per game.

17. $E = \$1\left(\dfrac{18}{38}\right) - \$1\left(\dfrac{20}{38}\right) \approx -\$0.053$. This means an expected loss of approximately \$0.053 per \$1.00 bet.

19. $E = \$499\left(\dfrac{1}{1000}\right) - \$1\left(\dfrac{999}{1000}\right) = -\$0.50$. This means an expected loss of \$0.50 per \$1.00 bet.

21-25. Answers will vary.

27. Let x = the charge for the policy. Note, the expected value, $E - \$60$.
    $\$60 = (x - \$200,000)(0.0005) + (x)(0.9995)$
    $\$60 = 0.0005x - \$100 + 0.9995x$
    $\$160 = x$
    The insurance company should charge \$160 for the policy.

Chapter 11 Review Exercises

1. Use the Fundamental Counting Principle with two groups of items. $20 \cdot 40 = 800$

2. Use the Fundamental Counting Principle with two groups of items. $4 \cdot 5 = 20$

3. Use the Fundamental Counting Principle with two groups of items. $100 \cdot 99 = 9900$

4. Use the Fundamental Counting Principle with three groups of items. $5 \cdot 5 \cdot 5 = 125$

5. Use the Fundamental Counting Principle with five groups of items. $3 \cdot 3 \cdot 3 \cdot 3 \cdot 3 = 243$

6. Use the Fundamental Counting Principle with four groups of items. $5 \cdot 2 \cdot 2 \cdot 3 = 60$

7. Use the Fundamental Counting Principle with six groups of items. $6 \cdot 5 \cdot 4 \cdot 3 \cdot 2 \cdot 1 - 720$

8. Use the Fundamental Counting Principle with five groups of items. $5 \cdot 4 \cdot 3 \cdot 2 \cdot 1 = 120$

9. Use the Fundamental Counting Principle with seven groups of items. $1 \cdot 5 \cdot 4 \cdot 3 \cdot 2 \cdot 1 \cdot 1 = 120$

10. $\dfrac{16!}{14!} = \dfrac{16 \cdot 15 \cdot 14!}{14!} = 240$

11. $\dfrac{800!}{799!} = \dfrac{800 \cdot 799!}{799!} = 800$

12. $5! - 3! = 5 \cdot 4 \cdot 3 \cdot 2 \cdot 1 - 3 \cdot 2 \cdot 1 = 120 - 6 = 114$

13. $\dfrac{11!}{(11-3)!} = \dfrac{11!}{8!} = \dfrac{11 \cdot 10 \cdot 9 \cdot 8!}{8!} = 990$

14. $_{10}P_6 = \dfrac{10!}{(10-6)!} = \dfrac{10!}{4!} = \dfrac{10 \cdot 9 \cdot 8 \cdot 7 \cdot 6 \cdot 5 \cdot 4!}{4!} = 151,200$

15. $_{100}P_2 = \dfrac{100!}{(100-2)!} = \dfrac{100!}{98!} = \dfrac{100 \cdot 99 \cdot 98!}{98!} = 9900$

16. $_{15}P_4 = \dfrac{15!}{(15-4)!} = \dfrac{15!}{11!} = \dfrac{15 \cdot 14 \cdot 13 \cdot 12 \cdot 11!}{11!} = 32,760$

17. $_{20}P_5 = \dfrac{20!}{(20-5)!} = \dfrac{20!}{15!} = \dfrac{20 \cdot 19 \cdot 18 \cdot 17 \cdot 16 \cdot 15!}{15!} = 1,860,480$

18. Order does not matter. This problem involves combinations.

19. Order matters. This problem involves permutations.

20. Order does not matter. This problem involves combinations.

21. $_{11}C_7 = \dfrac{11!}{(11-7)!7!} = \dfrac{11!}{4!7!} = \dfrac{11 \cdot 10 \cdot 9 \cdot 8 \cdot 7!}{4 \cdot 3 \cdot 2 \cdot 1 \cdot 7!} = 330$

22. $_{14}C_5 = \dfrac{14!}{(14-5)!5!} = \dfrac{14!}{9!5!} = \dfrac{14 \cdot 13 \cdot 12 \cdot 11 \cdot 10 \cdot 9!}{9! \cdot 5 \cdot 4 \cdot 3 \cdot 2 \cdot 1} = 2002$

23. $_{10}C_4 = \dfrac{10!}{(10-4)!4!} = \dfrac{10!}{6!4!} = \dfrac{10 \cdot 9 \cdot 8 \cdot 7 \cdot 6!}{6! 4 \cdot 3 \cdot 2 \cdot 1} = 210$

24. $_{13}C_5 = \dfrac{13!}{(13-5)!5!} = \dfrac{13!}{8!5!} = \dfrac{13 \cdot 12 \cdot 11 \cdot 10 \cdot 9 \cdot 8!}{8! 5 \cdot 4 \cdot 3 \cdot 2 \cdot 1} = 1287$

25. $_{20}C_3 = \dfrac{20!}{(20-3)!3!} = \dfrac{20!}{17!3!} = \dfrac{20 \cdot 19 \cdot 18 \cdot 17!}{17! 3 \cdot 2 \cdot 1} = 1140$

26. Choose the Republicans: $_{12}C_5 = \dfrac{12!}{(12-5)!5!} = \dfrac{12!}{7!5!} = \dfrac{12 \cdot 11 \cdot 10 \cdot 9 \cdot 8 \cdot 7!}{7! 5 \cdot 4 \cdot 3 \cdot 2 \cdot 1} = 792$

Choose the Democrats: $_8C_4 = \dfrac{8!}{(8-4)!4!} = \dfrac{8!}{4!4!} = \dfrac{8 \cdot 7 \cdot 6 \cdot 5 \cdot 4!}{4! 4 \cdot 3 \cdot 2 \cdot 1} = 70$

Multiply the choices: $792 \cdot 70 = 55,440$

27. $P(6) = \dfrac{\text{number of ways a 6 can occur}}{\text{total number of possible outcomes}} = \dfrac{1}{6}$

28. $P(\text{less than 5}) = \dfrac{\text{number of ways a number less than 5 can occur}}{\text{total number of possible outcomes}} = \dfrac{4}{6} = \dfrac{2}{3}$

29. $P(\text{less than } 7) = \dfrac{\text{number of ways a number less than 7 can occur}}{\text{total number of possible outcomes}} = \dfrac{6}{6} = 1$

30. $P(\text{greater than } 6) = \dfrac{\text{number of ways a number greater than 6 can occur}}{\text{total number of possible outcomes}} = \dfrac{0}{6} = 0$

31. $P(5) = \dfrac{\text{number of ways a 5 can occur}}{\text{total number of possible outcomes}} = \dfrac{4}{52} = \dfrac{1}{13}$

32. $P(\text{picture card}) = \dfrac{\text{number of ways a picture card can occur}}{\text{total number of possible outcomes}} = \dfrac{12}{52} = \dfrac{3}{13}$

33. $P(\text{greater than 4 and less than 8}) = \dfrac{\text{number of ways a card greater than 4 and less than 8 can occur}}{\text{total number of possible outcomes}} = \dfrac{12}{52} = \dfrac{3}{13}$

34. $P(\text{4 of diamonds}) = \dfrac{\text{number of ways a 4 of diamonds can occur}}{\text{total number of possible outcomes}} = \dfrac{1}{52}$

35. $P(\text{red ace}) = \dfrac{\text{number of ways a red ace can occur}}{\text{total number of possible outcomes}} = \dfrac{2}{52} = \dfrac{1}{26}$

36. $P(\text{chocolate}) = \dfrac{\text{number of ways a chocolate can occur}}{\text{total number of possible outcomes}} = \dfrac{15}{30} = \dfrac{1}{2}$

37. $P(\text{caramel}) = \dfrac{\text{number of ways a caramel can occur}}{\text{total number of possible outcomes}} = \dfrac{10}{30} = \dfrac{1}{3}$

38. $P(\text{peppermint}) = \dfrac{\text{number of ways a peppermint can occur}}{\text{total number of possible outcomes}} = \dfrac{5}{30} = \dfrac{1}{6}$

39. a. $P(\text{carrier without the disease}) = \dfrac{\text{number of ways to be a carrier without the disease}}{\text{total number of possible outcomes}} = \dfrac{2}{4} = \dfrac{1}{2}$

    b. $P(\text{disease}) = \dfrac{\text{number of ways to have the disease}}{\text{total number of possible outcomes}} = \dfrac{0}{4} = 0$

40. $P(\text{Californian is Hispanic}) = \dfrac{\text{number of Hispanics in California}}{\text{total population of California}} = \dfrac{10,112,986}{32,666,550} \approx 0.310$

41. $P(\text{Texan is Hispanic}) = \dfrac{\text{number of Hispanics in Texas}}{\text{total population of Texas}} = \dfrac{5,862,835}{19,759,614} \approx 0.297$

42. number of ways to visit in order D, B, A, C = 1
    total number of possible permutations = $4 \cdot 3 \cdot 2 \cdot 1 = 24$

    $P(D, B, A, C) = \dfrac{1}{24}$

43. number of permutations with C last $= 5 \cdot 4 \cdot 3 \cdot 2 \cdot 1 \cdot 1 = 120$
    total number of possible permutations $= 6 \cdot 5 \cdot 4 \cdot 3 \cdot 2 \cdot 1 = 720$

    $$P(\text{C last}) = \frac{120}{720} = \frac{1}{6}$$

44. number of permutations with B first and A last $= 1 \cdot 4 \cdot 3 \cdot 2 \cdot 1 \cdot 1 = 24$
    total number of possible permutations $= 6 \cdot 5 \cdot 4 \cdot 3 \cdot 2 \cdot 1 = 720$

    $$P(\text{B first and A last}) = \frac{24}{720} = \frac{1}{30}$$

45. number of permutations in order F, E, A, D, C, B $= 1$
    total number of possible permutations $= 6 \cdot 5 \cdot 4 \cdot 3 \cdot 2 \cdot 1 = 720$

    $$P(\text{F, E, A, D, C, B}) = \frac{1}{720}$$

46. number of permutations with A or C first $= 2 \cdot 5 \cdot 4 \cdot 3 \cdot 2 \cdot 1 = 240$
    total number of possible permutations $= 6 \cdot 5 \cdot 4 \cdot 3 \cdot 2 \cdot 1 = 720$

    $$P(\text{A or C first}) = \frac{240}{720} = \frac{1}{3}$$

47. **a.** number of ways to win $= 1$
    total number of possible combinations:

    $$_{20}C_5 = \frac{20!}{(20-5)!5!} = \frac{20!}{15!5!} = \frac{20 \cdot 19 \cdot 18 \cdot 17 \cdot 16 \cdot 15!}{15!5 \cdot 4 \cdot 3 \cdot 2 \cdot 1} = 15,504$$

    $$P(\text{winning with one ticket}) = \frac{1}{15,504} \approx 0.0000645$$

    b. number of ways to win $= 100$

    $$P(\text{winning with 100 different tickets}) = \frac{100}{15,504} \approx 0.00645$$

48. **a.** number of ways to select 4 Democrats: $_6C_4 = \dfrac{6!}{(6-4)!4!} = \dfrac{6!}{2!4!} = \dfrac{6 \cdot 5 \cdot 4!}{2 \cdot 1 \cdot 4!} = 15$

    total number of possible combinations: $_{10}C_4 = \dfrac{10!}{(10-4)!4!} = \dfrac{10!}{6!4!} = \dfrac{10 \cdot 9 \cdot 8 \cdot 7 \cdot 6!}{6!4 \cdot 3 \cdot 2 \cdot 1} = 210$

    $$P(\text{all Democrats}) = \frac{15}{210} = \frac{1}{14}$$

    b. number of ways to select 2 Democrats: $_6C_2 = \dfrac{6!}{(6-2)!2!} = \dfrac{6!}{4!2!} = \dfrac{6 \cdot 5 \cdot 4!}{4!2 \cdot 1} = 15$

    number of ways to select 2 Republicans: $_4C_2 = \dfrac{4!}{(4-2)!2!} = \dfrac{4!}{2!2!} = \dfrac{4 \cdot 3 \cdot 2!}{2!2 \cdot 1} = 6$

    number of ways to select 2 Democrats and 2 Republicans $= 15 \cdot 6 = 90$

    $$P(\text{2 Democrats and 2 Republicans}) = \frac{90}{210} = \frac{3}{7}$$

49. number of ways to get 2 picture cards: $_6C_2 = \dfrac{6!}{(6-2)!2!} = \dfrac{6!}{4!2!} = \dfrac{6 \cdot 5 \cdot 4!}{4!2 \cdot 1} = 15$

    number of ways to get one non-picture card $= 20$
    number of ways to get 2 picture cards and one non-picture card $= 15 \cdot 20 = 300$

total number of possible combinations: $_{26}C_3 = \dfrac{26!}{(26-3)!3!} = \dfrac{26!}{23!3!} = \dfrac{26 \cdot 25 \cdot 24 \cdot 23!}{23!3 \cdot 2 \cdot 1} = 2600$

$P(2 \text{ picture cards}) = \dfrac{300}{2600} = \dfrac{3}{26}$

50. $P(\text{not a } 5) = 1 - P(5) = 1 - \dfrac{1}{6} = \dfrac{5}{6}$

51. $P(\text{not less than } 4) = 1 - P(\text{less than } 4) = 1 - \dfrac{3}{6} = 1 - \dfrac{1}{2} = \dfrac{1}{2}$

52. $P(3 \text{ or } 5) = P(3) + P(5) = \dfrac{1}{6} + \dfrac{1}{6} = \dfrac{2}{6} = \dfrac{1}{3}$

53. $P(\text{less than } 3 \text{ or greater than } 4) = P(\text{less than } 3) + P(\text{greater than } 4) = \dfrac{2}{6} + \dfrac{2}{6} = \dfrac{1}{3} + \dfrac{1}{3} = \dfrac{2}{3}$

54. $P(\text{less than } 5 \text{ or greater than } 2) = P(\text{less than } 5) + P(\text{greater than } 2) - P(\text{less than } 5 \text{ and greater than } 2)$

$$= \dfrac{4}{6} + \dfrac{4}{6} - \dfrac{2}{6} = 1$$

55. $P(\text{not a picture card}) = 1 - P(\text{picture card}) = 1 - \dfrac{12}{52} = 1 - \dfrac{3}{13} = \dfrac{10}{13}$

56. $P(\text{not a diamond}) = 1 - P(\text{diamond}) = 1 - \dfrac{13}{52} = 1 - \dfrac{1}{4} = \dfrac{3}{4}$

57. $P(\text{ace or king}) = P(\text{ace}) + P(\text{king}) = \dfrac{4}{52} + \dfrac{4}{52} = \dfrac{1}{13} + \dfrac{1}{13} = \dfrac{2}{13}$

58. $P(\text{black } 6 \text{ or red } 7) = P(\text{black } 6) + P(\text{red } 7) = \dfrac{2}{52} + \dfrac{2}{52} = \dfrac{1}{26} + \dfrac{1}{26} = \dfrac{2}{26} = \dfrac{1}{13}$

59. $P(\text{queen or red card}) = P(\text{queen}) + P(\text{red card}) - P(\text{red queen}) = \dfrac{4}{52} + \dfrac{26}{52} - \dfrac{2}{52} = \dfrac{28}{52} = \dfrac{7}{13}$

60. $P(\text{club or picture card}) = P(\text{club}) + P(\text{picture card}) - P(\text{club and picture card}) = \dfrac{13}{52} + \dfrac{12}{52} - \dfrac{3}{52} = \dfrac{22}{52} = \dfrac{11}{26}$

61. $P(\text{not } 4) = 1 - P(4) = 1 - \dfrac{1}{6} = \dfrac{5}{6}$

62. $P(\text{not yellow}) = 1 - P(\text{yellow}) = 1 - \dfrac{1}{6} = \dfrac{5}{6}$

63. $P(\text{not red}) = 1 - P(\text{red}) = 1 - \dfrac{3}{6} = 1 - \dfrac{1}{2} = \dfrac{1}{2}$

64. $P(\text{red or yellow}) = P(\text{red}) + P(\text{yellow}) = \dfrac{3}{6} + \dfrac{1}{6} = \dfrac{4}{6} = \dfrac{2}{3}$

65. $P(\text{red or even}) = P(\text{red}) + P(\text{even}) - P(\text{red and even}) = \frac{3}{6} + \frac{3}{6} - \frac{0}{6} = 1$

66. $P(\text{red or greater than 3}) = P(\text{red}) + P(\text{greater than 3}) - P(\text{red and greater than 3}) = \frac{3}{6} + \frac{3}{6} - \frac{1}{6} = \frac{5}{6}$

67. $P(\text{not in Northeast}) = 1 - P(\text{Northeast}) = 1 - \frac{51,829,962}{272,690,813} = \frac{220,860,851}{272,690,813} \approx 0.810$

68. $P(\text{South or West}) = P(\text{South}) + P(\text{West}) = \frac{96,468,455}{272,690,813} + \frac{61,150,112}{272,690,813} = \frac{157,618,567}{272,690,813} \approx 0.578$

69. $P(\text{African American or male}) = P(\text{African American}) + P(\text{male}) - P(\text{African American male})$
$$= \frac{50+20}{200} + \frac{50+90}{200} - \frac{50}{200} = \frac{160}{200} = \frac{4}{5}$$

70. $P(\text{female or white}) = P(\text{female}) + P(\text{white}) - P(\text{white female}) = \frac{20+40}{200} + \frac{90+40}{200} - \frac{40}{200} = \frac{150}{200} = \frac{3}{4}$

71. number of favorable outcomes = 4, number of unfavorable outcomes = 48
    Odds in favor of getting a queen are 4:48, or 1:12. Odds against getting a queen are 12:1.

72. number of favorable outcomes = 20, number of unfavorable outcomes = 1980
    Odds against winning are 1980: 20, or 99:1.

73. $P(\text{win}) = \frac{3}{3+1} = \frac{3}{4}$

74. $P(\text{yellow then red}) = P(\text{yellow}) \cdot P(\text{red}) = \frac{2}{6} \cdot \frac{4}{6} = \frac{1}{3} \cdot \frac{2}{3} = \frac{2}{9}$

75. $P(1 \text{ then } 3) = P(1) \cdot P(3) = \frac{1}{6} \cdot \frac{1}{6} = \frac{1}{36}$

76. $P(\text{yellow both times}) = P(\text{yellow}) \cdot P(\text{yellow}) = \frac{2}{6} \cdot \frac{2}{6} = \frac{1}{3} \cdot \frac{1}{3} = \frac{1}{9}$

77. $P(\text{yellow then 4 then odd}) = P(\text{yellow}) \cdot P(4) \cdot P(\text{odd}) = \frac{2}{6} \cdot \frac{1}{6} \cdot \frac{3}{6} = \frac{1}{3} \cdot \frac{1}{6} \cdot \frac{1}{2} = \frac{1}{36}$

78. $P(\text{red every time}) = P(\text{red}) \cdot P(\text{red}) \cdot P(\text{red}) = \frac{4}{6} \cdot \frac{4}{6} \cdot \frac{4}{6} = \frac{2}{3} \cdot \frac{2}{3} \cdot \frac{2}{3} = \frac{8}{27}$

79. $P(\text{five boys in a row}) = P(\text{boy}) \cdot P(\text{boy}) \cdot P(\text{boy}) \cdot P(\text{boy}) \cdot P(\text{boy}) = \frac{1}{2} \cdot \frac{1}{2} \cdot \frac{1}{2} \cdot \frac{1}{2} \cdot \frac{1}{2} = \frac{1}{2^5} = \frac{1}{32}$

80. a. $P(\text{flood two years in a row}) = P(\text{flood}) \cdot P(\text{flood}) = (0.2)(0.2) = 0.04$

b. $P(\text{flood for three consecutive years}) = P(\text{flood}) \cdot P(\text{flood}) \cdot P(\text{flood}) = (0.2)(0.2)(0.2) = 0.008$

c. $P(\text{no flooding for four consecutive years}) = [1 - P(\text{flood})]^4 = (1 - 0.2)^4 = (0.8)^4 = 0.4096$

d. $P(\text{flood at least once in next four years}) = 1 - P(\text{no flooding for four consecutive years})$
$$= 1 - 0.4096 = 0.5904$$

81. $P(\text{music major then psychology major}) = P(\text{music major}) \cdot P\left(\begin{array}{c}\text{psychology major given} \\ \text{first was music major}\end{array}\right) = \dfrac{2}{9} \cdot \dfrac{4}{8} = \dfrac{2}{9} \cdot \dfrac{1}{2} = \dfrac{1}{9}$

82. $P(\text{two business majors}) = P(\text{bus. major}) \cdot P(\text{bus. major given first was bus. major}) = \dfrac{3}{9} \cdot \dfrac{2}{8} = \dfrac{1}{3} \cdot \dfrac{1}{4} = \dfrac{1}{12}$

83. $P(\text{solid then two cherry})$
$$= P(\text{solid}) \cdot P\left(\begin{array}{c}\text{cherry given} \\ \text{first was solid}\end{array}\right) \cdot P\left(\begin{array}{c}\text{cherry given first was solid} \\ \text{and second was cherry}\end{array}\right) = \dfrac{30}{50} \cdot \dfrac{5}{49} \cdot \dfrac{4}{48} = \dfrac{3}{5} \cdot \dfrac{5}{49} \cdot \dfrac{1}{12} = \dfrac{1}{196}$$

84. $P(5|\text{odd}) - \dfrac{1}{3}$

85. $P(\text{vowel}|\text{precedes the letter k}) = \dfrac{3}{10}$

86. a. $P(\text{odd}|\text{red}) = \dfrac{2}{4} = \dfrac{1}{2}$

b. $P(\text{yellow}|\text{at least 3}) = \dfrac{2}{7}$

87. $P(\text{freshman}|\text{female}) = \dfrac{\text{observed number of female freshmen}}{\text{observed number of females}} = \dfrac{368}{368 + 314 + 262 + 220} = \dfrac{368}{1164} = \dfrac{92}{291}$

88. $P(\text{woman}|\text{senior}) = \dfrac{\text{observed number of female seniors}}{\text{observed number of seniors}} = \dfrac{220}{192 + 220} = \dfrac{220}{412} = \dfrac{55}{103}$

89. $E = 1 \cdot \dfrac{1}{4} + 2 \cdot \dfrac{1}{8} + 3 \cdot \dfrac{1}{8} + 4 \cdot \dfrac{1}{4} + 5 \cdot \dfrac{1}{4} = 3.125$

90. a. $E = \$0(0.9999995) + (-\$1,000,000)(0.0000005) = -\$.50$
The insurance company spends an average of $0.50 per person insured.

b. charge $\$9.50 - (-\$0.50) = \$10.00$

91. $E = \$27,000\left(\dfrac{1}{4}\right) + (-\$3000)\left(\dfrac{3}{4}\right) = \$4500$. The expected gain is $4500 per bid.

92. $E - \$1\left(\dfrac{2}{4}\right) + \$1\left(\dfrac{1}{4}\right) + (-\$4)\left(\dfrac{1}{4}\right) = -\$0.25$. The expected loss is $0.25 per game.

Chapter 11: Counting Methods and Probability Theory

Chapter 11 Test

1. Use the Fundamental Counting Principle with five groups of items. $10 \cdot 2 \cdot 2 \cdot 2 \cdot 3 = 240$

2. Use the Fundamental Counting Principle with four groups of items. $4 \cdot 3 \cdot 2 \cdot 1 = 24$

3. Use the Fundamental Counting Principle with seven groups of items. $1 \cdot 6 \cdot 5 \cdot 4 \cdot 3 \cdot 2 \cdot 1 = 720$

4. $_{11}P_3 = \dfrac{11!}{(11-3)!} = \dfrac{11!}{8!} = \dfrac{11 \cdot 10 \cdot 9 \cdot 8!}{8!} = 990$

5. $_{10}C_4 = \dfrac{10!}{(10-4)!4!} = \dfrac{10!}{6!4!} = \dfrac{10 \cdot 9 \cdot 8 \cdot 7 \cdot 6!}{6!4 \cdot 3 \cdot 2 \cdot 1} = 210$

6. $P(\text{freshman}) = \dfrac{12}{50} = \dfrac{6}{25}$

7. $P(\text{not a sophomore}) = 1 - P(\text{sophomore}) = 1 - \dfrac{16}{50} = 1 - \dfrac{8}{25} = \dfrac{17}{25}$

8. $P(\text{junior or senior}) = P(\text{junior}) + P(\text{senior}) = \dfrac{20}{50} + \dfrac{2}{50} = \dfrac{22}{50} = \dfrac{11}{25}$

9. $P(\text{greater than 4 and less than 10}) = \dfrac{20}{52} = \dfrac{5}{13}$

10. $P(C \text{ first, } A \text{ next-to-last, } E \text{ last})$

    $= P(C) \cdot P(A \text{ given } C \text{ was first}) \cdot P(E \text{ given } C \text{ was first and } A \text{ was next-to-last}) = \dfrac{1}{7} \cdot \dfrac{1}{6} \cdot \dfrac{1}{5} = \dfrac{1}{210}$

11. total number of possible combinations: $_{15}C_6 = \dfrac{15!}{(15-6)!6!} = \dfrac{15!}{9!6!} = \dfrac{15 \cdot 14 \cdot 13 \cdot 12 \cdot 11 \cdot 10 \cdot 9!}{9!6 \cdot 5 \cdot 4 \cdot 3 \cdot 2 \cdot 1} = 5005$

    $P(\text{winning with 50 tickets}) = \dfrac{50}{5005} = \dfrac{10}{1001} \approx 0.00999$

12. $P(\text{red or blue}) = P(\text{red}) + P(\text{blue}) = \dfrac{2}{8} + \dfrac{2}{8} = \dfrac{4}{8} = \dfrac{1}{2}$

13. $P(\text{red then blue}) = P(\text{red}) \cdot P(\text{blue}) = \dfrac{2}{8} \cdot \dfrac{2}{8} = \dfrac{1}{4} \cdot \dfrac{1}{4} = \dfrac{1}{16}$

14. $P(\text{flooding for three consecutive years}) = P(\text{flood}) \cdot P(\text{flood}) \cdot P(\text{flood}) = \dfrac{1}{20} \cdot \dfrac{1}{20} \cdot \dfrac{1}{20} = \dfrac{1}{8000}$

15. $P(\text{black or picture card}) = P(\text{black}) + P(\text{picture card}) - P(\text{black picture card}) = \dfrac{26}{52} + \dfrac{12}{52} - \dfrac{6}{52} = \dfrac{32}{52} = \dfrac{8}{13}$

16. $P(\text{freshman or female}) = P(\text{freshman}) + P(\text{female}) - P(\text{female freshman}) = \dfrac{10+15}{50} + \dfrac{15+5}{50} - \dfrac{15}{50} = \dfrac{30}{50} = \dfrac{3}{5}$

17. $P(\text{both red}) = P(\text{red}) \cdot P(\text{red given first ball was red}) = \dfrac{5}{20} \cdot \dfrac{4}{19} = \dfrac{1}{4} \cdot \dfrac{4}{19} = \dfrac{1}{19}$

18. $P(\text{all correct}) = P(\text{correct}) \cdot P(\text{correct}) \cdot P(\text{correct}) \cdot P(\text{correct}) = \dfrac{1}{4} \cdot \dfrac{1}{4} \cdot \dfrac{1}{4} \cdot \dfrac{1}{4} = \left(\dfrac{1}{4}\right)^4 = \dfrac{1}{256}$

19. The populations of the 50 states are different.

20. number of favorable outcomes = 20, number of unfavorable outcomes = 15
Odds against being a man are 15:20, or 3:4.

21. a. Odds in favor are 4:1.

    b. $P(\text{win}) = \dfrac{4}{1+4} = \dfrac{4}{5}$

22. $P\left(\text{man} \middle| \text{blue eyes}\right) = \dfrac{\text{observed number of males with blue eyes}}{\text{observed number of people with blue eyes}} = \dfrac{18}{18+20} = \dfrac{18}{38} = \dfrac{9}{19}$

23. $E = \$65{,}000(0.2) + (-\$15{,}000)(0.8) = \$1000$. This means the expected gain is \$1000 for this bid.

24. $E = (-\$19) \cdot \dfrac{10}{20} + (-\$18) \cdot \dfrac{5}{20} + (-\$15) \cdot \dfrac{3}{20} + (-\$10) \cdot \dfrac{1}{20} + (\$80) \cdot \dfrac{1}{20}$

$\quad = \dfrac{-\$190 - \$90 - \$45 - \$10 + \$80}{20}$

$\quad = \dfrac{-\$255}{20}$

$\quad = -\$12.75$

This expected value of $-\$12.75$ means that a player will lose an average of \$12.75 per play in the long run.

Chapter 12
Statistics

Check Points 12.1

1. a. The population is the set containing all the of the city's homeless people.

   b. This is not a good idea. This sample of people currently in a shelter is more likely to hold opinions that favor required residence in city shelters than the population of all the city's homeless.

2. The sampling technique described in Check Point 1b does not produce a random sample because homeless people who do not go to shelters have no chance of being selected for the survey. In this instance, an appropriate method would be to randomly select neighborhoods of the city and then randomly survey homeless people within the selected neighborhood.

3.

Grade	Number of students
A	3
B	5
C	9
D	2
F	1
	20

4.

Exam Scores (class)	Tally	Number of students (frequency)
40 – 49	|	1
50 – 59	⊬	5
60 – 69	||||	4
70 – 79	⊬ ⊬ ⊬	15
80 – 89	⊬	5
90 – 99	⊬ ||	7
		37

5.

Stems	Leaves
4	1
5	8 2 8 0 7
6	8 2 9 9
7	3 5 9 9 7 5 5 3 3 6 7 1 7 1 5
8	7 3 9 9 1
9	4 6 9 7 5 8 0

Exercise Set 12.1

1. The population is the set containing all the college professors in the United States.

3. The population is the set containing all the residents of the United States.

5. The population is the set containing all high school students in Connecticut.

7. c

9.

Time Spent on Homework (in hours)	Number of students
15	4
16	5
17	6
18	5
19	4
20	2
21	2
22	0
23	0
24	2
	30

11.

Age	Number of Runners
10–19	8
20–29	13
30–39	7
40–49	6
50–59	4
60–69	2
	40

13.

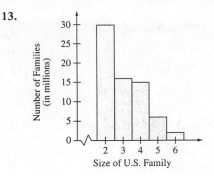

Histogram for Size of U.S. Family

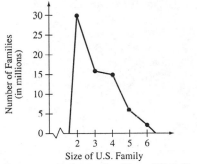

Frequency Polygon for Size of U.S. Family

15.

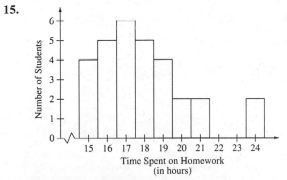

Histogram for Time Spent on Homework

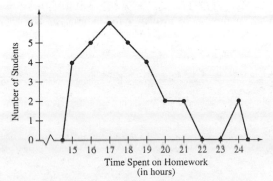

Frequency Polygon for Time Spent on Homework

17. c

19.

Stems	Leaves
2	8 8 9 5
3	8 7 0 1 2 7 6 4 0 5
4	8 2 2 1 4 5 4 6 2 0 8 2 7 9
5	9 4 1 9 1 0
6	3 2 3 6 6 3

21. Ages of the actors:
31, 32, 32, 33, 35, 37, 38, 39, 40, 40, 41, 42, 42, 43, 43, 44, 45, 47, 46, 47, 48, 48, 51, 55, 56, 56, 60, 61, 62, 76

Ages of the actresses:
21, 24, 26, 26, 26, 27, 30, 30, 31, 31, 33, 33, 34, 34, 34, 35, 35, 37, 37, 38, 41, 41, 41, 42, 49, 60, 61, 61, 74, 80
Explanations will vary.

23-37. Answers will vary.

39.

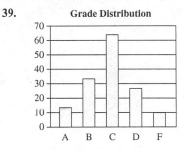

Check Points 12.2

1. a. $\dfrac{10+20+30+40+50}{5} = \dfrac{150}{5} = 30$

b. $\dfrac{3+10+10+10+117}{5} = \dfrac{150}{5} = 30$

2.

x	f	xf
30	3	$30 \cdot 3 = 90$
33	4	$33 \cdot 4 = 132$
40	4	$40 \cdot 4 = 160$
50	1	$50 \cdot 1 = 50$
	12	$\sum xf = 432$

$\text{Mean} = \dfrac{\sum xf}{n} = \dfrac{432}{12} = 36$

3. a. First arrange the data items from smallest to largest: 25, 28, <u>35</u>, 40, 42
        The number of data items is odd, so the median is the middle number. The median is 35.

    b. First arrange the data items from smallest to largest: 61, 72, <u>79</u>, <u>85</u>, 87, 93
        The number of data items is even, so the median is the mean of the two middle data items.
        The median is $\dfrac{79+85}{2} = \dfrac{164}{2} = 82$.

4. The data items are arranged from smallest to largest with $n = 19$, which gives $\dfrac{n+1}{2} = \dfrac{19+1}{2} = \dfrac{20}{2} = 10$

    The median is in the 10th position, which means the median is 5

5. The data items are arranged from smallest to largest with $n = 10$, which gives $\dfrac{n+1}{2} = \dfrac{10+1}{2} = \dfrac{11}{2} = 5.5$

    The median is in the 5.5 position, which means the median is the mean of the data items in positions 5 and 6.
    The median is $\dfrac{23+26}{2} = \dfrac{49}{2} = 24.5$.

6. The total frequency is $1+1+1+3+1+1+2+2+1+2+1+1 = 17$, therefore $n = 17$

    The median's position is $\dfrac{n+1}{2} = \dfrac{17+1}{2} = \dfrac{18}{2} = 9$.

    Counting through the frequency row identifies that the 9th data item is 55.

7. Mean $= \dfrac{17.1+11.8+5.0+4.5+4.3+4.2+3.1+2.1+2.0+1.0+1.0+0.8+0.7+0.3}{14} = \dfrac{57.9}{14} \approx \$4.1$ million

    The mean salary of \$4.1 million is substantially higher than the median salary of \$2.6 million. This is due to the fact that there are two players that are paid substantially more than the other twelve. These two extreme data items are greatly influencing the mean. These two extreme data items do not greatly influence the median.

8. The number 8 occurs more often than any other. The mode is 8.

9. lowest data value = \$60,000, highest data value = \$151,500

    Midrange $= \dfrac{\$60,000+\$151,500}{2} = \dfrac{\$211,500}{2} = \$105,750$

10. a. Mean $= \dfrac{173+191+182+190+172+147+146+138+175+136+179+153+107+195+135+140+138}{17}$

        $= \dfrac{2697}{17} = 158.6$ calories

    b. Order the data items: 107, 135, 136, 138, 138, 140, 146, 147, <u>153</u>, 172, 173, 175, 179, 182, 190, 191, 195
        The number of data items is odd, so the median is the middle number. The median is 153 calories.

    c. The number 138 occurs more often than any other. The mode is 138 calories.

    d. Midrange $= \dfrac{107+195}{2} = \dfrac{302}{2} = 151$ calories

Exercise Set 12.2

1. $\dfrac{7+4+3+2+8+5+1+3}{8} = \dfrac{33}{8} = 4.125$

3. $\dfrac{91+95+99+97+93+95}{6} = \dfrac{570}{6} = 95$

5. $\dfrac{100+40+70+40+60}{5} = \dfrac{310}{5} = 62$

7. $\dfrac{1.6+3.8+5.0+2.7+4.2+4.2+3.2+4.7+3.6+2.5+2.5}{11} = \dfrac{38}{11} \approx 3.45$

9.

x	f	xf
1	1	$1 \cdot 1 = 1$
2	3	$2 \cdot 3 = 6$
3	4	$3 \cdot 4 = 12$
4	4	$4 \cdot 4 = 16$
5	6	$5 \cdot 6 = 30$
6	5	$6 \cdot 5 = 30$
7	3	$7 \cdot 3 = 21$
8	2	$8 \cdot 2 = 16$
	28	$\sum xf = 132$

$$\text{Mean} = \frac{\sum xf}{n} = \frac{132}{28} \approx 4.71$$

11.

x	f	xf
1	1	$1 \cdot 1 = 1$
2	1	$2 \cdot 1 = 2$
3	2	$3 \cdot 2 = 6$
4	5	$4 \cdot 5 = 20$
5	7	$5 \cdot 7 = 35$
6	9	$6 \cdot 9 = 54$
7	8	$7 \cdot 8 = 56$
8	6	$8 \cdot 6 = 48$
9	4	$9 \cdot 4 = 36$
10	3	$10 \cdot 3 = 30$
	46	$\sum xf = 288$

$$\text{Mean} = \frac{\sum xf}{n} = \frac{288}{46} \approx 6.26$$

13. First arrange the data items from smallest to largest: 1, 2, 3, 3, 4, 5, 7, 8
The number of data items is even, so the median is the mean of the two middle data items. The median is 3.5.

15. First arrange the data items from smallest to largest: 91, 93, 95, 95, 97, 99
The number of data items is even, so the median is the mean of the two middle data items.
$$\text{Median} = \frac{95+95}{2} = 95$$

17. First arrange the data items from smallest to largest: 40, 40, 60, 70, 100
The number of data items is odd, so the median is the middle number. The median is 60.

19. First arrange the data items from smallest to largest: 1.6, 2.5, 2.5, 2.7, 3.2, 3.6, 3.8, 4.2, 4.2, 4.7, 5.0
The number of data items is odd, so the median is the middle number. The median is 3.6.

21. $n = 28$, thus $\dfrac{n+1}{2} = \dfrac{28+1}{2} = \dfrac{29}{2} = 14.5$

The median is in the 14.5 position, which means the median is the mean of the data items in positions 14 and 15. Counting down the frequency column, the 14th and 15th data items are both 5.

$\text{Median} = \dfrac{5+5}{2} = 5$

23. $n = 46$, thus $\dfrac{n+1}{2} = \dfrac{46+1}{2} = 23.5$

The median is in the 23.5 position, which means the median is the mean of the data items in positions 23 and 24. Counting down the frequency column, the 23rd and 24th data items are both 6.

$\text{Median} = \dfrac{6+6}{2} = 6$

25. The number 3 occurs more often than any other. The mode is 3.

27. The number 95 occurs more often than any other. The mode is 95.

29. The number 40 occurs more often than any other. The mode is 40.

31. No single value occurs most frequently. There is no mode.

33. The number 5 occurs more often than any other. The mode is 5.

35. The number 6 occurs more often than any other. The mode is 6.

37. lowest data value = 1, highest data value = 8, $\text{Midrange} = \dfrac{1+8}{2} = 4.5$

39. lowest data value = 91, highest data value = 99, $\text{Midrange} = \dfrac{91+99}{2} = 95$

41. lowest data value = 40, highest data value = 100, $\text{Midrange} = \dfrac{40+100}{2} = 70$

43. lowest data value = 1.6, highest data value = 5.0, $\text{Midrange} = \dfrac{1.6+5.0}{2} = 3.3$

45. $\text{Midrange} = \dfrac{1+8}{2} = 4.5$

47. $\text{Midrange} = \dfrac{1+10}{2} = 5.5$

49. a. $\text{Mean} = \dfrac{54+59+35+41+46+25+47+60+54+46+49+46+41+34+22}{15} = \dfrac{659}{15} \approx 43.9$

b. Order the data items: 22, 25, 34, 35, 41, 41, 46, <u>46</u>, 46, 47, 49, 54, 54, 59, 60
The number of data items is odd, so the median is the middle number. The median is 46.

c. The number 46 occurs more often than any other. The mode is 46.

d. $\text{Midrange} = \dfrac{22+60}{2} = \dfrac{82}{2} = 41$

51. a. $\text{Mean} = \dfrac{\$666,000 + \$562,500 + \$507,500 + \$453,500 + \$275,000}{5} = \dfrac{\$2,464,500}{5} = \$492,900$

b. The median is $507,500.

c. No single value occurs most frequently. There is no mode.

d. $\text{Midrange} = \dfrac{\$666,000 + \$275,000}{2} = \dfrac{\$941,000}{2} = \$470,500$

53. Reading of graph values will vary. Possible data readings (from 1981 to 2000):
480, 480, 490, 550, 620, 600, 650, 590, 380, 420, 550, 370, 420, 300, 420, 280, 290, 260, 390, and 410

$$\text{Mean} = \dfrac{480+480+490+550+620+600+650+590+380+420+550+370+420+300+420+280+290+260+390+410}{20}$$

$$= \dfrac{8950}{20} = 447.5$$

Put the data items in order to find median: 260, 280, 290, 300, 370, 380, 390, 410, 420, 420, 420, 480, 480, 490, 550, 550, 590, 600, 620, 650
The number of data items is even, so the median is the mean of the two middle data items.

$$\text{Median} = \dfrac{420 + 420}{2} = 420$$

Mode: The number 420 occurs more often than any other. The mode is 420.

$$\text{Midrange} = \dfrac{650 + 260}{2} = \dfrac{910}{2} = 455$$

55. $n = 40$, $\dfrac{n+1}{2} = \dfrac{40+1}{2} = \dfrac{41}{2} = 20.5$
The median is in the 20.5 position, which means the median is the mean of the data items in positions 20 and 21.
$$\text{Median} = \dfrac{175 + 175}{2} = 175 \text{ lb}$$

57. $\text{Midrange} = \dfrac{150 + 205}{2} = 177.5 \text{ lb}$

59. a.

Words per Minute	Number of People
600	1
650	2
700	2
750	2
800	3
850	3
900	3
950	2
1000	4
1050	1
1100	1
	24

$$\text{Mean} = \frac{1\cdot600+2\cdot650+2\cdot700+2\cdot750+3\cdot800+3\cdot850+3\cdot900+2\cdot950+4\cdot1000+1\cdot1050+1\cdot1100}{24}$$

≈ 854 words per minute

$\text{Median} = \dfrac{850+850}{2} = 850$ words per minute

$\text{Mode} = 1000$ words per minute

$\text{Midrange} = \dfrac{600+1100}{2} = 850$ words per minute

b. Mode

c. Answers will vary.

61-71. Answers will vary.

73. All 30 students had the same grade.

75. 61.6

Check Points 12.3

1. $\text{Range} = 11 - 2 = 9$

2. $\text{Mean} = \dfrac{2+4+7+11}{4} = \dfrac{24}{4} = 6$

Data item	Deviation: Data item – mean
2	$2 - 6 = -4$
4	$4 - 6 = -2$
7	$7 - 6 = 1$
11	$11 - 6 = 5$

3. $\text{Mean} = \dfrac{2+4+7+11}{4} = \dfrac{24}{4} = 6$

Data item	Deviation: Data item – mean	$(\text{Deviation})^2$: $(\text{Data item–mean})^2$
2	$2 - 6 = -4$	$(-4)^2 = 16$
4	$4 - 6 = -2$	$(-2)^2 = 4$
7	$7 - 6 = 1$	$1^2 = 1$
11	$11 - 6 = 5$	$5^2 = 25$

$\sum(\text{data item–mean})^2 = 46$

$\text{Standard deviation} = \sqrt{\dfrac{46}{4-1}} = \sqrt{\dfrac{46}{3}} \approx 3.92$

4. *Sample A*

$$\text{Mean} = \frac{73+75+77+79+81+83}{6} = \frac{468}{6} = 78$$

Data item	Deviation: Data item − mean	(Deviation)^2 : (Data item−mean)^2
73	$73-78 = -5$	$(-5)^2 = 25$
75	$75-78 = -3$	$(-3)^2 = 9$
77	$77-78 = -1$	$(-1)^2 = 1$
79	$79-78 = 1$	$1^2 = 1$
81	$81-78 = 3$	$3^2 = 9$
83	$83-78 = 5$	$5^2 = 25$
		$\sum(\text{data item−mean})^2 = 70$

$$\text{Standard deviation} = \sqrt{\frac{70}{6-1}} = \sqrt{\frac{70}{5}} \approx 3.74$$

Sample B

$$\text{Mean} = \frac{40+44+92+94+98+100}{6} = \frac{468}{6} = 78$$

Data item	Deviation: Data item − mean	(Deviation)^2 : (Data item−mean)^2
40	$40-78 = -38$	$(-38)^2 = 1444$
44	$44-78 = -34$	$(-34)^2 = 1156$
92	$92-78 = 14$	$14^2 = 196$
94	$94-78 = 16$	$16^2 = 256$
98	$98-78 = 20$	$20^2 = 400$
100	$100-78 = 22$	$22^2 = 484$
		$\sum(\text{data item−mean})^2 = 3936$

$$\text{Standard deviation} = \sqrt{\frac{3936}{6-1}} = \sqrt{\frac{3936}{5}} \approx 28.06$$

5. a. Stocks had a greater return on investment.

   b. Stocks have the greater risk. The high standard deviation indicates that stocks are more likely to lose money.

Exercise Set 12.3

1. $\text{Range} = 5-1 = 4$

3. $\text{Range} = 15-7 = 8$

5. $\text{Range} = 5-3 = 2$

7. Range $= 7 - 1 = 6$

9. Range $= 9 - 5 = 4$

11. **a.**

Data item	Deviation: Data item – mean
3	$3 - 12 = -9$
5	$5 - 12 = -7$
7	$7 - 12 = -5$
12	$12 - 12 = 0$
18	$18 - 12 = 6$
27	$27 - 12 = 15$

b. $-9 - 7 - 5 + 0 + 6 + 15 = 0$

13. **a.**

Data item	Deviation: Data item – mean
29	$29 - 49 = -20$
38	$38 - 49 = -11$
48	$48 - 49 = -1$
49	$49 - 49 = 0$
53	$53 - 49 = 4$
77	$77 - 49 = 28$

b. $-20 - 11 - 1 + 0 + 4 + 28 = 0$

15. **a.** Mean $= \dfrac{85 + 95 + 90 + 85 + 100}{5} = 91$

b.

Data item	Deviation: Data item – mean
85	$85 - 91 = -6$
95	$95 - 91 = 4$
90	$90 - 91 = -1$
85	$85 - 91 = -6$
100	$100 - 91 = 9$

c. $-6 + 4 - 1 - 6 + 9 = 0$

17. **a.** Mean $= \dfrac{146 + 153 + 155 + 160 + 161}{5} = 155$

b.

Data item	Deviation: Data item – mean
146	$146 - 155 = -9$
153	$153 - 155 = -2$
155	$155 - 155 = 0$
160	$160 - 155 = 5$
161	$161 - 155 = 6$

c. $-9 - 2 + 0 + 5 + 6 = 0$

19. a. $\text{Mean} = \dfrac{2.25 + 3.50 + 2.75 + 3.10 + 1.90}{5} = 2.70$

b.

Data item	Deviation: Data item – mean
2.25	$2.25 - 2.70 = -0.45$
3.50	$3.50 - 2.70 = 0.80$
2.75	$2.75 - 2.70 = 0.05$
3.10	$3.10 - 2.70 = 0.40$
1.90	$1.90 - 2.70 = -0.80$

c. $-0.45 + 0.80 + 0.05 + 0.40 - 0.80 = 0$

21. $\text{Mean} = \dfrac{1 + 2 + 3 + 4 + 5}{5} = 3$

Data item	Deviation: Data item – mean	$(\text{Deviation})^2$: $(\text{Data item–mean})^2$
1	$1 - 3 = -2$	$(-2)^2 = 4$
2	$2 - 3 = -1$	$(-1)^2 = 1$
3	$3 - 3 = 0$	$0^2 = 0$
4	$4 - 3 = 1$	$1^2 = 1$
5	$5 - 3 = 2$	$2^2 = 4$

$$\sum (\text{data item–mean})^2 = 10$$

$$\text{Standard deviation} = \sqrt{\dfrac{10}{5-1}} = \sqrt{\dfrac{10}{4}} \approx 1.58$$

23. $\text{Mean} = \dfrac{7 + 9 + 9 + 15}{4} = 10$

Data item	Deviation: Data item – mean	$(\text{Deviation})^2$: $(\text{Data item–mean})^2$
7	$7 - 10 = -3$	$(-3)^2 = 9$
9	$9 - 10 = -1$	$(-1)^2 = 1$
9	$9 - 10 = -1$	$(-1)^2 = 1$
15	$15 - 10 = 5$	$5^2 = 25$

$$\sum (\text{data item–mean})^2 = 36$$

$$\text{Standard deviation} = \sqrt{\dfrac{36}{4-1}} = \sqrt{\dfrac{36}{3}} \approx 3.46$$

25. $\text{Mean} = \dfrac{3+3+4+4+5+5}{6} = 4$

Data item	Deviation: Data item − mean	$(\text{Deviation})^2$: $(\text{Data item–mean})^2$
3	$3-4=-1$	$(-1)^2 = 1$
3	$3-4=-1$	$(-1)^2 = 1$
4	$4-4=0$	$0^2 = 0$
4	$4-4=0$	$0^2 = 0$
5	$5-4=1$	$1^2 = 1$
5	$5-4=1$	$1^2 = 1$

$$\sum (\text{data item–mean})^2 = 4$$

$$\text{Standard deviation} = \sqrt{\dfrac{4}{6-1}} = \sqrt{\dfrac{4}{5}} \approx 0.89$$

27. $\text{Mean} = \dfrac{1+1+1+4+7+7+7}{7} = 4$

Data item	Deviation: Data item − mean	$(\text{Deviation})^2$: $(\text{Data item–mean})^2$
1	$1-4=-3$	$(-3)^2 = 9$
1	$1-4=-3$	$(-3)^2 = 9$
1	$1-4=-3$	$(-3)^2 = 9$
4	$4-4=0$	$0^2 = 0$
7	$7-4=3$	$3^2 = 9$
7	$7-4=3$	$3^2 = 9$
7	$7-4=3$	$3^2 = 9$

$$\sum (\text{data item–mean})^2 = 54$$

$$\text{Standard deviation} = \sqrt{\dfrac{54}{7-1}} = \sqrt{\dfrac{54}{6}} = 3$$

29. Mean $= \dfrac{9+5+9+5+9+5+9+5}{8} = 7$

Data item	Deviation: Data item − mean	(Deviation)^2 : (Data item–mean)^2
9	$9 - 7 = 2$	$2^2 = 4$
5	$5 - 7 = -2$	$(-2)^2 = 4$
9	$9 - 7 = 2$	$2^2 = 4$
5	$5 - 7 = -2$	$(-2)^2 = 4$
9	$9 - 7 = 2$	$2^2 = 4$
5	$5 - 7 = -2$	$(-2)^2 = 4$
9	$9 - 7 = 2$	$2^2 = 4$
5	$5 - 7 = -2$	$(-2)^2 = 4$
		$\sum$ (data item–mean)$^2 = 32$

Standard deviation $= \sqrt{\dfrac{32}{8-1}} = \sqrt{\dfrac{32}{7}} \approx 2.14$

31. *Sample A*

Mean $= \dfrac{6+8+10+12+14+16+18}{7} = 12$

Range $= 18 - 6 = 12$

Data item	Deviation: Data item − mean	(Deviation)^2 : (Data item–mean)^2
6	$6 - 12 = -6$	$(-6)^2 = 36$
8	$8 - 12 = -4$	$(-4)^2 = 16$
10	$10 - 12 = -2$	$(-2)^2 = 4$
12	$12 - 12 = 0$	$0^2 = 0$
14	$14 - 12 = 2$	$2^2 = 4$
16	$16 - 12 = 4$	$4^2 = 16$
18	$18 - 12 = 6$	$6^2 = 36$
		$\sum$ (data item–mean)$^2 = 112$

Standard deviation $= \sqrt{\dfrac{112}{7-1}} = \sqrt{\dfrac{112}{6}} \approx 4.32$

Sample B

$$\text{Mean} = \frac{6 + 7 + 8 + 12 + 16 + 17 + 18}{7} = 12$$

$$\text{Range} = 18 - 6 = 12$$

Data item	Deviation: Data item − mean	$(\text{Deviation})^2$: $(\text{Data item–mean})^2$
6	$6 - 12 = -6$	$(-6)^2 = 36$
7	$7 - 12 = -5$	$(-5)^2 = 25$
8	$8 - 12 = -4$	$(-4)^2 = 16$
12	$12 - 12 = 0$	$0^2 = 0$
16	$16 - 12 = 4$	$4^2 = 16$
17	$17 - 12 = 5$	$5^2 = 25$
18	$18 - 12 = 6$	$6^2 = 36$
		$\sum (\text{data item–mean})^2 = 154$

$$\text{Standard deviation} = \sqrt{\frac{154}{7-1}} = \sqrt{\frac{154}{6}} \approx 5.07$$

Sample C

$$\text{Mean} = \frac{6 + 6 + 6 + 12 + 18 + 18 + 18}{7} = 12$$

$$\text{Range} = 18 - 6 = 12$$

Data item	Deviation: Data item − mean	$(\text{Deviation})^2$: $(\text{Data item–mean})^2$
6	$6 - 12 = -6$	$(-6)^2 = 36$
6	$6 - 12 = -6$	$(-6)^2 = 36$
6	$6 - 12 = -6$	$(-6)^2 = 36$
12	$12 - 12 = 0$	$0^2 = 0$
18	$18 - 12 = 6$	$6^2 = 36$
18	$18 - 12 = 6$	$6^2 = 36$
18	$18 - 12 = 6$	$6^2 = 36$
		$\sum (\text{data item–mean})^2 = 216$

$$\text{Standard deviation} = \sqrt{\frac{216}{7-1}} = \sqrt{\frac{216}{6}} = 6$$

The samples have the same mean and range, but different standard deviations.

33. a. $\text{Mean} = \dfrac{150 + 120 + 50 + 30 + 20}{5} = \dfrac{370}{5} = 74$

b.

Data item	Deviation: Data item – mean	(Deviation)^2 : (Data item–mean)^2
150	$150 - 74 = 76$	$(76)^2 = 5776$
120	$120 - 74 = 46$	$(46)^2 = 2116$
50	$50 - 74 = -24$	$(-24)^2 = 576$
30	$30 - 74 = -44$	$(-44)^2 = 1936$
20	$20 - 74 = -54$	$(-54)^2 = 2916$

$$\sum(\text{data item–mean})^2 = 13,320$$

$$\text{Standard deviation} = \sqrt{\dfrac{13320}{5-1}} = \sqrt{\dfrac{13320}{4}} \approx 57.71$$

35. a. $\text{Mean} = \dfrac{9 + 12 + 2 + 0 + 0 + 7 + 12 + 2 + 9 + 1 + 2}{11} = \dfrac{56}{11} \approx 5.09$

Data item	Deviation: Data item – mean	(Deviation)^2 : (Data item–mean)^2
9	$9 - 5.1 = 3.9$	$3.9^2 = 15.21$
12	$12 - 5.1 = 6.9$	$6.9^2 = 47.61$
2	$2 - 5.1 = -3.1$	$(-3.1)^2 = 9.61$
0	$0 - 5.1 = -5.1$	$(-5.1)^2 = 26.01$
0	$0 - 5.1 = -5.1$	$(-5.1)^2 = 26.01$
7	$7 - 5.1 = 1.9$	$1.9^2 = 3.61$
12	$12 - 5.1 = 6.9$	$6.9^2 = 47.61$
2	$2 - 5.1 = -3.1$	$(-3.1)^2 = 9.61$
9	$9 - 5.1 = 3.9$	$3.9^2 = 15.21$
1	$1 - 5.1 = -4.1$	$(-4.1)^2 = 16.81$
2	$2 - 5.1 = -3.1$	$(-3.1)^2 = 9.61$

$$\sum(\text{data item–mean})^2 = 226.91$$

b. $\text{Standard deviation} = \sqrt{\dfrac{226.91}{11-1}} = \sqrt{\dfrac{226.91}{10}} \approx 4.76$

37-43. Answers will vary.

45. a

47. Answers will vary.

49. Original data:

$$\text{Mean} = \frac{0+1+3+4+4+6}{6} = \frac{18}{6} = 3 \; ; \qquad \text{Standard deviation} = \sqrt{\frac{24}{6-1}} = \sqrt{\frac{24}{5}} \approx 2.19$$

Adjusted data:

$$\text{Mean} = \frac{2+3+5+6+6+8}{6} = \frac{30}{6} = 5 \; ; \qquad \text{Standard deviation} = \sqrt{\frac{24}{6-1}} = \sqrt{\frac{24}{5}} \approx 2.19$$

Adding 2 to each data item raises the mean by 2, but does not affect the standard deviation.

51. 30.28

Check Points 12.4

1. a. Height = $70 + 3 \cdot 2.5 = 70 + 7.5 = 77.5$ in.

   b. Height = $70 - 2 \cdot 2.5 = 70 - 5 = 65$ in.

2. a. 300 is 2 standard deviations below the mean. $500 - 2 \cdot 100 = 300$
     700 is 2 standard deviations above the mean. $500 + 2 \cdot 100 = 700$
     The 68-95-99.7 rule indicates that approximately 95% score between 300 and 700.

   b. 700 is 2 standard deviations above the mean. $500 + 2 \cdot 100 = 700$
     The 68-95-99.7 rule states that approximately 95% score between 2 standard deviations below and 2 standard deviations above the mean. Because the distribution is symmetrical, half of the 95%, or 47.5% score between 500 and 700.

   c. 600 is 1 standard deviation above the mean. $500 + 1 \cdot 100 = 600$
     The 68-95-99.7 rule states that approximately 68% score between 1 standard deviation below and 1 standard deviation above the mean. Therefore, $100\% - 68\% = 32\%$ are farther than 1 standard deviation from the mean. Because the distribution is symmetrical, half of the 32%, or 16% score above 600.

3. a. z-score for $342 = \dfrac{\text{data item} - \text{mean}}{\text{standard deviation}} = \dfrac{342 - 336}{3} = \dfrac{6}{3} = 2$

   b. z-score for $336 = \dfrac{\text{data item} - \text{mean}}{\text{standard deviation}} = \dfrac{336 - 336}{3} = \dfrac{0}{3} = 0$

   c. z-score for $333 = \dfrac{\text{data item} - \text{mean}}{\text{standard deviation}} = \dfrac{333 - 336}{3} = \dfrac{-3}{3} = -1$

4. Find the z-score for each test taken.

   SAT: z-score for $550 = \dfrac{\text{data item} - \text{mean}}{\text{standard deviation}} = \dfrac{550 - 500}{100} = \dfrac{50}{100} = 0.5$

   ACT: z-score for $24 = \dfrac{\text{data item} - \text{mean}}{\text{standard deviation}} = \dfrac{24 - 18}{6} = \dfrac{6}{6} = 1$

   You scored better on the ACT test because the score is 1 standard deviation above the mean. The SAT score is only half a standard deviation above the mean.

5. Score = mean + $2.5 \cdot$ standard deviation = $100 + 2.5(15) = 100 + 37.5 = 137.5$

6. Scoring in the 62nd percentile means that the student did better than about 62% of all those who took the SAT.

7. First, find the z-score. The z-score for $68 = \dfrac{\text{data item} - \text{mean}}{\text{standard deviation}} = \dfrac{68 - 65}{2.5} = \dfrac{3}{2.5} = 1.2$

   Looking up 1.2 in the z-score column of table 12.14 indicates that the percentile is 88.49.
   Therefore, approximately 88.49% of young women are shorter than 68 inches.

8. First, find the z-score. The z-score for $145 = \dfrac{\text{data item} - \text{mean}}{\text{standard deviation}} = \dfrac{145 - 110}{25} = \dfrac{35}{25} = 1.4$

   Looking up 1.4 in the z-score column of table 12.14 indicates that the percentile is 91.92.
   Therefore, approximately 100% − 91.92%, or 8.08% of all IQ scores for this age group are greater than 145.

9. Convert each given data item to a z-score and look up the corresponding percentile in table 12.14.

   z-score for $67 = \dfrac{\text{data item} - \text{mean}}{\text{standard deviation}} = \dfrac{67 - 70}{2.5} = \dfrac{-3}{2.5} = -1.2$, which corresponds to a percentile of 11.51.

   z-score for $74 = \dfrac{\text{data item} - \text{mean}}{\text{standard deviation}} = \dfrac{74 - 70}{2.5} = \dfrac{4}{2.5} = 1.6$, which corresponds to a percentile of 94.52.

   $94.52 - 11.51 = 83.01$. Therefore, approximately 83.01% of all young men have heights between 67 in. and 74 in.

10. a. Margin of error $= \pm \dfrac{1}{\sqrt{1082}} \approx \pm 0.03 = \pm 3.0\%$

    b. Note: $64\% - 3\% = 61\%$ and $64\% + 3\% = 67\%$. We can be 95% confident that between 61% and 67% of all adult Americans think that raising airplane ticket prices $50 to increase security is acceptable.

Exercise Set 12.4

1. Score $= 100 + 1 \cdot 20 = 100 + 20 = 120$

3. Score $= 100 + 3 \cdot 20 = 100 + 60 = 160$

5. Score $= 100 + 2.5(20) = 100 + 50 = 150$

7. Score $= 100 - 2 \cdot 20 = 100 - 40 = 60$

9. Score $= 100 - 0.5(20) = 100 - 10 = 90$

11. $16,500 is 1 standard deviation below the mean and $17,500 is 1 standard deviation above the mean. The Rule and the figure indicate that 68% of the buyers paid between $16,500 and $17,500.

13. $17,500 is 1 standard deviation above the mean. 68% of the buyers paid between $16,500 and $17,500.

    Because of symmetry, the percent that paid between $17,000 and $17,500 is $\dfrac{1}{2}(68\%) = 34\%$.

15. $16,000 is 2 standard deviations below the mean. 95% of the buyers paid between $16,000 and $18,000.

    Because of symmetry, the percent that paid between $16,000 and $17,000 is $\dfrac{1}{2}(95\%) = 47.5\%$.

17. $15,500 is 3 standard deviations below the mean. 99.7% of the buyers paid between $15,500 and $18,500.

    Because of symmetry, the percent that paid between $15,500 and $17,000 is $\dfrac{1}{2}(99.7\%) = 49.85\%$.

19. $17,500 is 1 standard deviation above the mean. Since 68% of the data items fall within 1 standard deviation of the mean, $100\% - 68\% = 32\%$ fall farther than 1 standard deviation from the mean.

    Because of symmetry, the percent that paid more than $17,500 is $\dfrac{1}{2}(32\%) = 16\%$.

21. $16,000 is 2 standard deviations below the mean. Since 95% of the data items fall within 2 standard deviations of the mean, 100% − 95% = 5% fall farther than 2 standard deviations from the mean. Because of symmetry, the percent that paid less than $16,000 is $\frac{1}{2}(5\%) = 2.5\%$.

23. $530 - 2 \cdot 128 = 274$
$530 + 2 \cdot 128 = 786$
95% score between 274 and 786.

25. From Exercise 23, we know 95% score between 274 and 786. By symmetry, $\frac{1}{2}(95\%) = 47.5\%$ score between 274 and 530.

27. $530 + 1 \cdot 128 = 658$
68% score within 1 standard deviation of the mean, so 100% − 68% = 32% of the scores are farther than 1 standard deviation from the mean. Because of symmetry, the percent that score above 658 is $\frac{1}{2}(32\%) = 16\%$.

29. From Exercise 23, we know 95% score between 274 and 786. That means that 100% − 95% = 5% score below 274 or above 786. Because of symmetry, the percent that score below 274 is $\frac{1}{2}(5\%) = 2.5\%$.

31. $530 + 3 \cdot 128 = 914$
99.7% score within 3 standard deviations of the mean, so 100% − 99.7% = 0.3% of the scores are farther than 3 standard deviations from the mean. Because of symmetry, the percent that score above 914 is $\frac{1}{2}(0.3\%) = 0.15\%$.

33. z-score for $68 = \dfrac{68 - 60}{8} = \dfrac{8}{8} = 1$

35. z-score for $84 = \dfrac{84 - 60}{8} = \dfrac{24}{8} = 3$

37. z-score for $64 = \dfrac{64 - 60}{8} = \dfrac{4}{8} = 0.5$

39. z-score for $74 = \dfrac{74 - 60}{8} = \dfrac{14}{8} = 1.75$

41. z-score for $60 = \dfrac{60 - 60}{8} = \dfrac{0}{8} = 0$

43. z-score for $52 = \dfrac{52 - 60}{8} = \dfrac{-8}{8} = -1$

45. z-score for $48 = \dfrac{48 - 60}{8} = \dfrac{-12}{8} = -1.5$

47. z-score for $34 = \dfrac{34 - 60}{8} = \dfrac{-26}{8} = -3.25$

49. z-score for $290 = \dfrac{290 - 266}{16} = \dfrac{24}{16} = 1.5$

51. z-score for $302 = \dfrac{302 - 266}{16} = \dfrac{36}{16} = 2.25$

53. z-score for $258 = \dfrac{258 - 266}{16} = \dfrac{-8}{16} = -0.5$

55. z-score for $242 = \dfrac{242 - 266}{16} = \dfrac{-24}{16} = -1.5$

57. math test:
z-score for $230 = \dfrac{230 - 200}{10} = \dfrac{30}{10} = 3$
reading test:
z-score for $540 = \dfrac{540 - 500}{15} = \dfrac{40}{15} \approx 2.7$
The student had the better score on the math test.

59. $2 \cdot 50 = 100$
The data item is 100 units above the mean.
$400 + 100 = 500$

61. $1.5(50) = 75$
The data item is 75 units above the mean.
$400 + 75 = 475$

63. $-3 \cdot 50 = -150$
The data item is 150 units below the mean.
$400 - 150 = 250$

65. $-2.5(50) = -125$
The data item is 125 units below the mean.
$400 - 125 = 275$

67. a. 72.57%

    b. 100% − 72.57% = 27.43%

69. a. 88.49%

b. 100% − 88.49% = 11.51%

71. a. 24.20%

b. 100% − 24.20% = 75.8%

73. a. 11.51%

b. 100% − 11.51% = 88.49%

75. $z = 0.2 \rightarrow 57.93\%$
$z = 1.4 \rightarrow 91.92\%$
$91.92\% - 57.93\% = 33.99\%$

77. $z = 1 \rightarrow 84.13\%$
$z = 3 \rightarrow 99.87\%$
$99.87\% - 84.13\% = 15.74\%$

79. $z = -1.5 \rightarrow 6.68\%$
$z = 1.5 \rightarrow 93.32\%$
$93.32\% - 6.68\% = 86.64\%$

81. $z = -2 \rightarrow 2.28\%$
$z = -0.5 \rightarrow 30.85\%$
$30.85\% - 2.28\% = 28.57\%$

83. z-score for $650 = \dfrac{650 - 500}{100} = \dfrac{150}{100} = 1.5$

$z = 1.5 \rightarrow 93.32\%$
93.32% score below 650.

85. z-score for $560 = \dfrac{560 - 500}{100} = \dfrac{60}{100} = 0.6$

$z = 0.6 \rightarrow 72.57\%$
$100\% - 72.57\% = 27.43\%$ score above 560.

87. z-score for $380 = \dfrac{380 - 500}{100} = \dfrac{-120}{100} = -1.2$

$z = -1.2 \rightarrow 11.51\%$
$100\% - 11.51\% = 88.49\%$ score above 380.

89. z-score for $640 = \dfrac{640 - 500}{100} = \dfrac{140}{100} = 1.4$

$z = 1.4 \rightarrow 91.92\%$

z-score for $710 = \dfrac{710 - 500}{100} = \dfrac{210}{100} = 2.1$

$z = 2.1 \rightarrow 98.21\%$
$98.21\% - 91.92\% = 6.29\%$ score between 640 and 710.

91. z-score for $440 = \dfrac{440 - 500}{100} = \dfrac{-60}{100} = -0.6$

$z = -0.6 \rightarrow 27.43\%$

z-score for $560 = \dfrac{560 - 500}{100} = \dfrac{60}{100} = 0.6$

$z = 0.6 \rightarrow 72.57\%$
$72.57\% - 27.43\% = 45.14\%$ score between 440 and 560.

93. z-score for $25.8 = \dfrac{25.8 - 22.5}{2.2} = 1.5$

$z = 1.5 \rightarrow 93.32\%$
$100\% - 93.32\% = 6.68\%$ weigh more than 25.8 pounds.

95. z-score for $19.2 = \dfrac{19.2 - 22.5}{2.2} = -1.5$

$z = -1.5 \rightarrow 6.68\%$

z-score for $21.4 = \dfrac{21.4 - 22.5}{2.2} = -0.5$

$z = -0.5 \rightarrow 30.85\%$
$30.85\% - 6.68\% = 24.17\%$ weigh between 19.2 and 21.4 pounds.

97. a. margin of error $= \pm \dfrac{1}{\sqrt{397}} \approx \pm 0.050 \approx \pm 5.0\%$

b. 26% − 5% = 21%
26% + 5% = 31%
We can be 95% confident that between 21% and 31% of all parents feel that crime is a bad thing about being a kid.

99. a. margin of error $= \pm \dfrac{1}{\sqrt{4000}} \approx \pm 0.016 \approx \pm 1.6\%$

b. 60.2% − 1.6% = 58.6%
60.2% + 1.6% = 61.8%
We can be 95% confident that between 58.6% and 61.8% of all TV households watched the final episode of M*A*S*H.

101. new margin of error $= \pm \dfrac{1}{\sqrt{5000}} \approx \pm 0.014 \approx \pm 1.4\%$

improvement = 1.6% − 1.4% = 0.2%

103. No, explanations will vary.

105-117. Answers will vary.

119. $\dfrac{1}{400} = 0.0025 = 0.25\%$

A z-score of -2.8 has 0.26% of the data items below it. So find the height corresponding to $z = -2.8$.
$69 - 2.8(2.5) = 62$ inches. Therefore, the woman is 62 inches tall.

Check Points 12.5

1. 0.51 would indicate a moderate correlation between the two.

2.

x	y	xy	x^2	y^2
2.5	211	527.5	6.25	44,521
3.9	167	651.3	15.21	27,889
2.9	131	379.9	8.41	17,161
2.4	191	458.4	5.76	36,481
2.9	220	638	8.41	48,400
0.8	297	237.6	0.64	88,209
9.1	71	646.1	82.81	5041
0.8	211	168.8	0.64	44,521
0.7	300	210	0.49	90,000
7.9	107	845.3	62.41	11,449
1.8	167	300.6	3.24	27,889
1.9	266	505.4	3.61	70,756
0.8	227	181.6	0.64	51,529
6.5	86	559	42.25	7396
1.6	207	331.2	2.56	42,849
5.8	115	667	33.64	13,225
1.3	285	370.5	1.69	81,225
1.2	199	238.8	1.44	39,601
2.7	172	464.4	7.29	29,584

$$\sum x = 57.5 \qquad \sum y = 3630 \qquad \sum xy = 8381.4 \qquad \sum x^2 = 287.39 \qquad \sum y^2 = 777,726$$

$$\left(\sum x\right)^2 = (57.5)^2 = 3306.25 \text{ and } \left(\sum y\right)^2 = (3630)^2 = 13,176,900$$

$$r = \dfrac{19(8381.4) - (57.5)(3630)}{\sqrt{19(287.39) - 3306.25}\sqrt{19(777,726) - 13,176,900}} = \dfrac{-49,478.4}{\sqrt{2154.16}\sqrt{1599894}} \approx -0.84$$

This value for r is fairly close to -1 and indicates a strong negative correlation. This means the wins a person drinks, the less likely the person is to die from heart disease.

3. $m = \dfrac{19(8381.4) - (57.5)(3630)}{19(287.39) - 3306.25} = \dfrac{-49,478.4}{2154.16} \approx -22.97$

$b = \dfrac{3630 - (-22.97)(57.5)}{19} = \dfrac{4950.775}{19} \approx 260.56$

The equation of the regression line is $y = -22.97x + 260.56$.

The predicted heart disease death rate in a country where adults average 10 liters of alcohol per person per year can be found by substituting 10 for x.
$y = -22.97x + 260.56 = -22.97(10) + 260.56 = 30.86$

4. Yes, $|r| = 0.84$. Since $0.84 > 0.456$ and 0.575 (using table 12.16), we may conclude that a correlation does exist.

Exercise Set 12.5

1. There appears to be a positive correlation.

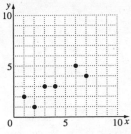

3. There appears to be a negative correlation.

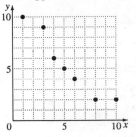

5. There appears to be a positive correlation.

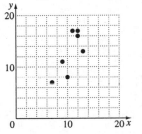

7. There appears to be a positive correlation.

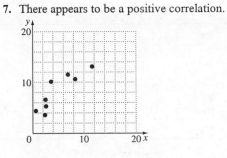

9. False; the correlation is negative.

11. True

13. True

15. False; see for example, Syria and Vietnam.

17. False; there is some positive correlation.

19. True

21. False; very few of the points lie on the regression line.

23. False

25. False; both correlations are negative.

27. True

29. a

31. d

33.

x	y	xy	x^2	y^2
1	2	2	1	4
6	5	30	36	25
4	3	12	16	9
3	3	9	9	9
7	4	28	49	16
2	1	2	4	1
$\sum x = 23$	$\sum y = 18$	$\sum xy = 83$	$\sum x^2 = 115$	$\sum y^2 = 64$

$$\left(\sum x\right)^2 = (23)^2 = 529 \text{ and } \left(\sum y\right)^2 = (18)^2 = 324$$

$$r = \frac{6(83) - (23)(18)}{\sqrt{6(115) - 529}\sqrt{6(64) - (324)}}$$

$$= \frac{84}{\sqrt{161}\sqrt{60}}$$

$$\approx 0.855$$

35.

x	y	xy	x^2	y^2
8	2	16	64	4
6	4	24	36	16
1	10	10	1	100
5	5	25	25	25
4	6	24	16	36
10	2	20	100	4
3	9	27	9	81

$\sum x = 37 \quad \sum y = 38 \quad \sum xy = 146 \quad \sum x^2 = 251 \quad \sum y^2 = 266$

$\left(\sum x\right)^2 = (37)^2 = 1369$ and $\left(\sum y\right)^2 = (38)^2 = 1444$

$r = \dfrac{7(146) - (37)(38)}{\sqrt{7(251) - 1369}\sqrt{7(266) - 1444}}$

$= \dfrac{-384}{\sqrt{388}\sqrt{418}}$

≈ -0.954

37. a.

x	y	xy	x^2	y^2
13	13	169	169	169
9	11	99	81	121
7	7	49	49	49
12	16	192	144	256
12	17	204	144	289
10	8	80	100	64
11	17	187	121	289

$\sum x = 74 \quad \sum y = 89 \quad \sum xy = 980 \quad \sum x^2 = 808 \quad \sum y^2 = 1237$

$\left(\sum x\right) = (74)^2 = 5476$ and $\left(\sum y\right)^2 = (89)^2 = 7921$

$r = \dfrac{7(980) - (74)(89)}{\sqrt{7(808) - 5476}\sqrt{7(1237) - 7921}}$

$= \dfrac{274}{\sqrt{180}\sqrt{738}}$

≈ 0.75

b. $m = \dfrac{7(980) - (74)(89)}{7(808) - 5476} = \dfrac{274}{180} \approx 1.52$

$b = \dfrac{89 - 1.52(74)}{7} = \dfrac{-23.48}{7} \approx -3.38$

$y = 1.52x - 3.38$

c. $x = 16$

$y = 1.52(16) - 3.38 = 20.98$

21 years

39. a.

x	y	xy	x^2	y^2
11.6	13.1	151.96	134.56	171.61
8.3	10.6	87.98	68.89	112.36
6.9	11.5	79.35	47.61	132.25
3.6	10.1	36.36	12.96	102.01
2.6	5.3	13.78	6.76	28.09
2.5	6.6	16.50	6.25	43.56
2.4	3.6	8.64	5.76	12.96
0.6	4.4	2.64	0.36	19.36

$\sum x = 38.5$ $\sum y = 65.2$ $\sum xy = 397.21$ $\sum x^2 = 283.15$ $\sum y^2 = 622.2$

$\left(\sum x\right)^2 = (38.5)^2 = 1482.25$ and $\left(\sum y\right)^2 = (65.2)^2 = 4251.04$

$$r = \frac{8(397.21) - (38.5)(65.2)}{\sqrt{8(283.15) - 1482.25}\sqrt{8(622.2) - 4251.04}}$$

$$= \frac{667.48}{\sqrt{782.95}\sqrt{726.56}}$$

$$\approx 0.885$$

b. $m = \dfrac{8(397.21) - (38.5)(65.2)}{8(283.15) - 1482.25} = \dfrac{667.48}{782.95} \approx 0.8525 \approx 0.85$

$b = \dfrac{65.2 - 0.8525(38.5)}{8} = \dfrac{32.378}{8} \approx 4.05$

$y = 0.85x + 4.05$

c. $x = 14$

$y = 0.85(14) + 4.05 = 15.95$

16 murders per 100,000 people

41. $|r| = 0.5$;   Since $0.5 > 0.444$, conclude that a correlation does exist.

43. $|r| = 0.5$;   Since $0.5 < 0.576$, conclude that a correlation does not exist.

45. $|r| = 0.351$;   Since $0.351 > 0.232$, conclude that a correlation does exist.

47. $|r| = 0.37$;   Since $0.37 < 0.444$, conclude that a correlation does not exist.

49-59. Answers will vary.

61. Answers will vary.

63.

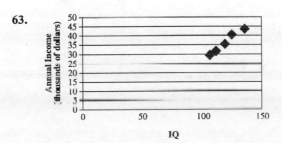

Chapter 12 Review Exercises

1. The population is the set containing all the American adults earning $100,000 or more each year.

2. The population is the set containing all the rental apartments in San Francisco.

3. a

4.

Time Spent on Homework (in hours)	Number of students
6	1
7	3
8	3
9	2
10	1
	10

5.

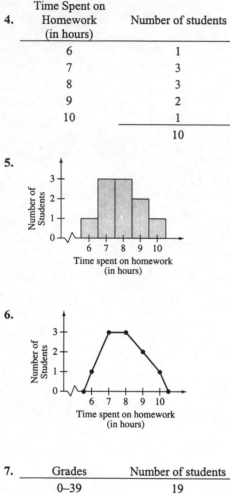

6.

7.

Grades	Number of students
0–39	19
40–49	8
50–59	6
60–69	6
70–79	5
80–89	3
90–100	3
	50

8.

Stems	Leaves
1	3 4 1 3 7 8
2	4 9 6 9 2 7
3	4 9 6 5 1 1 1
4	4 0 2 7 9 1 2 5
5	7 9 6 4 0 1
6	3 3 7 0 8 9
7	2 3 4 0 5
8	7 1 6
9	5 1 0

9. Answers will vary.

10. Mean $= \dfrac{84+90+95+89+98}{5}$

$= \dfrac{456}{5}$

$= 91.2$

11. Mean $= \dfrac{33+27+9+10+6+7+11+23+27}{9}$

$= \dfrac{153}{9}$

$= 17$

12. Mean $= \dfrac{1\cdot 2 + 2\cdot 4 + 3\cdot 3 + 4\cdot 1}{10}$

$= \dfrac{2+8+9+4}{10}$

$= \dfrac{23}{10}$

$= 2.3$

13. First arrange the data items from smallest to largest.
6, 7, 9, 10, <u>11</u>, 23, 27, 27, 33
There is an odd number of data items, so the median is the middle number. The median is 11.

14. First arrange the data items from smallest to largest.
16, 22, <u>28</u>, 28, 34
There is an odd number of data items, so the median is the middle number. The median is 28.

15. The median is the value in the
$\dfrac{n+1}{2} = \dfrac{10+1}{2} = \dfrac{11}{2} = 5.5$ position, which means the median is the mean of the 5th and 6th values. The 5th and 6th values are both 2, therefore the median is 2.

16. The number 27 occurs most frequently, so the mode is 27.

17. No single value occurs most frequently.
There is no mode.

18. The number 2 occurs most frequently,
so the mode is 2.

19. lowest data value = 84,
highest data value = 98
Midrange $= \dfrac{84+98}{2} = \dfrac{182}{2} = 91$

20. lowest data value = 6,
highest data value = 33
Midrange $= \dfrac{6+33}{2} = \dfrac{39}{2} = 19.5$

21. lowest data value = 1,
highest data value = 4
Midrange $= \dfrac{1+4}{2} = \dfrac{5}{2} = 2.5$

22-23. Answers will vary.

24. a.

Age at first inauguration	Number of Presidents
42	1
43	1
44	0
45	0
46	2
47	1
48	1
49	2
50	1
51	5
52	2
53	0
54	4
55	4
56	3
57	4
58	1
59	0
60	1
61	3
62	1
63	0
64	2
65	1
66	0
67	0
68	1
69	1
	42

b. $\text{Mean} = \dfrac{\left(\begin{array}{l}42\cdot1+43\cdot1+46\cdot2+47\cdot1+48\cdot1+49\cdot2+50\cdot1+51\cdot5+52\cdot2+54\cdot4+55\cdot4 \\ +56\cdot3+57\cdot4+58\cdot1+60\cdot1+61\cdot3+62\cdot1+64\cdot2+65\cdot1+68\cdot1+69\cdot1\end{array}\right)}{42} = \dfrac{2304}{42} \approx 54.86 \text{ years}$

The median is the value in the $\dfrac{n+1}{2} = \dfrac{42+1}{2} = \dfrac{43}{2} = 21.5$ position, which means the median is the mean of the 21st and 22nd values.

$\text{Median} = \dfrac{55+55}{2} = \dfrac{110}{2} = 55 \text{ years}$

The model age is 51 years.

$\text{Midrange} = \dfrac{42+69}{2} = 55.5 \text{ years}$

25. Range $= 34 - 16 = 18$

26. Range $= 783 - 219 = 564$

27. a.

Data item	Deviation: Data item – mean
29	$29 - 35 = -6$
9	$9 - 35 = -26$
8	$8 - 35 = -27$
22	$22 - 35 = -13$
46	$46 - 35 = 11$
51	$51 - 35 = 16$
48	$48 - 35 = 13$
42	$42 - 35 = 7$
53	$53 - 35 = 18$
42	$42 - 35 = 7$

b. $-6 - 26 - 27 - 13 + 11 + 16 + 13 + 7 + 18 + 7 = 0$

28. a. $\text{Mean} = \dfrac{36+26+24+90+74}{5} = \dfrac{250}{5} = 50$

b.

Data item	Deviation: Data item – mean
36	$36 - 50 = -14$
26	$26 - 50 = -24$
24	$24 - 50 = -26$
90	$90 - 50 = 40$
74	$74 - 50 = 24$

c. $-14 - 24 - 26 + 40 + 24 = 0$

29. Mean $= \dfrac{3+3+5+8+10+13}{6} = \dfrac{42}{6} = 7$

Data item	Deviation: Data item − mean	(Deviation)^2: (Data item−mean)^2
3	$3 - 7 = -4$	$(-4)^2 = 16$
3	$3 - 7 = -4$	$(-4)^2 = 16$
5	$5 - 7 = -2$	$(-2)^2 = 4$
8	$8 - 7 = 1$	$1^2 = 1$
10	$10 - 7 = 3$	$3^2 = 9$
13	$13 - 7 = 6$	$6^2 = 36$

$\sum(\text{data item−mean})^2 = 82$

Standard deviation $= \sqrt{\dfrac{82}{6-1}} = \sqrt{\dfrac{82}{5}} \approx 4.05$

30. Mean $= \dfrac{20+27+23+26+28+32+33+35}{8} = \dfrac{224}{8} = 28$

Data item	Deviation: Data item − mean	(Deviation)^2: (Data item−mean)^2
20	$20 - 28 = -8$	$(-8)^2 = 64$
27	$27 - 28 = -1$	$(-1)^2 = 1$
23	$23 - 28 = -5$	$(-5)^2 = 25$
26	$26 - 28 = -2$	$(-2)^2 = 4$
28	$28 - 28 = 0$	$0^2 = 0$
32	$32 - 28 = 4$	$4^2 = 16$
33	$33 - 28 = 5$	$5^2 = 25$
35	$35 - 28 = 7$	$7^2 = 49$

$\sum(\text{data item−mean})^2 = 184$

Standard deviation $= \sqrt{\dfrac{184}{8-1}} = \sqrt{\dfrac{184}{7}} \approx 5.13$

31. Mean $= \dfrac{10 + 30 + 37 + 40 + 43 + 44 + 45 + 69 + 86 + 86}{10} = \dfrac{490}{10} = 49$

Range $= 86 - 10 = 76$

Data item	Deviation: Data item − mean	(Deviation)^2: (Data item−mean)^2
10	$10 - 49 = -39$	$(-39)^2 = 1521$
30	$30 - 49 = -19$	$(-19)^2 = 361$
37	$37 - 49 = -12$	$(-12)^2 = 144$
40	$40 - 49 = -9$	$(-9)^2 = 81$
43	$43 - 49 = -6$	$(-6)^2 = 36$
44	$44 - 49 = -5$	$(-5)^2 = 25$
45	$45 - 49 = -4$	$(-4)^2 = 16$
69	$69 - 49 = 20$	$20^2 = 400$
86	$86 - 49 = 37$	$37^2 = 1369$
86	$86 - 49 = 37$	$37^2 = 1369$

$$\sum (\text{data item−mean})^2 = 5322$$

Standard deviation $= \sqrt{\dfrac{5322}{10-1}} = \sqrt{\dfrac{5322}{9}} \approx 24.32$

32. Set A:

Mean $= \dfrac{80 + 80 + 80 + 80}{4} = \dfrac{320}{4} = 80$

Data item	Deviation: Data item − mean	(Deviation)^2: (Data item−mean)^2
80	$80 - 80 = 0$	$0^2 = 0$
80	$80 - 80 = 0$	$0^2 = 0$
80	$80 - 80 = 0$	$0^2 = 0$
80	$80 - 80 = 0$	$0^2 = 0$

$$\sum (\text{data item−mean})^2 = 0$$

Standard deviation $= \sqrt{\dfrac{0}{4-1}} = \sqrt{\dfrac{0}{3}} = 0$

Set B:

$$\text{Mean} = \frac{70 + 70 + 90 + 90}{4} = \frac{320}{4} = 80$$

Data item	Deviation: Data item – mean	(Deviation)^2: (Data item–mean)^2
70	$70 - 80 = -10$	$(-10)^2 = 100$
70	$70 - 80 = -10$	$(-10)^2 = 100$
90	$90 - 80 = 10$	$10^2 = 100$
90	$90 - 80 = 10$	$10^2 = 100$

$$\sum (\text{data item–mean})^2 = 400$$

$$\text{Standard deviation} = \sqrt{\frac{400}{4-1}} = \sqrt{\frac{400}{3}} \approx 11.55$$

Written descriptions of the similarities and differences between the two sets of data will vary.

33. Answers will vary.

34. $70 + 2 \cdot 8 = 70 + 16 = 86$

35. $70 + 3.5(8) = 70 + 28 = 98$

36. $70 - 1.25(8) = 70 - 10 = 60$

37. 64 is one standard deviation below the mean and 72 is one standard deviation above the mean, so 68% of the people in the retirement community are between 64 and 72 years old.

38. 60 is two standard deviations below the mean and 76 is two standard deviations above the mean, so 95% of the people in the retirement community are between 60 and 76 years old.

39. 68 is the mean and 72 is one standard deviation above the mean, so half of 68%, or 34% of the people in the retirement community are between 68 and 72 years old.

40. 56 is three standard deviations below the mean and 80 is three standard deviations above the mean, so 99.7% of the people in the retirement community are between 56 and 80 years old.

41. 72 is one standard deviation above the mean, so 16% of the people in the retirement community are over 72 years old. (Note: 100% – 68% – 32%, half of 32% is 16%).

42. 72 is one standard deviation above the mean, so 84% of the people in the retirement community are under 72 years old. (Note: Question #41 showed that 16% is above 72, 100% – 16% = 84%)

43. 76 is two standard deviations above the mean, so 2.5% of the people in the retirement community are over 76 years old. (Note: 100% – 95% = 5%, half of 5% is 2.5%).

44. z-score for $50 = \frac{50 - 50}{5} = \frac{0}{5} = 0$

45. z-score for $60 = \frac{60 - 50}{5} = \frac{10}{5} = 2$

46. z-score for $58 = \dfrac{58-50}{5} = \dfrac{8}{5} = 1.6$

47. z-score for $35 = \dfrac{35-50}{5} = \dfrac{-15}{5} = -3$

48. z-score for $44 = \dfrac{44-50}{5} = \dfrac{-6}{5} = -1.2$

49. vocabulary test:

z-score for $60 = \dfrac{60-50}{5} = \dfrac{10}{5} = 2$

grammar test:

z-score for $80 = \dfrac{80-72}{6} = \dfrac{8}{6} \approx 1.3$

The student scored better on the vocabulary test.

50. $1.5(4000) = 6000$
$32{,}000 + 6000 = 38{,}000$ miles

51. $2.25(4000) = 9000$
$32{,}000 + 9000 = 41{,}000$ miles

52. $-2.5(4000) = -10{,}000$
$32{,}000 - 10{,}000 = 22{,}000$ miles

53. z-score for $221 = \dfrac{221-200}{15} = \dfrac{21}{15} = 1.4$

$z = 1.4 \rightarrow 91.92\%$
91.92% have cholesterol less than 221.

54. z-score for $173 = \dfrac{173-200}{15} = \dfrac{-27}{15} = -1.8$

$z = -1.8 \rightarrow 3.59\%$
$100\% - 3.59\% = 96.41\%$ have cholesterol greater than 173.

55. z-score for $173 = \dfrac{173-200}{15} = \dfrac{-27}{15} = -1.8$

$z = -1.8 \rightarrow 3.59\%$

z-score for $221 = \dfrac{221-200}{15} = \dfrac{21}{15} = 1.4$

$z = 1.4 \rightarrow 91.92\%$
$91.92\% - 3.59\% = 88.33\%$ have cholesterol between 173 and 221.

56. z-score for $164 = \dfrac{164-200}{15} = \dfrac{-36}{15} = -2.4$

z-score $\rightarrow 0.82\%$

z-score for $182 = \dfrac{182-200}{15} = \dfrac{-18}{15} = -1.2$

$z = -1.2 \rightarrow 11.51\%$
$11.51\% - 0.82\% = 10.69\%$ have cholesterol between 164 and 182.

57. 75%

58. 100% − 86% = 14%

59. 86% − 75% = 11%

60. a. margin of error $= \pm\dfrac{1}{\sqrt{5000}}$

$$\approx \pm 0.014$$
$$\approx \pm 1.4\%$$

b. We can be 95% confident that between 8.4% and 11.2% of all TV households watch *Jeopardy*.

61. There appears to be a positive correlation.

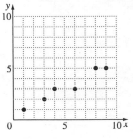

62. There appears to be a negative correlation.

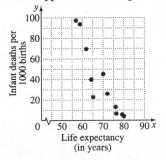

63. False; the correlation is only moderate.

64. True

65. False

66. False; data points that are vertically aligned dispute this statement.

67. True

68. False; there is a moderate negative correlation.

69. True

70. c

71. a.

x	y	xy	x^2	y^2
1	1	1	1	1
3	2	6	9	4
4	3	12	16	9
6	3	18	36	9
8	5	40	64	25
9	5	45	81	25

$\sum x = 31$ $\sum y = 19$ $\sum xy = 122$ $\sum x^2 = 207$ $\sum y^2 = 73$

$$\left(\sum x\right)^2 = (31)^2 = 961 \text{ and } \left(\sum y\right)^2 = (19)^2 = 361$$

$$r = \frac{6(122) - (31)(19)}{\sqrt{6(207) - 961}\sqrt{6(73) - 361}} = \frac{143}{\sqrt{281}\sqrt{77}} \approx 0.972$$

b.

$$m = \frac{6(122) - (31)(19)}{6(207) - 961} = \frac{143}{281} \approx 0.509$$

$$b = \frac{19 - (0.509)(31)}{6} = \frac{3.221}{6} \approx 0.537$$

$$y = 0.509x + 0.537$$

72. a.

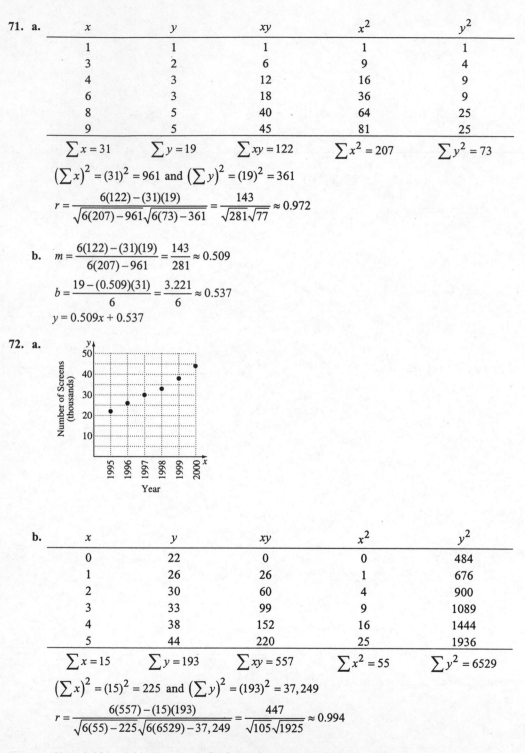

b.

x	y	xy	x^2	y^2
0	22	0	0	484
1	26	26	1	676
2	30	60	4	900
3	33	99	9	1089
4	38	152	16	1444
5	44	220	25	1936

$\sum x = 15$ $\sum y = 193$ $\sum xy = 557$ $\sum x^2 = 55$ $\sum y^2 = 6529$

$$\left(\sum x\right)^2 = (15)^2 = 225 \text{ and } \left(\sum y\right)^2 = (193)^2 = 37,249$$

$$r = \frac{6(557) - (15)(193)}{\sqrt{6(55) - 225}\sqrt{6(6529) - 37,249}} = \frac{447}{\sqrt{105}\sqrt{1925}} \approx 0.994$$

c. Since $0.994 > 0.811$, we can, indeed, conclude that there is a correlation.

Chapter 12 Test

1. d

2.

Score	Frequency
3	1
4	2
5	3
6	2
7	2
8	3
9	2
10	1
	16

3.

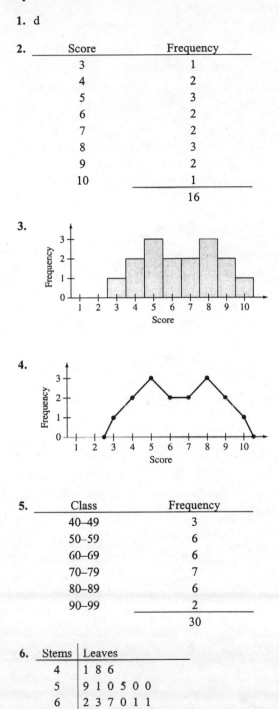

4.

5.

Class	Frequency
40–49	3
50–59	6
60–69	6
70–79	7
80–89	6
90–99	2
	30

6.

Stems	Leaves
4	1 8 6
5	9 1 0 5 0 0
6	2 3 7 0 1 1
7	9 3 1 5 8 9 1
8	8 9 9 1 3 0
9	0 3

7. Mean $= \dfrac{3+6+2+1+7+3}{6} = \dfrac{22}{6} \approx 3.67$

8. First arrange the numbers from smallest to largest.
   1, 2, 3, 3, 6, 7
   There is an even number of data items, so the median is the mean of the middle two data values.

   Median $= \dfrac{3+3}{2} = \dfrac{6}{2} = 3$

9. lowest data value $= 1$
   highest data value $= 7$

   Midrange $= \dfrac{1+7}{2} = \dfrac{8}{2} = 4$

10.

Data item	Deviation: Data item – mean	(Deviation)^2 : (Data item–mean)^2
3	$3 - 3.7 = -0.7$	$(-0.7)^2 = 0.49$
6	$6 - 3.7 = 2.3$	$(2.3)^2 = 5.29$
2	$2 - 3.7 = -1.7$	$(-1.7)^2 = 2.89$
1	$1 - 3.7 = -2.7$	$(-2.7)^2 = 7.29$
7	$7 - 3.7 = 3.3$	$(3.3)^2 = 10.89$
3	$3 - 3.7 = -0.7$	$(-0.7)^2 = 0.49$

$$\sum(\text{data item–mean})^2 = 27.34$$

Standard deviation $= \sqrt{\dfrac{27.34}{6-1}} = \sqrt{\dfrac{27.34}{5}} \approx 2.34$

11. Mean $= \dfrac{1 \cdot 3 + 2 \cdot 5 + 3 \cdot 2 + 4 \cdot 2}{12}$

    $= \dfrac{3 + 10 + 6 + 8}{12}$

    $= \dfrac{27}{12}$

    $= 2.25$

12. The median is in the $\dfrac{n+1}{2} = \dfrac{12+1}{2} = \dfrac{13}{2} = 6.5$ position, which means the median is the mean of the values in the 6th and 7th positions.

    Median $= \dfrac{2+2}{2} = \dfrac{4}{2} = 2$

13. Mode $= 2$

14. Answers will vary.

15. $7 + 1(5.3) = 12.3$

68% of the data values are within 1 standard deviation of the mean. Because of symmetry, $\frac{1}{2}(68\%) = 34\%$ of college freshmen study between 7 and 12.3 hours per week.

16. $7 + 2(5.3) = 17.6$
95% of the data values are within 2 standard deviations of the mean. $100\% - 95\% = 5\%$ of the values are farther than 2 standard deviations from the mean. Because of symmetry, $\frac{1}{2}(5\%) = 2.5\%$ of college freshmen study more than 17.6 hours per week.

17. student:
$$z\text{-score for } 120 = \frac{120 - 100}{10} = \frac{20}{10} = 2$$
professor:
$$z\text{-score for } 128 = \frac{128 - 100}{15} = \frac{28}{15} \approx 1.9$$
The student scored better, because the student's z-score is higher.

18. $z\text{-score for } 88 = \dfrac{88 - 74}{10} = \dfrac{14}{10} = 1.4$
$z = 1.4 \rightarrow 91.92\%$
$100\% - 91.92\% = 8.08\%$ of the scores are above 88.

19. $49\% - 8\% = 41\%$

20. a. margin of error $= \pm \dfrac{1}{\sqrt{n}}$
$$= \pm \frac{1}{\sqrt{100}}$$
$$= \pm 0.1$$
$$= \pm 10\%$$

b. We can be 95% confident that between 50% and 70% of all students are very satisfied with their professors.

21. There appears to be a strong negative correlation.

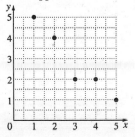

22. False; Though the data shows that there is a <u>correlation</u>, it does not prove <u>causation</u>.

23. False

24. True

25. Answers will vary.

Chapter 13
Mathematical Systems

Check Points 13.1

1. $O + O = E$. This means that the sum of two odd numbers is an even number.

2. No, the set is not closed under addition.
   Example: $2 + 2 = 4$; 4 is not an element of the set.

3. Yes, the natural numbers are closed under multiplication.

4. No, the natural numbers are not closed under division.
   Example: 2 divided by 3 is not a natural number.

5. We must show that $(1 \oplus 3) \oplus 2 = 1 \oplus (3 \oplus 2)$.
   $$(1 \oplus 3) \oplus 2 = 1 \oplus (3 \oplus 2)$$
   $$0 \oplus 2 = 1 \oplus 1$$
   $$2 = 2$$

6. The identity element is g, because it does not change anything.

7. The inverse is -12. (because $12 + (-12) = 0$)

8. The inverse is $\dfrac{1}{12}$. (because $12 \cdot \dfrac{1}{12} = 1$)

9. **a.** The identity element is k, because it does not change anything.

   b. The inverse of j is l because $j \circ l = k$
   The inverse of k is k because $k \circ k = k$
   The inverse of l is j because $l \circ j = k$
   The inverse of m is m because $m \circ m = k$

Exercise Set 13.1

1. $\{e, a, b, c\}$

3. $a \circ b = c$

5. $b \circ c = a$

7. $e \circ e = e$

9. $b \circ e = b$

11. $a \circ a = e$

13. $c \circ c = e$

15. No. For example, $4 + 1 = 5$, and 5 is not in the set.

17. Yes. The answer to any possible combination of two elements of the set is an element that is in the set.

19. No. For example, $1 - 2 = -1$, and -1 is not in the set.

21. No, the system is not closed under $*$. $b * a = c$, and c is not in the set.

23. $2 \oplus 4 = 1$
    $4 \oplus 2 = 1$
    So $2 \oplus 4 = 4 \oplus 2$

25. $4 \oplus 1 = 0$
    $1 \oplus 4 = 0$
    So $4 \oplus 1 = 1 \oplus 4$

27. The Commutative Property; since table entries are mirror images of each other across the main diagonal, $\oplus$ is commutative.

29. $(4 \oplus 3) \oplus 2 = 2 \oplus 2 = 4$
    $4 \oplus (3 \oplus 2) = 4 \oplus 0 = 4$
    $(4 \oplus 3) \oplus 2 = 4 \oplus (3 \oplus 2)$

31. $(4 \oplus 4) \oplus 2 = 3 \oplus 2 = 0$
    $4 \oplus (4 \oplus 2) = 4 \oplus 1 = 0$
    $(4 \oplus 4) \oplus 2 = 4 \oplus (4 \oplus 2)$

33. **a.** $(b \circ c) \circ b = c \circ b = b$

    b. $b \circ (c \circ b) = b \circ b = a$

    c. No. $(b \circ c) \circ b \neq b \circ (c \circ b)$

35. Answers will vary.

37. $29 + (-29) = 0$
    -29 is the inverse of 29 under the operation of addition.

39. $29 \cdot \dfrac{1}{29} = 1$
    $\dfrac{1}{29}$ is the inverse of 29 under the operation of multiplication.

41. $a \circ a = a$

43. $a \circ c = c$

45. $a \circ e = e$

47. $c \circ a = c$

49. $e \circ a = e$

51. a

53. d

55. b

57. 0 is the identity element.

59. $1 \oplus 4 = 0$
4 is the inverse for 1.

61. $3 \oplus 2 = 0$
2 is the inverse for 3.

63. a. d is the identity element.

   b.
element	inverse
a	none
b	none
c	none
d	d

65. Answers will vary.

67–71.
$\times$	E	O
E	E	E
O	E	O

67. Yes. The answer to any possible combination of two elements of the set is an element that is in the set.

69. $(O \times E) \times O = E$ and $O \times (E \times O) = E$

71. E does not have an inverse.

73-79. Answers will vary.

81. a.
$$\begin{bmatrix} 2 & 3 \\ 4 & 7 \end{bmatrix} \times \begin{bmatrix} 0 & 1 \\ 5 & 6 \end{bmatrix}$$
$$= \begin{bmatrix} 2 \cdot 0 + 3 \cdot 5 & 2 \cdot 1 + 3 \cdot 6 \\ 4 \cdot 0 + 7 \cdot 5 & 4 \cdot 1 + 7 \cdot 6 \end{bmatrix}$$
$$= \begin{bmatrix} 0 + 15 & 2 + 18 \\ 0 + 35 & 4 + 42 \end{bmatrix}$$
$$= \begin{bmatrix} 15 & 20 \\ 35 & 46 \end{bmatrix}$$

b.
$$\begin{bmatrix} 0 & 1 \\ 5 & 6 \end{bmatrix} \times \begin{bmatrix} 2 & 3 \\ 4 & 7 \end{bmatrix}$$
$$= \begin{bmatrix} 0 \cdot 2 + 1 \cdot 4 & 0 \cdot 3 + 1 \cdot 7 \\ 5 \cdot 2 + 6 \cdot 4 & 5 \cdot 3 + 6 \cdot 7 \end{bmatrix}$$
$$= \begin{bmatrix} 0 + 4 & 0 + 7 \\ 10 + 24 & 15 + 42 \end{bmatrix}$$
$$= \begin{bmatrix} 4 & 7 \\ 34 & 57 \end{bmatrix}$$

c. Matrix multiplication is not commutative.

Check Points 13.2

1. a. 1. The set is closed under the binary operation because the entries in the body of the table are all elements of the set.

   2. Associative Property: For example,
   $(E + O) + O = E + (O + O)$
   $$O + O = E + E$$
   $$E = E$$
   and
   $(O + E) + O = O + (E + O)$
   $$O + O = O + O$$
   $$E = E$$

   3. E is the identity element.

   4.
element	inverse
E	E
O	O

   Each element has an inverse.

   b. The Commutative Property holds for this group (as can be seen by the symmetry along the diagonal from the upper left to lower right). Therefore, this system is a commutative group.

2. a. $(8+5)+11 = 8+(5+11)$

$1+11 = 8+4$

$0 = 0$

b. Locating 9 on the left and 4 across the top indicates that $9+4=1$.

Locating 4 on the left and 9 across the top indicates that $4+9=1$.

3. a. true; $61 \equiv 5 \pmod{7}$ because

$61 \div 7 = 8$, remainder 5.

b. true; $36 \equiv 0 \pmod{6}$ because

$36 \div 6 = 6$, remainder 0.

c. false; $57 \equiv 2 \pmod{11}$ because

$57 \div 11 = 5$, remainder 2 (not 3).

4. a. $(1+3)(\bmod\,5) \equiv 4(\bmod\,5)$

b. $(5+4)(\bmod\,7) \equiv 9(\bmod\,7) \equiv 2(\bmod\,7)$

c. $(8+10)(\bmod\,13) \equiv 18(\bmod\,13) \equiv 5(\bmod\,13)$

5. $97 \equiv 6(\bmod\,7)$ thus, the desired day of the week is 6 days past Wednesday, or Tuesday.

Exercise Set 13.2

1. 8-fold rotational symmetry.

3. 18-fold rotational symmetry.

5. For any 2 elements in the set, the result is in the set.

7. $(r \circ t) \circ q = p \circ q = e$

$r \circ (t \circ q) = r \circ r = e$

So $(r \circ t) \circ q = r \circ (t \circ q)$

9. e is the identity element.

11. $p \circ q = e$

q is the inverse of p.

13. $r \circ r = e$

r is the inverse of r.

15. $t \circ t = e$

t is the inverse of t.

17. $r \circ p = t$

19. This mathematical system is not commutative.

21. Most elements do not have an inverse. For example, no natural number will satisfy the expression $5 \times ? = 1$.

23. a.

+	0	1	2	3	4	5
0	0	1	2	3	4	5
1	1	2	3	4	5	0
2	2	3	4	5	0	1
3	3	4	5	0	1	2
4	4	5	0	1	2	3
5	5	0	1	2	3	4

b. 1. The set is closed under the operation of clock addition because the entries in the body of the table are all elements of the set.

2. Associative Property: For example,

$(2+3)+4 = 2+(3+4)$

$5+4 = 2+1$

$3 = 3$

and

$(3+4)+0 = 3+(4+0)$

$1+0 = 3+4$

$1 = 1$

3. 0 is the identity element.

4.

element	inverse
0	0
1	5
2	4
3	3
4	2
5	1

Each element has an inverse.

5. The table is symmetric, so the Commutative Property holds. Therefore, this system is a commutative group.

25. $7 \equiv 2 \pmod{5}$

$7 \div 5 = 1$, remainder 2

True

27. $41 \equiv 6 \pmod 7$

$41 \div 7 = 5$, remainder 6

True

29. $84 \equiv 1 \pmod 7$

$84 \div 7 = 12$, remainder 0

False

A true statement is $84 \equiv 0 \pmod 7$

31. $23 \equiv 2 \pmod 4$

$23 \div 4 = 5$, remainder 3

False

A true statement is $23 \equiv 3 \pmod 4$

33. $55 \equiv 0 \pmod{11}$

$55 \div 11 = 5$, remainder 0

True

35. $(3 + 2)\pmod 6$

$3+2 = 5$, $5 < 6$

$3+2 \equiv 5 \pmod 6$

37. $(4 + 5) \pmod 6$

$4+5 = 9$, $9 > 6$

$9 \div 6 = 1$, remainder 3

$4+5 \equiv 3 \pmod 6$

39. $(6 + 5) \pmod 7$

$6+5 = 11$, $11 > 7$

$11 \div 7 = 1$, remainder 4

$6+5 \equiv 4 \pmod 7$

41. $(49 + 49) \pmod 5$

$49 + 49 = 98$, $98 > 50$

$98 \div 50 = 1$, remainder 48

$49+49 \equiv 48 \pmod{50}$

43. $(1200 + 0600) \pmod{2400}$

$1200 + 0600 = 1800$, $1800 < 2400$

$1200 + 0600 \equiv 1800 \pmod{2400}$

45. $(0830 + 1550) \pmod{2400}$

$0830 + 1550 = 2380$,

23 hr 80 min $\equiv 24$ hr 20 min

$2420 \equiv 0020 \pmod{2400}$

47. $67 \div 7 = 9$, remainder 4

$67 \equiv 4 \pmod 7$

Thus, the desired day of the week is 4 days past Wednesday, or Sunday.

49-55. Answers will vary.

57. Answers will vary.

59. $99,999,999 \div 24 = 4,166,666$, remainder 15

$99,999,999 \equiv 15 \pmod{24}$

It will be 15 hours past 5:00 P.M., or 8:00 A.M.

Chapter 13 Review Exercises

1. $\{e, c, f, r\}$

2. Yes. Any possible combination of two elements of the set is an element of the set.

3. $c \circ f = r$

4. $r \circ r = f$

5. $e \circ c = c$

6. $c \circ r = e$

$r \circ c = e$

7. $f \circ r = c$

$r \circ f = c$

8. $f \circ e = f$

$e \circ f = f$

9. The Commutative Property

10. Since the table entries are symmetric about the main diagonal, the operation is commutative.

11. $(c \circ r) \circ f = e \circ f = f$

$c \circ (r \circ f) = c \circ c = f$

12. $(r \circ e) \circ c = r \circ c = e$

$r \circ (e \circ c) = r \circ c = e$

13. The Associative Property

14. e is the identity element.

15. $e \circ e = e$

e is the inverse for e.

16. $c \circ r = e$
r is the inverse for c.

17. $f \circ f = e$
f is the inverse for f.

18. $r \circ c = e$
c is the inverse for r.

19. No. $1 + 1 = 2$, and 2 is not in the set.

20. Yes. Any possible combination of two elements of the set is an element of the set.

21. No. For example, $1 \div 2 = \dfrac{1}{2}$, and $\dfrac{1}{2}$ is not in the set.

22. $123 + (-123) = 0$
-123 is the additive inverse of 123.

23. $123 \cdot \dfrac{1}{123} = 1$

$\dfrac{1}{123}$ is the multiplicative inverse of 123.

24. a.

+	0	1	2	3	4
0	0	1	2	3	4
1	1	1	2	3	4
2	2	2	2	3	4
3	3	3	3	3	4
4	4	4	4	4	4

b. Zero is the identity element.

c. No. There is no element in the set such that $2 \circ ? = 0$.

25. 3-fold rotational symmetry

26. 18-fold rotational symmetry

27. a.

+	0	1	2	3	4
0	0	1	2	3	4
1	1	2	3	4	0
2	2	3	4	0	1
3	3	4	0	1	2
4	4	0	1	2	3

b. 1. The set is closed under the operation of clock addition because the entries in the body of the table are all elements of the set.

2. Associative Property: For example,
$(1+2)+3 = 1+(2+3)$
$3+3 = 1+0$
$1 = 1$

3. 0 is the identity element.

4.

element	inverse
0	0
1	4
2	3
3	2
4	1

Each element has an inverse.

5. The table is symmetric, so the Commutative Property holds. Therefore, this system is a commutative group.

28. $17 \equiv 2 \ (\text{mod}\,8)$
$17 \div 8 = 2$, remainder 1
False
A true statement is $17 \equiv 1 \ (\text{mod}\,8)$.

29. $37 \equiv 3 \ (\text{mod}\,5)$
$37 \div 5 = 7$, remainder 2
False
A true statement is $37 \equiv 2 \ (\text{mod}\,5)$.

30. $60 \equiv 0 \ (\text{mod}\,10)$
$60 \div 10 = 6$, remainder 0
True

31. $(4+3) \pmod 6$

$4+3 = 7,\ 7 > 6$

$7 \div 6 = 1,\ \text{remainder } 1$

$4+3 \equiv 1 \pmod 6$

32. $(7+7) \pmod 8$

$7+7 = 14,\ 14 > 8$

$14 \div 8 = 1,\ \text{remainder } 6$

$7+7 \equiv 6 \pmod 8$

33. $(4+3) \pmod 9$

$4+3 = 7,\ 7 < 9$

$4+3 = 7 \pmod 9$

34. $(3+18) \pmod{20}$

$3+18 = 21,\ 21 > 20$

$21 \div 20 = 1,\ \text{remainder } 1$

$3+18 \equiv 1 \pmod{20}$

Chapter 13 Test

1. Yes. Any possible combination of two elements of the set is an element of the set.

2. $z \circ y = x$

$y \circ z = x$

This illustrates the Commutative Property.

3. $(x \circ z) \circ z = z \circ z = y$

$x \circ (z \circ z) = x \circ y = y$

This illustrates the Associative Property.

4. x is the identity element.

5.

element	inverse
x	x
y	z
z	y

6. No. For example, $1 + 1 = 2$, and 2 is not in the set.

7. $5 \cdot \dfrac{1}{5} = 1$

$\dfrac{1}{5}$ is the multiplicative inverse of 5.

8. 6-fold symmetry; Answers will vary.

9.

+	0	1	2	3
0	0	1	2	3
1	1	2	3	0
2	2	3	0	1
3	3	0	1	2

10. 1. The set is closed under the operation of clock addition because the entries in the body of the table are all elements of the set.

2. Associative Property:
For example,
$(1+2)+3 = 1+(2+3)$
$3+3 = 1+1$
$2 = 2$

3. 0 is the identity element.

4.

element	inverse
0	0
1	3
2	2
3	1

Each element has an inverse.

5. The table is symmetric, so the Commutative Property holds. Therefore, this system is a commutative group.

11. $39 \equiv 3 \pmod 6$

$39 \div 6 = 6,\ \text{remainder } 3$

True

12. $14 = 2 \pmod 7$

$14 \div 7 = 2,\ \text{remainder } 0$

False

A true statement is $14 \equiv 0 \pmod 7$

13. $(9+1) \pmod{11}$

$9+1 = 10,\ 10 < 11$

$9+1 = 10 \pmod{11}$

14. $(9+6) \pmod{10}$

$9+6 = 15,\ 15 > 10$

$15 \div 10 = 1,\ \text{remainder } 5$

$9+6 \equiv 5 \pmod{10}$

Chapter 14
Voting and Apportionment

14.1 Voting Methods

Check Point 1.

a. We find the number of people who voted in the election by adding the numbers in the row labeled Number of Votes: $2100 + 1305 + 765 + 40 = 4210$. Thus, 4210 people voted in the election.

b. We find how many people selected the candidates in the order B, S, A, C by referring to the fourth column of letters in the preference table. Above this column is the number 40. Thus, 40 people voted in the order B, S, A, C.

c. We find the number of people who selected S as their first choice by reading across the row that says First Choice: $2100 + 765 = 2865$. Thus, 2865 students selected S (Samir) as their first choice for student body president.

Check Point 2.

The candidate with the most first-place votes is the winner. When using Table 14.2, it is only necessary to look at the row which indicates the number of first-place votes. This indicates that A (Antonio) gets 130 first-place votes, C (Carmen) gets 150 first-place votes, and D (Donna) gets $120 + 100 = 220$ first-place votes. Thus Donna is declared the winner using the plurality method.

Check Point 3.

Because there are four candidates, a first-place vote is worth 4 points, a second-place vote is worth 3 points, a third-place vote is worth 2 points, and a fourth-place vote is worth 1 point. We show the points produced by the votes in the preference table.

Number of Votes	130	120	100	150
First Choice: 4 points	A: $130 \times 4 = 520$ pts	D: $120 \times 4 = 480$ pts	D: $100 \times 4 = 400$ pts	C: $150 \times 4 = 600$ pts
Second Choice: 3 points	B: $130 \times 3 = 390$ pts	B: $120 \times 3 = 360$ pts	B: $100 \times 3 = 300$ pts	B: $150 \times 3 = 450$ pts
Third Choice: 2 points	C: $130 \times 2 = 260$ pts	C: $120 \times 2 = 240$ pts	A: $100 \times 2 = 200$ pts	A: $150 \times 2 = 300$ pts
Fourth Choice: 1 point	D: $130 \times 1 = 130$ pts	A: $120 \times 1 = 120$ pts	C: $100 \times 1 = 100$ pts	D: $150 \times 1 = 150$ pts

Now we read down each column and total the points for each candidate separately.

A gets $520 + 120 + 200 + 300 = 1140$ points
B gets $390 + 360 + 300 + 450 = 1500$ points
C gets $260 + 240 + 100 + 600 = 1200$ points
D gets $130 + 480 + 400 + 150 = 1160$ points

Because B (Bob) has received the most points, he is the winner and the new mayor of Smallville.

Check Point 4.

There are $130 + 120 + 100 + 150$, or 500, people voting. In order to receive a majority, a candidate must receive more than 50% of the votes, meaning more than 250 votes. The number of first-place votes for each candidate is
A (Antonio) = 130   B (Bob) = 0   C (Carmen) = 150   D (Donna) = 220

We see that no candidate receives a majority of first-place votes. Because Bob received the fewest first-place votes, he is eliminated in the next round. We construct a new preference table in which B is removed. Each candidate below B moves up one place, while the positions of candidates above B remain unchanged.

Number of Votes	130	120	100	150
First Choice	A	D	D	C
Second Choice	C	C	A	A
Third Choice	D	A	C	D

The number of first-place votes for each candidate is now
A (Antonio) = 130   C (Carmen) = 150   D (Donna) = 220

No candidate receives a majority of first-place votes. Because Antonio received the fewest first-place votes, he is eliminated in the next round.

Number of Votes	130	120	100	150
First Choice	C	D	D	C
Second Choice	D	C	C	D

The number of first-place votes for each candidate is now
C (Carmen) = 280   D (Donna) = 220

Because Carmen has received the majority of first-place votes, she is the winner and the new mayor of Smallville.

Check Point 5.

A vs. B

130	120	100	150
A	D	D	C
B	B	B	B
C	C	A	A
D	A	C	D

130 voters prefer A to B.
120 + 100 + 150 = 370 voters prefer B to A.

Conclusion: B wins this comparison and gets one point.

A vs. C

130	120	100	150
A	D	D	C
B	B	B	B
C	C	A	A
D	A	C	D

130 + 100 = 230 voters prefer A to C.
120 + 150 = 270 voters prefer C to A.

Conclusion: C wins this comparison and gets one point.

A vs. D

130	120	100	150
A	**D**	**D**	C
B	B	B	B
C	C	*A*	**A**
D	*A*	C	*D*

130 + 150 = 280 voters prefer A to D.
120 + 100 = 220 voters prefer D to A.

Conclusion: A wins this comparison and gets one point.

B vs. C

130	120	100	150
A	D	D	**C**
B	**B**	**B**	*B*
C	*C*	A	A
D	A	*C*	D

130 + 120 + 100 = 350 voters prefer B to C.
150 voters prefer C to B.

Conclusion: B wins this comparison and gets one point.

B vs. D

130	120	100	150
A	**D**	**D**	C
B	*B*	*B*	**B**
C	C	A	A
D	A	C	*D*

130 + 150 = 280 voters prefer B to D.
120 + 100 = 220 voters prefer D to B.

Conclusion: B wins this comparison and gets one point.

C vs. D

130	120	100	150
A	**D**	**D**	**C**
B	B	B	B
C	*C*	A	A
D	A	*C*	*D*

130 + 150 = 280 voters prefer C to D.
120 + 100 = 220 voters prefer D to C.

Conclusion: C wins this comparison and gets one point.

We now use each of the six conclusions and add points for the six comparisons.

    A gets 1 point.
    B gets $1 + 1 + 1 = 3$ points.
    C gets $1 + 1 = 2$ points.

After all comparisons have been made, the candidate receiving the most points is B (Bob). He is the winner and the new mayor of Smallville.

Exercise Set 14.1

1.

Number of Votes	7	5	4
First Choice	A	B	C
Second Choice	B	C	B
Third Choice	C	A	A

3.

Number of Votes	5	1	4	2
First Choice	A	B	C	C
Second Choice	B	D	B	B
Third Choice	C	C	D	A
Fourth Choice	D	A	A	D

5. a. $14 + 8 + 3 + 1 = 126$

   b. 8

   c. $14 + 8 = 22$

   d. 3

7. "Musical" received 12 first-place votes, "comedy" received 10 first-place votes, and "drama" received 8 first-place votes, so the type of play selected is a musical.

9. Darwin received 30 first-place votes, Einstein received 22 first-place votes, Freud received 20 first-place votes, and Hawking received 14 first-place votes, so the professor declared chair is Darwin.

11.

Number of Votes	10	6	6	4	2	2
First Choice: 3 points	M: $10 \times 3 = 30$	C: $6 \times 3 = 18$	D: $6 \times 3 = 18$	C: $4 \times 3 = 12$	D: $2 \times 3 = 6$	M: $2 \times 3 = 6$
Second Choice: 2 points	C: $10 \times 2 = 20$	M: $6 \times 2 = 12$	C: $6 \times 2 = 12$	D: $4 \times 2 = 8$	M: $2 \times 2 = 4$	D: $2 \times 2 = 4$
Third Choice: 1 point	D: $10 \times 1 = 10$	D: $6 \times 1 = 6$	M: $6 \times 1 = 6$	M: $4 \times 1 = 4$	C: $2 \times 1 = 2$	C: $2 \times 1 = 2$

C gets $20 + 18 + 12 + 12 + 2 + 2 = 66$ points.
D gets $10 + 6 + 18 + 8 + 6 + 4 = 52$ points.
M gets $30 + 12 + 6 + 4 + 4 + 6 = 62$ points.

C (Comedy) receives the most points, and is selected.

13.

Number of Votes	30	22	20	12	2
First Choice: 4 points	D: $30 \times 4 = 120$	E: $22 \times 4 = 88$	F: $20 \times 4 = 80$	H: $12 \times 4 = 48$	H: $2 \times 4 = 8$
Second Choice: 3 points	H: $30 \times 3 = 90$	F: $22 \times 3 = 66$	E: $20 \times 3 = 60$	E: $12 \times 3 = 36$	F: $2 \times 3 = 6$
Third Choice: 2 points	F: $30 \times 2 = 60$	H: $22 \times 2 = 44$	H: $20 \times 2 = 40$	F: $12 \times 2 = 24$	D: $2 \times 2 = 4$
Fourth Choice: 1 point	E: $30 \times 1 = 30$	D: $22 \times 1 = 22$	D: $20 \times 1 = 20$	D: $12 \times 1 = 12$	E: $2 \times 1 = 2$

D gets $120 + 22 + 20 + 12 + 4 = 178$ points.
E gets $30 + 88 + 60 + 36 + 2 = 216$ points.
F gets $60 + 66 + 80 + 24 + 6 = 236$ points.
H gets $90 + 44 + 40 + 48 + 8 = 230$ points.

F (Freud) receives the most points and is declared the new department chair.

15. There are 30 people voting, so the winner needs more than 15 votes for a majority.
The number of first-place votes for each candidate is

C (Comedy) = 10   D (Drama) = 8   M (Musical) = 12

No candidate has a majority. Drama received the fewest first-place votes, so we eliminate it in the next round.

Number of Votes	10	6	6	4	2	2
First Choice	M	C	C	C	M	M
Second Choice	C	M	M	M	C	C

The number of first-place votes for each candidate is now

C (Comedy) = 16   M (Musical) = 14

C (Comedy) has 16 votes, which is a majority, so "Comedy" is selected.

17. There are 86 people voting, so the winner needs more than 43 votes for a majority. The number of first-place votes for each candidate is

D (Darwin) = 30   E (Einstein) = 22
F (Freud) = 20   H (Hawking) = 14

No candidate has a majority. Hawking received the fewest first-place votes, so we eliminate him in the next round.

Number of Votes	30	22	20	12	2
First Choice	D	E	F	E	F
Second Choice	F	F	E	F	D
Third Choice	E	D	D	D	E

The number of first-place votes for each candidate is now

D (Darwin) = 30 E (Einstein) = 34 F (Freud) = 22

No candidate has a majority. Freud received the fewest first-place votes, so we eliminate him in the next round:

Number of Votes	30	22	20	12	2
First Choice	D	E	E	E	D
Second Choice	E	D	D	D	E

The number of first-place votes for each candidate is now

D (Darwin) = 32 E (Einstein) = 54

E (Einstein) has 54 votes, which is a majority, so Einstein is declared the new department chair.

19. With $n = 5$, there are $\dfrac{5(5-1)}{2} = 10$ comparisons.

21. With $n = 8$, there are $\dfrac{8(8-1)}{2} = 28$ comparisons.

23.

10	6	6	4	2	2
M	C	D	C	D	M
C	M	C	D	M	D
D	D	M	M	C	C

C vs. D
10 + 6 + 4 = 20 voters prefer C to D.
6 + 2 + 2 = 10 voters prefer D to C.
C wins this comparison and gets one point.

C vs. M
6 + 6 + 4 = 16 voters prefer C to M.
10 + 2 + 2 = 14 voters prefer M to C.
C wins this comparison and gets one point.

D vs. M
6 + 4 + 2 = 12 voters prefer D to M.
10 + 6 + 2 = 18 voters prefer M to D.
M wins this comparison and gets one point.

Adding points for the three comparisons:
C gets 1 + 1 = 2 points.
D gets 0 points.
M gets 1 point.

C (Comedy) receives the most points, so a comedy is selected.

25.

30	22	20	12	2
D	E	F	H	H
H	F	E	E	F
F	H	H	F	D
E	D	D	D	E

D vs. E
30 + 2 = 32 voters prefer D to E.
22 + 20 + 12 = 54 voters prefer E to D.
E wins the comparison and gets one point.

D vs. F
30 voters prefer D to F.
22 + 20 + 12 + 2 = 56 voters prefer F to D.
F wins this comparison and gets one point.

D vs. H
30 voters prefer D to H.
22 + 20 + 12 + 2 = 56 voters prefer H to D.
H wins this comparison and gets one point.

E vs. F
22 + 12 = 34 voters prefer E to F.
30 + 20 + 2 = 52 voters prefer F to E.
F wins this comparison and gets one point.

E vs. H
22 + 20 = 42 voters prefer E to H.
30 + 12 + 2 = 44 voters prefer H to E.
H wins this comparison and gets one point.

F vs. H
22 + 20 = 42 voters prefer F to H.
30 + 12 + 2 = 44 voters prefer H to F.
H wins this comparison and gets one point.

Adding points for the six comparisons:
D gets 0 points.
E gets 1 point.
F gets 1 + 1 = 2 points.
H gets 1 + 1 + 1 = 3 points.

H (Hawking) receives the most points, so Hawking is declared the new department chair.

27. A received 34 first-place votes, B received 30 first-place votes, C received 6 first-place votes, and D received 2 first-place votes, so A is the winner.

29. There are 72 people voting, so the winner needs more than 36 votes for a majority. The number of first-place votes for each candidate is:

A = 34   B = 30   C = 6   D = 2

No candidate has a majority. D received the fewest first-place votes, so we eliminate it in the next round.

Number of Voters	34	30	6	2
First Choice	A	B	C	B
Second Choice	B	C	B	C
Third Choice	C	A	A	A

The number of first-place votes for each candidate is now

A = 34   B = 32   C = 6

No candidate has a majority. C received the fewest first-place votes, so we eliminate it in the next round.

Number of Voters	34	30	6	2
First Choice	A	B	B	B
Second Choice	B	A	A	A

The number of first-place votes for each candidate is now

A = 34   B = 38

B has 38 votes, which is a majority, so B is selected.

31.

Number of Votes	5	5	4	3	3	2
First choice: 5 points	C: $5 \times 5 = 25$	S: $5 \times 5 = 25$	C: $4 \times 5 = 20$	W: $3 \times 5 = 15$	W: $3 \times 5 = 15$	P: $2 \times 5 = 10$
Second choice: 4 points	R: $5 \times 4 = 20$	R: $5 \times 4 = 20$	P: $4 \times 4 = 16$	P: $3 \times 4 = 12$	R: $3 \times 4 = 12$	S: $2 \times 4 = 8$
Third choice: 3 points	P: $5 \times 3 = 15$	W: $5 \times 3 = 15$	R: $4 \times 3 = 12$	R: $3 \times 3 = 9$	S: $3 \times 3 = 9$	C: $2 \times 3 = 6$
Fourth choice: 2 points	W: $5 \times 2 = 10$	P: $5 \times 2 = 10$	S: $4 \times 2 = 8$	S: $3 \times 2 = 6$	C: $3 \times 2 = 6$	R: $2 \times 2 = 4$
Fifth choice: 1 point	S: $5 \times 1 = 5$	C: $5 \times 1 = 5$	W: $4 \times 1 = 4$	C: $3 \times 1 = 3$	P: $3 \times 1 = 3$	W: $2 \times 1 = 2$

C gets $25 + 5 + 20 + 3 + 6 + 6 = 65$ points.
P gets $15 + 10 + 16 + 12 + 3 + 10 = 66$ points.
R gets $20 + 20 + 12 + 9 + 12 + 4 = 77$ points.
S gets $5 + 25 + 8 + 6 + 9 + 8 = 61$ points.
W gets $10 + 15 + 4 + 15 + 15 + 2 = 61$ points.

R (Rent) receives the most points and is selected.

33.

5	5	4	3	3	2
C	S	C	W	W	P
R	R	P	P	R	S
P	W	R	R	S	C
W	P	S	S	C	R
S	C	W	C	P	W

C vs. P

5 + 4 + 3 = 12 voters prefer C to P.
5 + 3 + 2 = 10 voters prefer P to C.
C wins this comparison and gets one point.

C vs. R

5 + 4 + 2 = 11 voters prefer C to R.
5 + 3 + 3 = 11 voters prefer R to C.

C and R are tied. Each gets $\frac{1}{2}$ point.

C vs. S

5 + 4 = 9 voters prefer C to S.
5 + 3 + 3 + 2 = 13 voters prefer S to C.
S wins this comparison and gets one point.

C vs. W

5 + 4 + 2 = 11 voters prefer C to W.
5 + 3 + 3 = 11 voters prefer W to C.

C and W are tied. Each gets $\frac{1}{2}$ point.

P vs. R

4 + 3 + 2 = 9 voters prefer P to R.
5 + 5 + 3 = 13 voters prefer R to P.
R wins this comparison and gets one point.

P vs. S

5 + 4 + 3 + 2 = 14 voters prefer P to S.
5 + 3 = 8 voters prefer S to P.
P wins this comparison and gets one point.

P vs. W

5 + 4 + 2 = 11 voters prefer P to W.
5 + 3 + 3 = 11 voters prefer W to P.

P and W are tied. Each gets $\frac{1}{2}$ point.

R vs. S

5 + 4 + 3 + 3 = 15 voters prefer R to S.
5 + 2 = 7 voters prefer S to R.
R wins this comparison and gets one point.

R vs. W

5 + 5 + 4 + 2 = 16 voters prefer R to W.
3 + 3 = 6 voters prefer W to R.
R wins this comparison and gets one point.

S vs. W

5 + 4 + 2 = 11 voters prefer S to W.
5 + 3 + 3 = 11 voters prefer W to S.

S and W are tied. Each gets $\frac{1}{2}$ point.

Adding points for 10 comparisons:

C gets $1 + \frac{1}{2} + \frac{1}{2} = 2$ points.

P gets $1 + \frac{1}{2} = 1\frac{1}{2}$ points.

R gets $\frac{1}{2} + 1 + 1 + 1 = 3\frac{1}{2}$ points.

S gets $1 + \frac{1}{2} = 1\frac{1}{2}$ points.

W gets $\frac{1}{2} + \frac{1}{2} + \frac{1}{2} = 1\frac{1}{2}$ points.

R (Rent) receives the most points, so Rent is the winner.

35. a.

Number of Votes	5	5	3	3	3	2
First Choice: 5 points	A: $5 \times 5 = 25$	C: $5 \times 5 = 25$	D: $3 \times 5 = 15$	A: $3 \times 5 = 15$	B: $3 \times 5 = 15$	D: $2 \times 5 = 10$
Second Choice: 4 points	B: $5 \times 4 = 20$	E: $5 \times 4 = 20$	C: $3 \times 4 = 12$	D: $3 \times 4 = 12$	E: $3 \times 4 = 12$	C: $2 \times 4 = 8$
Third Choice: 3 points	C: $5 \times 3 = 15$	D: $5 \times 3 = 15$	B: $3 \times 3 = 9$	B: $3 \times 3 = 9$	A: $3 \times 3 = 9$	B: $2 \times 3 = 6$
Fourth Choice: 2 points	D: $5 \times 2 = 10$	A: $5 \times 2 = 10$	E: $3 \times 2 = 6$	C: $3 \times 2 = 6$	C: $3 \times 2 = 6$	A: $2 \times 2 = 4$
Fifth Choice: 1 point	E: $5 \times 1 = 5$	B: $5 \times 1 = 5$	A: $3 \times 1 = 3$	E: $3 \times 1 = 3$	D: $3 \times 1 = 3$	E: $2 \times 1 = 2$

A gets $25 + 10 + 3 + 15 + 9 + 4 = 66$ points.
B gets $20 + 5 + 9 + 9 + 15 + 6 = 64$ points.
C gets $15 + 25 + 12 + 6 + 6 + 8 = 72$ points.
D gets $10 + 15 + 15 + 12 + 3 + 10 = 65$ points.
E gets $5 + 20 + 6 + 3 + 12 + 2 = 48$ points.

C receives the most points and is the winner.

b.

Number of Votes	5	5	3	3	3	2
First Choice: 4 points	A: $5 \times 4 = 20$	C: $5 \times 4 = 20$	D: $3 \times 4 = 12$	A: $3 \times 4 = 12$	B: $3 \times 4 = 12$	D: $2 \times 4 = 8$
Second Choice: 3 points	B: $5 \times 3 = 15$	D: $5 \times 3 = 15$	C: $3 \times 3 = 9$	D: $3 \times 3 = 9$	A: $3 \times 3 = 9$	C: $2 \times 3 = 6$
Third Choice: 2 points	C: $5 \times 2 = 10$	A: $5 \times 2 = 10$	B: $3 \times 2 = 6$	B: $3 \times 2 = 6$	C: $3 \times 2 = 6$	B: $2 \times 2 = 4$
Fourth Choice: 1 points	D: $5 \times 1 = 5$	B: $5 \times 1 = 5$	A: $3 \times 1 = 3$	C: $3 \times 1 = 3$	D: $3 \times 1 = 3$	A: $2 \times 1 = 2$

A gets $20 + 10 + 3 + 12 + 9 + 2 = 56$ points.
B gets $15 + 5 + 6 + 6 + 12 + 4 = 48$ points.
C gets $10 + 20 + 9 + 3 + 6 + 6 = 54$ points.
D gets $5 + 15 + 12 + 9 + 3 + 8 = 52$ points.

A receives the most points and is the winner.

37. First use the plurality method: C receives 12,000 first-place votes, and A receives 12,000 first-place votes. This results in a tie, so we use the Borda count method.

Number of Votes	12,000	7500	4500
First Choice: 3 points	C: $12,000 \times 3 = 36,000$	A: $7500 \times 3 = 22,500$	A: $4500 \times 3 = 13,500$
Second Choice: 2 points	B: $12,000 \times 2 = 24,000$	B: $7500 \times 2 = 15,000$	C: $4500 \times 3 = 9000$
Third Choice: 1 points	A: $12,000 \times 1 = 12,000$	C: $7500 \times 1 = 7500$	B: $4500 \times 1 = 4500$

A gets $12,000 + 22,500 + 13,500 = 48,000$ points.
B gets $24,000 + 15,000 + 4500 = 43,500$ points.
C gets $36,000 + 7500 + 9000 = 52,500$ points.
C receives the most points and is the winner.

39. Answers will vary.

41. Answers will vary.

43. Answers will vary.

45. Answers will vary.

47. Answers will vary.

49. a, b, c , and d are all false.

51. Answers will vary.

53. Answers will vary.

14.2 Flaws of Voting Methods

Check Point 1.

a. There are 14 first-place votes. A candidate with more than half of these receives a majority. The first-choice row shows that candidate A received 8 first-place votes. Thus, candidate A has a majority of first-place votes.

b. Using the Borda count method with four candidates, a first-place vote is worth 4 points, a second-place vote is worth 3 points, a third-place vote is worth 2 points, and a fourth-place vote is worth 1 point.

Number of Votes	6	4	2	2
First Choice: 4 points	A: $6 \times 4 = 24$ pts	B: $4 \times 4 = 16$ pts	B: $2 \times 4 = 8$ pts	A: $2 \times 4 = 8$ pts
Second Choice: 3 points	B: $6 \times 3 = 18$ pts	C: $4 \times 3 = 12$ pts	D: $2 \times 3 = 6$ pts	B: $2 \times 3 = 6$ pts
Third Choice: 2 points	C: $6 \times 2 = 12$ pts	D: $4 \times 2 = 8$ pts	C: $2 \times 2 = 4$ pts	D: $2 \times 2 = 4$ pts
Fourth Choice: 1 point	D: $6 \times 1 = 6$ pts	A: $4 \times 1 = 4$ pts	A: $2 \times 1 = 2$ pts	C: $2 \times 1 = 2$ pts

Now we read down the columns and total the points for each candidate.
A gets $24 + 4 + 2 + 8 = 38$ points.
B gets $18 + 16 + 8 + 6 = 48$ points.
C gets $12 + 12 + 4 + 2 = 30$ points.
D gets $6 + 8 + 6 + 4 = 24$ points.

Because candidate B has received the most points, candidate B is declared the new principal using the Borda count method.

Check Point 2.

a. We begin by comparing A and B. A is favored over B in column 1, giving A 3 votes. B is favored over A in columns 2 and 3, giving B $2 + 2$, or 4, votes. Thus, B is favored when compared to A.

Now we compare B to C. B is favored over C in columns 1 and 2, giving B $3 + 2$, or 5, votes. C is favored over B in column 3, giving C 2 votes. Thus, B is favored when compared to C.

We see that B is favored over both A and C using a head-to-head comparison.

b. Using the plurality method, the brand with the most first-place votes is the winner. In the row indicating first choice, A received 3 votes, B received 2 votes, and C received 2 votes. A wins using the plurality method.

Check Point 3.

a. There are 120 people voting. No candidate initially receives more than 60 votes. Because C receives the fewest first-place votes, C is eliminated in the next round. The new preference table is

Number of Votes	42	34	28	16
First Choice	A	A	B	B
Second Choice	B	B	A	A

Because A has received a majority of first-place votes, A is the winner of the straw poll.

b. No candidate initially receives more than 60 votes. Because B receives the fewest first-place votes, B is eliminated in the next round. The new preference table is

Number of Votes	54	34	28	4
First Choice	A	C	C	A
Second Choice	C	A	A	C

Because C has received a majority of first-place votes, C is the winner of the second election.

c. A won the first election. A then gained additional support with the 12 voters who changed their ballots to make A their first choice. A lost the second election. This violates the monotonicity criterion.

Check Point 4.

a. Because there are 4 candidates, $n = 4$ and the number of comparisons we must make is $\frac{n(n-1)}{2} = \frac{4(4-1)}{2} = \frac{4 \cdot 3}{2} = \frac{12}{2} = 6$.

The following table shows the results of these 6 comparisons.

Comparison	Vote Results	Conclusion
A vs. B	270 voters prefer A to B. 90 voters prefer B to A.	A wins and gets 1 point.
A vs. C	270 voters prefer A to C. 90 voters prefer C to A.	A wins and gets 1 point.
A vs. D	150 voters prefer A to D. 210 voters prefer D to A.	D wins and gets 1 point.
B vs. C	180 voters prefer B to C. 180 voters prefer C to B.	B and C tie. Each gets $\frac{1}{2}$ point.
B vs. D	240 voters prefer B to D. 120 voters prefer D to B.	B wins and gets 1 point.
C vs. D	240 voters prefer C to D. 120 voters prefer D to C.	C wins and gets 1 point.

Thus A gets 2 points, B gets $1\frac{1}{2}$ points, C gets $1\frac{1}{2}$ points, and D gets 1 point. Therefore A is the winner.

b. After B and C withdraw, there is a new preference table:

Number of Votes	150	90	90	30
First Choice	A	D	D	D
Second Choice	D	A	A	A

Using the pairwise comparison test with 2 candidates, there is only one comparison to make namely A vs. D.

150 voters prefer A to D, and 210 voters prefer D to A. D gets 1 point, A gets 0 points, and D wins the election.

c. The first election count produced A as the winner. The removal of B and C from the ballots produced D as the winner. This violates the irrelevant alternatives criterion.

Exercise Set 14.2

1. a. D has 300 first-place votes, which is more than half of the 570 total votes, so D has a majority of first-place votes.

b.

Number of Votes	300	120	90	60
First Choice: 4 points	D: $300 \times 4 = 1200$	C: $120 \times 4 = 480$	C: $90 \times 4 = 360$	A: $60 \times 4 = 240$
Second Choice: 3 points	A: $300 \times 3 = 900$	A: $120 \times 3 = 360$	A: $90 \times 3 = 270$	D: $60 \times 3 = 180$
Third Choice: 2 points	B: $300 \times 2 = 600$	B: $120 \times 2 = 240$	D: $90 \times 2 = 180$	B: $60 \times 2 = 120$
Fourth Choice: 1 point	C: $300 \times 1 = 300$	D: $120 \times 1 = 120$	B: $90 \times 1 = 90$	C: $60 \times 1 = 60$

A gets $900 + 360 + 270 + 240 = 1770$ points.
B gets $600 + 240 + 90 + 120 = 1050$ points.
C gets $300 + 480 + 360 + 60 = 1200$ points.
D gets $1200 + 120 + 180 + 180 = 1680$ points.

A receives the most points, so A is the chosen design.

c. No. D receives a majority of first-place votes, but A is chosen by the Borda count method.

3. a. A is favored over R in columns 1 and 3, giving A $12 + 4$, or 16, votes. R is favored over A in columns 2 and 4, giving R $9 + 4$, or 13, votes. Thus, A is favored when compared to R.

A is favored over V in columns 1 and 4, giving A $12 + 4$, or 16, votes. V is favored over A in columns 2 and 3, giving V $9 + 4$, or 13, votes. Thus, A is favored when compared to V.

We see that A is favored over the other two cities using a head-to-head comparison.

b. A gets 12 first-place votes, V gets 13 first-place votes, and R gets 4 first-place votes, so V wins using the plurality method.

c. No. A wins the head-to-head comparison, but V wins the election.

5. a. A is favored over B in columns 1 and 4, giving A $120 + 30$, or 150, votes. B is favored over A in columns 2, 3, and 5, giving B $60 + 30 + 30$, or 120 votes. Thus, A is favored when compared to B.

A is favored over C in columns 1 and 3, giving A $120 + 30$, or 150 votes. C is favored over A in columns 2, 4, and 5, giving C $60 + 30 + 30$, or 120, votes. Thus, A is favored when compared to C.

We see that A is favored over the other two options using a head-to-head comparison.

b.

Number of Votes	120	60	30	30	30
First Choice: 3 points	A: $120 \times 3 = 360$	C: $60 \times 3 = 180$	B: $30 \times 3 = 90$	C: $30 \times 3 = 90$	B: $30 \times 3 = 90$
Second Choice: 2 points	C: $120 \times 2 = 240$	B: $60 \times 2 = 120$	A: $30 \times 2 = 60$	A: $30 \times 2 = 60$	C: $30 \times 2 = 60$
Third Choice: 1 point	B: $120 \times 1 = 120$	A: $60 \times 1 = 60$	C: $30 \times 1 = 30$	B: $30 \times 1 = 30$	A: $30 \times 1 = 30$

A gets $360 + 60 + 60 + 60 + 30 = 570$ points.
B gets $120 + 120 + 90 + 30 + 90 = 450$ points.
C gets $240 + 180 + 30 + 90 + 60 = 600$ points.

C receives the most points, so C is the winner.

c. No. A wins the head-to-head comparison, but C wins the election.

7. a.   There are 29 people voting. No one receives the 15 first-place votes needed for a majority.
B receives the fewest first-place votes and is eliminated in the next round.

Number of Votes	18	11
First Choice	C	A
Second Choice	A	C

C receives the majority of first-place votes, so C is the winner.

b.   With the voting change, a new preference table results.

Number of Votes	14	8	7
First Choice	C	B	A
Second Choice	A	C	B
Third Choice	B	A	C

No one receives a majority of first-place votes. A receives the fewest first-place votes, and is eliminated in the next
round.

Number of Votes	14	15
First Choice	C	B
Second Choice	B	C

B receives the majority of first-place votes, so B is the winner.

c.   No. C wins the straw vote, and the only change increases the number of first-place votes for C, but B wins the
election.

9. a.   There are 3 candidates, so $n = 3$ and the number of comparisons we must make is $\frac{n(n-1)}{2} = \frac{3(2)}{2} = 3$.

Comparison	Vote Results	Conclusion
H vs. L	10 voters prefer H to L. 13 voters prefer L to H.	L wins and gets one point.
H vs. S	10 voters prefer H to S. 13 voters prefer S to H.	S wins and gets one point.
L vs. S	8 voters prefer L to S. 15 voters prefer S to L.	S wins and gets one point.

Thus, L gets 1 point and S gets 2 points. Therefore, S is the winner when candidates H and L are included.

b.   New preference table:

Number of Votes	15	8
First Choice	S	L
Second Choice	L	S

With only two candidates, we can only make one comparison. We see that S wins, defeating L by 15 votes to 8
votes. Thus S gets 1 point, L gets 0 points, and S is the winner.

c.   Yes. S wins whether or not H withdraws.

11. a.

Number of Votes	20	16	10	4
First Choice: 4 points	D: $20 \times 4 = 80$	C: $16 \times 4 = 64$	C: $10 \times 4 = 40$	A: $4 \times 4 = 16$
Second Choice: 3 points	A: $20 \times 3 = 60$	A: $16 \times 3 = 48$	B: $10 \times 3 = 30$	B: $4 \times 3 = 12$
Third Choice: 2 points	B: $20 \times 2 = 40$	B: $16 \times 2 = 32$	D: $10 \times 2 = 20$	D: $4 \times 2 = 8$
Fourth Choice: 1 point	C: $20 \times 1 = 20$	D: $16 \times 1 = 16$	A: $10 \times 1 = 10$	C: $4 \times 1 = 4$

A gets $60 + 48 + 10 + 16 = 134$ points.
B gets $40 + 32 + 30 + 12 = 114$ points.
C gets $20 + 64 + 40 + 4 = 128$ points.
D gets $80 + 16 + 20 + 8 = 124$ points.

A receives the most points, so A is the winner.

b. No. A has only 4 first-place votes, out of 50 total votes. C has 26 first-place votes, which is a majority, but A wins the election.

13. a. There are 70 people voting. No one receives the 36 first-place votes needed for a majority. B receives the fewest first-place votes and is eliminated in the next round.

Number of Votes	24	20	10	8	8
First Choice	D	C	A	A	C
Second Choice	A	A	D	C	D
Third Choice	C	D	C	D	A

No one receives a majority of first-place votes. A receives the fewest first-place votes and is eliminated in the next round.

Number of Votes	34	36
First Choice	D	C
Second Choice	C	D

C receives 36 first-place votes, which is a majority, so C is the winner.

b. No. When compared individually to B, A wins with 60 votes to 10. Compared with C, A wins with 42 votes to 28. Compared with D, A wins with 38 votes to 32. So A is favored in all head-to-head contests but C wins the election.

15. a.

Number of Votes	14	8	4
First Choice: 4 points	A: $14 \times 4 = 56$	B: $8 \times 4 = 32$	D: $4 \times 4 = 16$
Second Choice: 3 points	B: $14 \times 3 = 42$	D: $8 \times 3 = 24$	A: $4 \times 3 = 12$
Third Choice: 2 points	C: $14 \times 2 = 28$	C: $8 \times 2 = 16$	C: $4 \times 2 = 8$
Fourth Choice: 1 point	D: $14 \times 1 = 14$	A: $8 \times 1 = 8$	B: $4 \times 1 = 4$

A gets $56 + 8 + 12 = 76$ points.
B gets $42 + 32 + 4 = 78$ points.
C gets $28 + 16 + 8 = 52$ points.
D gets $14 + 24 + 16 = 54$ points.

B receives the most points, so B is the winner.

b. No. A receives the majority of first-place votes, but B wins the election.

c. No. A wins all head-to-head comparisons, but B wins the election.

d. Using the Borda count method with C removed:

Number of Votes	14	8	4
First Choice: 3 points	A: $14 \times 3 = 42$	B: $8 \times 3 = 24$	D: $4 \times 3 = 12$
Second Choice: 2 points	B: $14 \times 2 = 28$	D: $8 \times 2 = 16$	A: $4 \times 2 = 8$
Third Choice: 1 point	D: $14 \times 1 = 14$	A: $8 \times 1 = 8$	B: $4 \times 1 = 4$

A gets $42 + 8 + 8 = 58$ points.
B gets $28 + 24 + 4 = 56$ points.
D gets $14 + 16 + 12 = 42$ points.

A receives the most points, and wins the election.

The irrelevant alternatives criterion is not satisfied. Candidate C's dropping out changed the outcome of the election.

17. a.

Number of Votes	16	14	12	4	2
First Choice: 5 points	A: $16 \times 5 = 80$	D: $14 \times 5 = 70$	D: $12 \times 5 = 60$	C: $4 \times 5 = 20$	E: $2 \times 5 = 10$
Second Choice: 4 points	B: $16 \times 4 = 64$	B: $14 \times 4 = 56$	B: $12 \times 4 = 48$	A: $4 \times 4 = 16$	A: $2 \times 4 = 8$
Third Choice: 3 points	C: $16 \times 3 = 48$	A: $14 \times 3 = 42$	E: $12 \times 3 = 36$	B: $4 \times 3 = 12$	D: $2 \times 3 = 6$
Fourth Choice: 2 points	D: $16 \times 2 = 32$	C: $14 \times 2 = 28$	C: $12 \times 2 = 24$	D: $4 \times 2 = 8$	B: $2 \times 2 = 4$
Fifth Choice: 1 point	E: $16 \times 1 = 16$	E: $14 \times 1 = 14$	A: $12 \times 1 = 12$	E: $4 \times 1 = 4$	C: $2 \times 1 = 2$

A gets $80 + 42 + 12 + 16 + 8 = 158$ points.
B gets $64 + 56 + 48 + 12 + 4 = 184$ points.
C gets $48 + 28 + 24 + 20 + 2 = 122$ points.
D gets $32 + 70 + 60 + 8 + 6 = 176$ points.
E gets $16 + 14 + 36 + 4 + 10 = 80$ points.

B receives the most points, so B is the winner.

b. No. D gets a majority of first-place votes, but B wins the election.

c. No. D wins all head-to-head comparisons, but B wins the election.

19. a. A receives the most first-place votes, and is the winner.

b. Yes. A has a majority of the first-place votes, and wins.

c. Yes. A wins in comparisons to B and C.

d. New preference table:

Number of Votes	7	3	2
First Choice	A	B	A
Second Choice	B	C	C
Third Choice	C	A	B

A has the majority of first-place votes, and wins using the plurality method.

e. Yes. A still receives the most first-place votes, and wins.

f. No. The fact that all four criteria are satisfied in a particular case does not mean that the method used always satisfies all four criteria.

21. Answers will vary.

23. Answers will vary.

25. Answers will vary.

27. Answers will vary.

29. Answers will vary.

31. Answers will vary.

14.3 Apportionment Methods

Check Point 1.

a. Standard divisor $= \dfrac{\text{total population}}{\text{number of allocated items}} = \dfrac{10,000}{200} = 50$

b. Standard quota for state A $= \dfrac{\text{population of state A}}{\text{standard divisor}} = \dfrac{1112}{50} = 22.24$

Standard quota for state B $= \dfrac{\text{population of state B}}{\text{standard divisor}} = \dfrac{1118}{50} = 22.36$

Standard quota for state C $= \dfrac{\text{population of state C}}{\text{standard divisor}} = \dfrac{1320}{50} = 26.4$

Standard quota for state D $= \dfrac{\text{population of state D}}{\text{standard divisor}} = \dfrac{1515}{50} = 30.3$

Standard quota for state E $= \dfrac{\text{population of state E}}{\text{standard divisor}} = \dfrac{4935}{50} = 98.7$

Table 14.27   **Population of Amador by State**

State	A	B	C	D	E	Total
Population (in thousands)	1112	1118	1320	1515	4935	10,000
Standard quota	22.24	22.36	26.4	30.3	98.7	200

Check Point 2.

State	Population (in thousands)	Standard Quota	Lower Quota	Fractional Part	Surplus	Final Apportionment
A	1112	22.24	22	0.24		22
B	1118	22.36	22	0.36		22
C	1320	26.4	26	0.4 (next largest)	1	27
D	1515	30.3	30	0.3		30
E	4935	98.7	98	0.7 (largest)	1	99
Total	10,000	200	198			200

Check Point 3.

State	Population (in thousands)	Modified Quota (using $d = 49.3$)	Modified Lower Quota	Final Apportionment
A	1112	22.56	22	22
B	1118	22.68	22	22
C	1320	26.77	26	26
D	1515	30.73	30	30
E	4935	100.10	100	100
Total	10,000		200	200

Check Point 4.

State	Population (in thousands)	Modified Quota (using $d = 50.5$)	Modified Upper Quota
A	1112	22.02	23
B	1118	22.14	23
C	1320	26.14	27
D	1515	30	30
E	4935	97.72	98
Total	10,000		201

This sum should be 200, not 201.

State	Population (in thousands)	Modified Quota (using $d = 50.6$)	Modified Upper Quota	Final Apportionment
A	1112	21.98	22	22
B	1118	22.09	23	23
C	1320	26.09	27	27
D	1515	29.94	30	30
E	4935	97.53	98	98
Total	10,000		200	200

Check Point 5.

State	Population (in thousands)	Modified Quota (using $d = 49.8$)	Modified Rounded Quota
A	1112	22.33	22
B	1118	22.45	22
C	1320	26.51	27
D	1515	30.42	30
E	4935	99.10	99
Total	10,000		200

Exercise Set 14.3

1. **a.** Standard divisor = $\frac{1600}{80} = 20$. There are 20,000 people for each seat in congress.

b–c.

State	A	B	C	D
Standard quota	$\frac{138}{20} = 6.9$	$\frac{266}{20} = 13.3$	$\frac{534}{20} = 26.7$	$\frac{662}{20} = 33.1$
Lower quota	6	13	26	33
Upper Quota	7	14	27	34

3.

State	Population (in thousands)	Standard Quota	Lower Quota	Fractional Part	Surplus	Final Apportionment
A	138	6.9	6	0.9	1	7
B	266	13.3	13	0.3		13
C	534	26.7	26	0.7	1	27
D	662	33.1	33	0.1		33
Total	1600	80	78			80

5.

School	Enrollment	Standard Quota	Lower Quota	Fractional Part	Surplus	Final Apportionment
Humanities	1050	30.26	30	0.26		30
Social Science	1410	40.63	40	0.63	1	41
Engineering	1830	52.74	52	0.74	1	53
Business	2540	73.20	73	0.20		73
Education	3580	103.17	103	0.17		103
Total	10,410	300	298			300

We use $\frac{10,410}{300} = 34.7$ as the standard divisor.

7.

State	Population	Modified Quota ($d = 32,920$)	Modified Lower Quota	Final Apportionment
A	126,316	3.84	3	3
B	196,492	5.97	5	5
C	425,264	12.92	12	12
D	526,664	15.998	15	15
E	725,264	22.03	22	22
Total	2,000,000		57	57

9. There are 15,000 patients. The standard divisor is $\frac{15,000}{150}$, or 100. Try a modified divisor of 98.

Clinic	Average Weekly Patient Load	Modified Quota	Modified Lower Quota	Final Apportionment
A	1714	17.49	17	17
B	5460	55.71	55	55
C	2440	24.90	24	24
D	5386	54.96	54	54
Total	15,000		150	150

11.

Precinct	Crimes	Modified Quota ($d = 16$)	Modified Upper Quota	Final Apportionment
A	446	27.88	28	28
B	526	32.88	33	33
C	835	52.19	53	53
D	227	14.19	15	15
E	338	21.13	22	22
F	456	28.5	29	29
Total	2828		180	180

13. There is a total of $2025 to be invested. The standard divisor is $\frac{2025}{30}$, or 67.5. Try a modified divisor of 72.

Person	Amount	Modified Quota	Modified Upper Quota	Final Apportionment
A	795	11.04	12	12
B	705	9.79	10	10
C	525	7.29	8	8
Total	2025		30	30

15.

Course	Enrollment	Modified Quota ($d = 29.6$)	Modified Rounded Quota	Final Apportionment
Introductory Algebra	130	4.39	4	4
Intermediate Algebra	282	9.53	10	10
Liberal Arts Math	188	6.35	6	6
Total	600		20	20

17. The total number of passengers is 11,060. The standard divisor is $\frac{11,060}{200}$ or 55.3. Try a modified divisor of 55.5.

Route	Average Number of Passengers	Modified Quota	Modified Rounded Quota	Final Apportionment
A	1087	19.59	20	20
B	1323	23.84	24	24
C	1592	28.68	29	29
D	1596	28.76	29	29
E	5462	98.41	98	98
Total	11,060		200	200

19. The total number of patients is 2000. The standard divisor is $\frac{2000}{250}$, or 8. Use Hamilton's method.

Shift	Average Number of Patients	Standard Quota	Lower Quota	Fractional Part	Surplus	Final Apportionment
A	453	56.625	56	0.625	1	57
B	650	81.25	81	0.25		81
C	547	68.375	68	0.375		68
D	350	43.75	43	0.75	1	44
Total	2000	250	248			250

21. Try a modified divisor of 8.06. Use Adams' method.

Shift	Average Number of Patients	Modified Quota	Modified Upper Quota	Final Apportionment
A	453	56.20	57	57
B	650	80.65	81	81
C	547	67.87	68	68
D	350	43.42	44	44
Total	2000		250	250

23. The total population is 3,615,920. The standard divisor is $\frac{3,615,920}{105}$, or 34,437.333. Use Hamilton's method.

State	Population	Standard Quota	Lower Quota	Fractional Part	Surplus	Final Apportionment
Connecticut	236,841	6.88	6	0.88	1	7
Delaware	55,540	1.61	1	0.61	1	2
Georgia	70,835	2.06	2	0.06		2
Kentucky	68,705	1.995	1	0.995	1	2
Maryland	278,514	8.09	8	0.09		8
Massachusetts	475,327	13.80	13	0.80	1	14
New Hampshire	141,822	4.12	4	0.12		4
New Jersey	179,570	5.21	5	0.21		5
New York	331,589	9.63	9	0.63	1	10
North Carolina	353,523	10.27	10	0.27		10
Pennsylvania	432,879	12.57	12	0.57	1	13
Rhode Island	68,446	1.99	1	0.99	1	2
South Carolina	206,236	5.99	5	0.99	1	6
Vermont	85,533	2.48	2	0.48		2
Virginia	630,560	18.31	18	0.31		18
Total	3,615,920	105.005	97			105

25. Use Adams' method with $d = 36,100$.

State	Population	Modified Quota	Modified Upper Quota	Final Apportionment
Connecticut	236,841	6.56	7	7
Delaware	55,540	1.54	2	2
Georgia	70,835	1.96	2	2
Kentucky	68,705	1.90	2	2
Maryland	278,514	7.72	8	8
Massachusetts	475,327	13.17	14	14
New Hampshire	141,822	3.93	4	4
New Jersey	179,570	4.97	5	5
New York	331,589	9.19	10	10
North Carolina	353,523	9.79	10	10
Pennsylvania	432,879	11.99	12	12
Rhode Island	68,446	1.90	2	2
South Carolina	206,236	5.71	6	6
Vermont	85,533	2.37	3	3
Virginia	630,560	17.47	18	18
Total	3,615,920		105	105

27. Answers will vary.

29. Answers will vary.

31. Answers will vary.

33. Answers will vary.

35. Answers will vary.

37. Answers will vary.

39. Answers will vary.

41. Answers will vary.

43. Answers will vary.

45. Answers will vary.

14.4   Flaws of Apportionment Methods

Check Point 1.

We begin with 99 seats in the Congress. First we compute the standard divisor.

$$\text{Standard divisor} = \frac{\text{total population}}{\text{number of allocated items}} = \frac{20,000}{99} = 202.02$$

Using this value, make a table showing apportionment using Hamilton's method.

State	Population	Standard Quota	Lower Quota	Fractional Part	Surplus Seats	Final Apportionment
A	2060	10.20	10	0.20		10
B	2080	10.30	10	0.30	1	11
C	7730	38.26	38	0.26		38
D	8130	40.24	40	0.24		40
Total	20,000	99	98			99

Now let's see what happens with 100 seats in Congress.

First we compute the standard divisor.

$$\text{Standard divisor} \frac{\text{total population}}{\text{number of allocated items}} = \frac{20,000}{100} = 200.$$

Using this value, make a table showing apportionment using Hamilton's method.

State	Population	Standard Quota	Lower Quota	Fractional Part	Surplus Seats	Final Apportionment
A	2060	10.3	10	0.3		10
B	2080	10.4	10	0.4		10
C	7730	38.65	38	0.65	1	39
D	8130	40.65	40	0.65	1	41
Total	20,000	100	98			100

The final apportionments are summarized in the following table.

State	Apportionment with 99 seats	Apportionment with 100 seats
A	10	10
B	11	10
C	38	39
D	40	41

When the number of seats increased from 99 to 100, B's apportionment decreased from 11 to 10.

Check Point 2.

a. We use Hamilton's method to find the apportionment for each state with its original population. First we compute the standard divisor.

$$\text{Standard divisor} = \frac{\text{total population}}{\text{number of allocated items}} = \frac{200,000}{100} = 2000$$

Using this value, we show the apportionment in the following table.

State	Original Population	Standard Quota	Lower Quota	Fractional Part	Surplus Seats	Final Apportionment
A	19,110	9.56	9	0.56	1	10
B	39,090	19.55	19	0.55		19
C	141,800	70.9	70	0.9	1	71
Total	200,000	100.01	98			100

b. The fraction for percent increase is the amount of increase divided by the original amount. The percent increase in the population of each state is determined as follows.

$$\text{State A: } \frac{19,302 - 19,110}{19,110} = \frac{192}{19,110} \approx 0.01005 = 1.005\%$$

$$\text{State B: } \frac{39,480 - 39,090}{39,090} = \frac{390}{39,090} \approx 0.00998 = 0.998\%$$

State A is increasing at a rate of 1.005%. This is faster than State B, which is increasing at a rate of 0.998%.

c. We use Hamilton's method to find the apportionment for each state with its new population. First we compute the standard divisor.

$$\text{Standard divisor} = \frac{\text{total population}}{\text{number of allocated items}} = \frac{200,582}{100} = 2005.82$$

Using this value, we show the apportionment in the following table.

State	New Population	Standard Quota	Lower Quota	Fractional Part	Surplus Seats	Final Apportionment
A	19,302	9.62	9	0.62		9
B	39,480	19.68	19	0.68	1	20
C	141,800	70.69	70	0.69	1	71
Total	200,582	99.99	98			100

The final apportionments are summarized in the following table.

State	Growth Rate	Original Apportionment	New Apportionment
A	1.005%	10	9
B	0.998%	19	20
C	0%	71	71

State A loses a seat to State B, even though the population of State A is increasing at a faster rate. This is an example of the population paradox.

Check Point 3.

a. We use Hamilton's method to find the apportionment for each school.
    First we compute the standard divisor.

$$\text{Standard divisor} = \frac{\text{total population}}{\text{number of allocated items}} = \frac{12,000}{100} = 120$$

Using this value, we show the apportionment in the following table.

School	Enrollment	Standard Quota	Lower Quota	Fractional Part	Surplus	Final Apportionment
East High	2574	21.45	21	0.45		21
West High	9426	78.55	78	0.55	1	79
Total	12,000	100	99			100

b. Again we use Hamilton's method.

$$\text{Standard divisor} = \frac{\text{total population}}{\text{number of allocated items}} = \frac{12,750}{106} = 120.28$$

Using this value, we show the apportionment in the following table

School	Enrollment	Standard Quota	Lower Quota	Fractional Part	Surplus	Final Apportionment
East High	2574	21.40	21	0.40	1	22
West High	9426	78.37	78	0.37		78
North High	750	6.24	6	0.24		6
Total	12,750	106.01	105			106

West High has lost a counselor to East High.

Exercise Set 14.4

1. a. The standard divisor is $\frac{1800}{30}$, or 60.

Course	Enrollment	Standard Quota	Lower Quota	Fractional Part	Surplus	Final Apportionment
College Algebra	978	16.30	16	0.30		16
Statistics	500	8.33	8	0.33		8
Liberal Arts Math	322	5.37	5	0.37	1	6
Total	1800	30	29			30

b. The standard divisor is $\frac{1800}{31}$, or 58.06.

Course	Enrollment	Standard Quota	Lower Quota	Fractional Part	Surplus	Final Apportionment
College Algebra	978	16.84	16	0.84	1	17
Statistics	500	8.61	8	0.61	1	9
Liberal Arts Math	322	5.55	5	0.55		5
Total	1800	31	29			31

Liberal Arts Math loses a teaching assistant when the total number of teaching assistants is raised from 30 to 31. This is an example of the Alabama paradox.

3. Standard divisor with 40 seats: $\frac{20,000}{40} = 500$. Use Hamilton's method.

State	Population	Standard Quota	Lower Quota	Fractional Part	Surplus	Final Apportionment
A	680	1.36	1	0.36	1	2
B	9150	18.30	18	0.30		18
C	10,170	20.34	20	0.34		20
Total	20,000	40	39			40

Standard divisor with 41 seats: $\frac{20,000}{41} = 487.8$. Use Hamilton's method.

State	Population	Standard Quota	Lower Quota	Fractional Part	Surplus	Final Apportionment
A	680	1.39	1	0.39		1
B	9150	18.76	18	0.76	1	19
C	10,170	20.85	20	0.85	1	21
Total	20,000	41	39			41

State A loses a seat when the total number of seats increases from 40 to 41.

5. a. Standard divisor: $\frac{3760}{24} = 156.7$. Use Hamilton's method.

State	Original Population	Standard Quota	Lower Quota	Fractional Part	Surplus	Final Apportionment
A	530	3.38	3	0.38	1	4
B	990	6.32	6	0.32		6
C	2240	14.30	14	0.30		14
Total	3760	24	23			24

b. Percent increase for state A: $\frac{680 - 530}{530} \approx 0.283 = 28.3\%$

Percent increase for state B: $\frac{1250 - 990}{990} \approx 0.263 = 26.3\%$

Percent increase for state C: $\frac{2570 - 2240}{2240} \approx 0.147 = 14.7\%$

c. Standard divisor: $\frac{4500}{24} = 187.5$. Use Hamilton's method.

State	New Population	Standard Quota	Lower Quota	Fractional Part	Surplus	Final Apportionment
A	680	3.63	3	0.63		3
B	1250	6.67	6	0.67	1	7
C	2570	13.71	13	0.71	1	14
Total	4500	24.01	22			24

A loses a seat while B gains, even though A has a faster increasing population. The population paradox does occur.

7. Original standard divisor: $\frac{8880}{40} = 222$

District	Original Population	Standard Quota	Lower Quota	Fractional Part	Surplus	Final Apportionment
A	1188	5.35	5	0.35		5
B	1424	6.41	6	0.41		6
C	2538	11.43	11	0.43	1	12
D	3730	16.80	16	0.80	1	17
Total	8880	39.99	38			40

New standard divisor: $\frac{9000}{40} = 225$

District	New Population	Standard Quota	Lower Quota	Fractional Part	Surplus	Final Apportionment
A	1188	5.28	5	0.28		5
B	1420	6.311	6	0.311	1	7
C	2544	11.307	11	0.307		11
D	3848	17.10	17	0.10		17
Total	9000	39.998	39			40

Percent increase by state:

A: 0% (no change)

B: $\frac{1420 - 1424}{1424} \approx -0.0028 = -0.28\%$

C: $\frac{2544 - 2538}{2538} \approx 0.0024 = 0.24\%$

D: $\frac{3848 - 3730}{3730} \approx 0.032 = 3.2\%$

C loses a truck to B even though C increased in population faster than B. This shows the population paradox occurs.

9. a. Standard divisor: $\frac{10,000}{100} = 100$

Branch	Employees	Standard Quota	Lower Quota	Fractional Part	Surplus	Final Apportionment
A	1045	10.45	10	0.45		10
B	8955	89.55	89	0.55	1	90
Total	10,000	100	99			100

b. New standard divisor: $\dfrac{10,525}{105} = 100.238$

Branch	Employees	Standard Quota	Lower Quota	Fractional Part	Surplus	Final Apportionment
A	1045	10.43	10	0.43	1	11
B	8955	89.34	89	0.34		89
C	525	5.24	5	0.24		5
Total	10,525	105.01	104			105

Branch B loses a promotion when branch C is added. This means the new-states paradox has occurred.

11. a. Standard divisor: $\dfrac{9450 + 90,550}{100} = 1000$

State	Population	Standard Quota	Lower Quota	Fractional Part	Surplus	Final Apportionment
A	9450	9.45	9	0.45		9
B	90,550	90.55	90	0.55	1	91
Total	100,000	100	99			100

b. New standard divisor: $\dfrac{100,000 + 10,400}{110} = 1003.64$

State	Population	Standard Quota	Lower Quota	Fractional Part	Surplus	Final Apportionment
A	9450	9.42	9	0.42	1	10
B	90,550	90.22	90	0.22		90
C	10,400	10.36	10	0.36		10
Total	110,400	110	109			110

State B loses a seat when state C is added.

13. a.

State	Population	Modified Quota	Modified Lower Quota	Final Apportionment
A	99,000	6.39	6	6
B	214,000	13.81	13	13
C	487,000	31.42	31	31
Total	800,000		50	50

b.

State	Population	Modified Quota	Modified Lower Quota	Final Apportionment
A	99,000	6.39	6	6
B	214,000	13.81	13	13
C	487,000	31.42	31	37
D	116,000	7.48	7	7
Total	916,000		57	57

The new-states paradox does not occur. As long as the modified divisor, d, remains the same, adding a new state cannot change the number of seats held by existing states.

15. Answers will vary.

17. Answers will vary.

19. The only true statement is d.

Review Exercises

1.

Number of Votes	4	3	3	2
First Choice	A	B	C	C
Second Choice	B	D	B	B
Third Choice	C	C	D	A
Fourth Choice	D	A	A	D

2. $9 + 5 + 4 + 2 + 2 + 1 = 23$

3. 4

4. $9 + 5 + 2 = 16$

5. $9 + 5 = 14$

6. M receives 12 first-choice votes, compared to 10 for C and 2 for D, so M (Musical) is selected.

7.

Number of Votes	10	8	4	2
First Choice: 3 points	C: $10 \times 3 = 30$	M: $8 \times 3 = 24$	M: $4 \times 3 = 12$	D: $2 \times 3 = 6$
Second Choice: 2 points	D: $10 \times 2 = 20$	C: $8 \times 2 = 16$	D: $4 \times 2 = 8$	M: $2 \times 2 = 4$
Third Choice: 1 point	M: $10 \times 1 = 10$	D: $8 \times 1 = 8$	C: $4 \times 1 = 4$	C: $2 \times 1 = 2$

C gets $30 + 16 + 4 + 2 = 52$ points.
D gets $20 + 8 + 8 + 6 = 42$ points.
M gets $10 + 24 + 12 + 4 = 50$ points.

C (Comedy) gets the most points and is chosen.

8. There are 24 voters, so 13 votes are needed for a majority. None of the candidates has 13 first-place votes. D has the fewest first-place votes and is eliminated in the next round.

Number of Votes	10	14
First Choice	C	M
Second Choice	M	C

M (Musical) has 14 first-place votes, a majority, so a musical is selected.

9. There are 3 choices so we make $\frac{3(3-1)}{2} = 3$ comparisons.

Comparison	Vote Results	Conclusion
C vs. D	18 voters prefer C to D. 6 voters prefer D to C.	C wins and gets 1 point.
C vs. M	10 voters prefer C to M. 14 voters prefer M to C.	M wins and gets 1 point.
D vs. M	12 voters prefer D to M. 12 voters prefer M to D.	D and M tie. Each gets $\frac{1}{2}$ point.

C gets 1 point, D gets $\frac{1}{2}$ point, and M gets $1\frac{1}{2}$ points. So M (Musical) wins, and is selected.

10. A receives 40 first-place votes, compared to 30 for B, 6 for C, and 2 for D. So A wins.

11.

Number of Votes	40	30	6	2
First Choice: 4 points	A: $40 \times 4 = 160$	B: $30 \times 4 = 120$	C: $6 \times 4 = 24$	D: $2 \times 4 = 8$
Second Choice: 3 points	B: $40 \times 3 = 120$	C: $30 \times 3 = 90$	D: $6 \times 3 = 18$	B: $2 \times 3 = 6$
Third Choice: 2 points	C: $40 \times 2 = 80$	D: $30 \times 2 = 60$	B: $6 \times 2 = 12$	C: $2 \times 2 = 4$
Fourth Choice: 1 point	D: $40 \times 1 = 40$	A: $30 \times 1 = 30$	A: $6 \times 1 = 6$	A: $2 \times 1 = 2$

A gets $160 + 30 + 6 + 2 = 198$ points.
B gets $120 + 120 + 12 + 6 = 258$ points.
C gets $80 + 90 + 24 + 4 = 198$ points.
D gets $40 + 60 + 18 + 8 = 126$ points.

B receives the most points, and wins.

12. There are 78 voters, so 40 first-place votes are needed for a majority. A has 40 first-place votes, and wins.

13. There are 4 candidates, so $\frac{4(4-1)}{2} = 6$ comparisons are needed.

Comparison	Vote Results	Conclusion
A vs. B	40 voters prefer A to B. 38 voters prefer B to A.	A wins and gets 1 point.
A vs. C	40 voters prefer A to C. 38 voters prefer C to A.	A wins and gets 1 point.
A vs. D	40 voters prefer A to D. 38 voters prefer D to A.	A wins and gets 1 point.
B vs. C	72 voters prefer B to C. 6 voters prefer C to B.	B wins and gets 1 point.
B vs. D	70 voters prefer B to D. 8 voters prefer D to B.	B wins and gets 1 point.
C vs. D	76 voters prefer C to D. 2 voters prefer D to C.	C wins and gets 1 point.

A gets 3 points, B gets 2 points, C gets 1 point, and D gets 0 points. So A wins.

14.

Number of Votes	1500	600	300
First Choice: 4 points	A: $1500 \times 4 = 6000$	B: $600 \times 4 = 2400$	C: $300 \times 4 = 1200$
Second Choice: 3 points	B: $1500 \times 3 = 4500$	D: $600 \times 3 = 1800$	B: $300 \times 3 = 900$
Third Choice: 2 points	C: $1500 \times 2 = 3000$	C: $600 \times 2 = 1200$	D: $300 \times 2 = 600$
Fourth Choice: 1 point	D: $1500 \times 1 = 1500$	A: $600 \times 1 = 600$	A: $300 \times 1 = 300$

A gets $6000 + 600 + 300 = 6900$ points.
B gets $4500 + 2400 + 900 = 7800$ points.
C gets $3000 + 1200 + 1200 = 5400$ points.
D gets $1500 + 1800 + 600 = 3900$ points.

B receives the most points, and wins.

15. A has a majority of first-place votes. In Exercise 14, B wins and so the majority criterion is not satisfied.

16. A is favored above all others using a head-to-head comparison. This is automatically true, since A has a majority of first place votes. In Exercise 14, B wins and so the head to head criterion is not satisfied.

17. There are 2500 voters. 1251 first-place votes are needed for a majority. B has 1500 first-place votes, and is the winner.

18. B is favored above all others using a head-to-head comparison. This is automatically true, since B has a majority of first-place votes. In Exercise 17, B wins and so the head-to-head criterion is satisfied.

19. A receives 180 first-place votes, compared with 100 for B, 30 for C, and 40 for D. Therefore A wins.

20.

Number of Votes	180	100	40	30
First Choice: 4 points	A: $180 \times 4 = 720$	B: $100 \times 4 = 400$	D: $40 \times 4 = 160$	C: $30 \times 4 = 120$
Second Choice: 3 points	B: $180 \times 3 = 540$	D: $100 \times 3 = 300$	B: $40 \times 3 = 120$	B: $30 \times 3 = 90$
Third Choice: 2 points	C: $180 \times 2 = 360$	A: $100 \times 2 = 200$	C: $40 \times 2 = 80$	A: $30 \times 2 = 60$
Fourth Choice: 1 point	D: $180 \times 1 = 180$	C: $100 \times 1 = 100$	A: $40 \times 1 = 40$	D: $30 \times 1 = 30$

A gets $720 + 200 + 40 + 60 = 1020$ points.
B gets $540 + 400 + 120 + 90 = 1150$ points.
C gets $360 + 100 + 80 + 120 = 660$ points.
D gets $180 + 300 + 160 + 30 = 670$ points.

B gets the most points, and wins.

21. There are 350 voters. 176 first-place votes are needed for a majority. A has 180 votes, a majority, and wins.

22. There are 4 candidates, and therefore $\frac{4(4-1)}{2} = 6$ comparisons.

Comparison	Vote Results	Conclusion
A vs. B	180 voters prefer A to B. 170 voters prefer B to A.	A wins and gets 1 point.
A vs. C	280 voters prefer A to C. 70 voters prefer C to A.	A wins and gets 1 point.
A vs. D	210 voters prefer A to D. 140 voters prefer D to A.	A wins and gets 1 point.
B vs. C	320 voters prefer B to C. 30 voters prefer C to B.	B wins and gets 1 point.
B vs. D	310 voters prefer B to D. 40 voters prefer D to B.	B wins and gets 1 point.
C vs. D	210 voters prefer C to D. 140 voters prefer D to C.	C wins and gets 1 point.

A gets 3 points, B gets 2 points, C gets 1 point, and D gets 0 points. Therefore A wins.

23. A has a majority of first-place votes. Based on Exercises 19–22, only the Borda count method violates the majority criterion. B wins by the Borda count method.

24. There are 1450 voters. 726 first-place votes are needed for a majority. No candidate has a majority. A has the fewest first-place votes and is eliminated in the next round.

Number of Votes	900	550
First Choice	B	C
Second Choice	C	B

B has the majority of first-place votes, and wins.

25. There is a new preference table:

Number of Votes	700	400	350
First Choice	B	A	C
Second Choice	C	B	A
Third Choice	A	C	B

No candidate has a majority of first-place votes. C has the fewest first-place votes, and is eliminated in the next round.

Number of Votes	700	750
First Choice	B	A
Second Choice	A	B

A has a majority of first-place votes, and wins. This does not satisfy the monotonicity criterion, since the only change gave B more first-place votes, but after the change B lost the election.

26. A has 400 first-place votes, compared to 200 for B and 250 for C. Therefore A wins.

27.

Number of Votes	400	450
First Choice	A	C
Second Choice	C	A

C has the majority of first-place votes, and wins this election. The irrelevant alternatives criterion is not satisfied, because removing B changes the winner from A to C.

28.

Number of Votes	400	250	200
First Choice: 3 points	A: $400 \times 3 = 1200$	C: $250 \times 3 = 750$	B: $200 \times 3 = 600$
Second Choice: 2 points	B: $400 \times 2 = 800$	B: $250 \times 2 = 500$	C: $200 \times 2 = 400$
Third Choice: 1 point	C: $400 \times 1 = 400$	A: $250 \times 1 = 250$	A: $200 \times 1 = 200$

A gets $1200 + 250 + 200 = 1650$ points.
B gets $800 + 500 + 600 = 1900$ points.
C gets $400 + 750 + 400 = 1550$ points.

B gets the most points, and wins.

29.

Number of Votes	400	450
First Choice: 2 points	A: $400 \times 2 = 800$	B: $450 \times 2 = 900$
Second Choice: 1 point	B: $400 \times 1 = 400$	A: $450 \times 1 = 450$

A gets $800 + 450 = 1250$ points.
B gets $400 + 900 = 1300$ points.

B still gets the most points, and wins. The same thing happens if A drops out instead of C, and so the irrelevant alternatives criterion is satisfied.

30. $\dfrac{275 + 392 + 611 + 724}{40} = \dfrac{2002}{40} = 50.05$

31. With a standard divisor of 50.05:

Clinic	A	B	C	D
Average weekly patient load	275	392	611	724
Standard Quota	5.49	7.83	12.21	14.47

32. Using the results of Exercise 31:

Clinic	Standard Quota	Lower Quota	Upper Quota
A	5.49	5	6
B	7.83	7	8
C	12.21	12	13
D	14.47	14	15

33.

Clinic	Standard Quota	Lower Quota	Fractional Part	Surplus	Final Apportionment
A	5.49	5	0.49	1	6
B	7.83	7	0.83	1	8
C	12.21	12	0.21		12
D	14.47	14	0.47		14
Total	40	38			40

34.

Clinic	Average Weekly Patient Load	Modified Quota ($d = 48$)	Modified Lower Quota	Final Apportionment
A	275	5.73	5	5
B	392	8.17	8	8
C	611	12.73	12	12
D	724	15.08	15	15
Total	2002		40	40

35.

Clinic	Average Weekly Patient Load	Modified Quota ($d = 52$)	Modified Upper Quota	Final Apportionment
A	275	5.29	6	6
B	392	7.54	8	8
C	611	11.75	12	12
D	724	13.92	14	14
Total	2002		40	40

36.

Clinic	Average Weekly Patient Load	Modified Quota ($d = 49.95$)	Modified Rounded Quota	Final Apportionment
A	275	5.51	6	6
B	392	7.85	8	8
C	611	12.23	12	12
D	724	14.49	14	14
Total	2002		40	40

37. Standard divisor: $\dfrac{3320 + 10,060 + 15,020 + 19,600}{200} = \dfrac{48,000}{200} = 240$

State	Population	Standard Quota	Lower Quota	Fractional Part	Surplus	Final Apportionment
A	3320	13.83	13	0.83	1	14
B	10,060	41.92	41	0.92	1	42
C	15,020	62.58	62	0.58		62
D	19,600	81.67	81	0.67	1	82
Total	48,000	200	197			200

38. Try modified divisor $d = 238$.

State	Population	Modified Quota	Modified Lower Quota	Final Apportionment
A	3320	13.95	13	13
B	10,060	42.27	42	42
C	15,020	63.11	63	63
D	19,600	82.35	82	82
Total	48,000		200	200

39. Try modified divisor $d = 242$.

State	Population	Modified Quota	Modified Upper Quota	Final Apportionment
A	3320	13.72	14	14
B	10,060	41.57	42	42
C	15,020	62.07	63	63
D	19,600	80.99	81	81
Total	48,000		200	200

40. Try modified divisor $d = 240.4$.

State	Population	Modified Quota	Modified Rounded Quota	Final Apportionment
A	3320	13.81	14	14
B	10,060	41.85	42	42
C	15,020	62.48	62	62
D	19,600	81.53	82	82
Total	48,000		200	200

41. a. Standard divisor: $\dfrac{7500}{150} = 50$

School	Enrollment	Standard Quota	Lower Quota	Fractional Part	Surplus	Final Apportionment
A	370	7.4	7	0.4	1	8
B	3365	67.3	67	0.3		67
C	3765	75.3	75	0.0		75
Total	7500	150	149			150

b. Standard divisor: $\dfrac{7500}{151} = 49.67$

School	Enrollment	Standard Quota	Lower Quota	Fractional Part	Surplus	Final Apportionment
A	370	7.45	7	0.45		7
B	3365	67.75	67	0.75	1	68
C	3765	75.80	75	0.80	1	76
Total	7500	151	149			151

The Alabama paradox occurs. A loses a laptop when the overall number of laptops changes from 150 to 151.

42. a. Standard divisor: $\dfrac{200,000}{100} = 2000$

School	Original Population	Standard Quota	Lower Quota	Fractional Part	Surplus	Final Apportionment
A	143,796	71.90	71	0.90	1	72
B	41,090	20.55	20	0.55		20
C	15,114	7.56	7	0.56	1	8
Total	200,000	100.01	98			100

b. Percent increase of B: $\dfrac{41,420 - 41,090}{41,090} \approx 0.0080 = 0.8\%$

Percent increase of C: $\dfrac{15,304 - 15,114}{15,114} \approx 0.0126 \approx 1.3\%$

c. Standard divisor: $\dfrac{200,520}{100} = 2005.2$

School	New Population	Standard Quota	Lower Quota	Fractional Part	Surplus	Final Apportionment
A	143,796	71.71	71	0.71	1	72
B	41,420	20.66	20	0.66	1	21
C	15,304	7.63	7	0.63		7
Total	200,520	100	98			100

The population paradox occurs. C loses a seat to B, even though C is growing faster.

43. a. Standard divisor: $\dfrac{1650}{33} = 50$

Branch	Employees	Standard Quota	Lower Quota	Fractional Part	Surplus	Final Apportionment
A	372	7.44	7	0.44		7
B	1278	25.56	25	0.56	1	26
Total	1650	33	32			33

b. Standard divisor: $\dfrac{2005}{40} = 50.125$

Branch	Employees	Standard Quota	Lower Quota	Fractional Part	Surplus	Final Apportionment
A	372	7.42	7	0.42		7
B	1278	25.50	25	0.50	1	26
C	355	7.08	7	0.08		7
Total	2005	40	39			40

The new-states paradox does not occur. Neither branch A nor branch B loses any promotions.

44. False. Answers will vary.

Chapter 14 Test

1. $1200 + 900 + 900 + 600 = 3600$

2. 600

3. $900 + 600 = 1500$

4. $900 + 600 = 1500$

5. A received 1200 first-place votes, B received 1500, and C received 900. Therefore B wins.

6.

Number of Votes	1200	900	900	600
First Choice: 3 points	A: $1200 \times 3 = 3600$	C: $900 \times 3 = 2700$	B: $900 \times 3 = 2700$	B: $600 \times 3 = 1800$
Second Choice: 2 points	B: $1200 \times 2 = 2400$	A: $900 \times 2 = 1800$	C: $900 \times 2 = 1800$	A: $600 \times 2 = 1200$
Third Choice: 1 point	C: $1200 \times 1 = 1200$	B: $900 \times 1 = 900$	A: $900 \times 1 = 900$	C: $600 \times 1 = 600$

A gets $3600 + 1800 + 900 + 1200 = 7500$ points.
B gets $2400 + 900 + 2700 + 1800 = 7800$ points.
C gets $1200 + 2700 + 1800 + 600 = 6300$ points.

B receives the most points and is the winner.

7. There are 3600 voters. 1801 first-place votes are needed for a majority. No candidate has a majority. C receives the fewest first-place votes and is eliminated in the next round.

Number of Votes	2100	1500
First Choice	A	B
Second Choice	B	A

A receives the majority of first-place votes, and wins.

8. There are 3 candidates. The number of comparisons is $\frac{3(3-1)}{2}$, or 3.

Comparison	Vote Results	Conclusion
A vs. B	2100 voters prefer A to B. 1500 voters prefer B to A.	A wins and gets 1 point.
A vs. C	1800 voters prefer A to C. 1800 voters prefer C to A.	A and C tie. Each gets $\frac{1}{2}$ point.
B vs. C	2700 voters prefer B to C. 900 voters prefer C to B.	B wins and gets 1 point.

A gets $1\frac{1}{2}$ points, B gets 1 point, and C gets $\frac{1}{2}$ point. Therefore A wins.

9.

Number of Votes	240	160	60
First Choice: 4 points	A: $240 \times 4 = 960$	C: $160 \times 4 = 640$	D: $60 \times 4 = 240$
Second Choice: 3 points	B: $240 \times 3 = 720$	B: $160 \times 3 = 480$	A: $60 \times 3 = 180$
Third Choice: 2 points	C: $240 \times 2 = 480$	D: $160 \times 2 = 320$	C: $60 \times 2 = 120$
Fourth Choice: 1 point	D: $240 \times 1 = 240$	A: $160 \times 1 = 160$	B: $60 \times 1 = 60$

A gets $960 + 160 + 180 = 1300$ points.
B gets $720 + 480 + 60 = 1260$ points.
C gets $480 + 640 + 120 = 1240$ points.
D gets $240 + 320 + 240 = 800$ points.

A gets the most points, and wins.

10. A has the majority of first-place votes. Based on Exercise 9, the majority criterion is satisfied.

11. A has 1500 first-place votes, whereas B and C have 1000 each. Therefore A wins.

12. B is favored when compared to A, by 2000 votes to 1500. B is favored when compared to C, by 2500 votes to 1000. So B is favored in each head-to-head comparison. Based on Exercise 11, the head-to-head criterion is not satisfied, because A wins the election.

13. There are 210 voters. 106 votes are needed for a majority. No candidate has a majority. B receives the fewest first-place votes and is eliminated in the next round.

Number of Votes	130	80
First Choice	C	A
Second Choice	A	C

C receives a majority of votes, and wins.

14. New preference table:

Number of Votes	100	60	50
First Choice	C	B	A
Second Choice	A	C	B
Third Choice	B	A	C

No candidate has a majority. A has the fewest first-place votes and is eliminated in the next round.

Number of Votes	100	110
First Choice	C	B
Second Choice	B	C

B has the majority of first-place votes, and wins. The monotonicity criterion is not satisfied, because the only change gave more first-place votes to C, but C lost the second election.

15. B has 90 first-place votes, C has 75, and A has 45. Therefore B wins. If C drops out, there is a new preference table:

Number of Votes	90	120
First Choice	B	A
Second Choice	A	B

A has a majority of first-place votes, and wins. This changed outcome shows that the irrelevant alternatives criterion is not satisfied.

16. $\dfrac{119 + 165 + 216}{10} = \dfrac{500}{10} = 50$

17. A: $\dfrac{119}{50} = 2.38$ B: $\dfrac{165}{50} = 3.3$ C: $\dfrac{216}{50} = 4.32$

18. A: 2, 3; B: 3, 4; C: 4, 5

19.

Clinic	Average Weekly Patient Load	Standard Quota	Lower Quota	Fractional Part	Surplus	Final Apportionment
A	119	2.38	2	0.38	1	3
B	165	3.3	3	0.3		3
C	216	4.32	4	0.32		4
Total	500	10	9			10

20.

Clinic	Average Weekly Patient Load	Modified Quota ($d = 42$)	Modified Lower Quota	Final Apportionment
A	119	2.83	2	2
B	165	3.93	3	3
C	216	5.14	5	5
Total	500		10	10

21.

Clinic	Average Weekly Patient Load	Modified Quota ($d = 56$)	Modified Upper Quota	Final Apportionment
A	119	2.13	3	3
B	165	2.95	3	3
C	216	3.86	4	4
Total	500		10	10

22.

Clinic	Average Weekly Patient Load	Modified Quota ($d = 47.7$)	Modified Rounded Quota	Final Apportionment
A	119	2.49	2	2
B	165	3.46	3	3
C	216	4.52	5	5
Total	500		10	10

23. New standard divisor: $\dfrac{500}{11} = 45.45$

Clinic	Average Weekly Patient Load	Standard Quota	Lower Quota	Fractional Part	Surplus	Final Apportionment
A	119	2.62	2	0.62		2
B	165	3.63	3	0.63	1	4
C	216	4.75	4	0.75	1	5
Total	500	11	9			11

The Alabama paradox occurs. Clinic A loses one doctor when the total number of doctors is raised from 10 to 11.

24. New standard divisor: $\dfrac{500 + 110}{12} = \dfrac{610}{12} = 50.83$

Clinic	Average Weekly Patient Load	Standard Quota	Lower Quota	Fractional Part	Surplus	Final Apportionment
A	119	2.34	2	0.34	1	3
B	165	3.25	3	0.25		3
C	216	4.25	4	0.25		4
D	110	2.16	2	0.16		2
Total	610	12	11			12

The new-states paradox does not occur. No clinic loses doctors when a new clinic is added.

25. Answers will vary.

Chapter 15
Graph Theory

15.1   Graphs, Paths, and Circuits

Check Point 1.

Graphs (a) and (b) both have vertices *A, B, C, D,* and *E.* Also, both graphs have edges *AB*, *AC*, *BD*, *BE*, *CD*, *CE*, and *DE*.

Because the two graphs have the same number of vertices connected to each other in the same way, they are the same. And in fact graph (b) is just graph (a) rotated clockwise and bent out of shape.

Check Point 2.

Draw points for the five land masses and label them *N, S, A, B,* and *C.*

There is one bridge that connects North Metroville to Island A, so one edge is drawn connecting vertex *N* to vertex *A.* Similarly, one edge connects vertex *A* with vertex *B,* and one edge connects vertex *B* with vertex *C.* Since there are two bridges connecting Island C to South Metroville, two edges connect vertex *C* with vertex *S.*

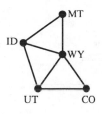

Check Point 3.

We use the abbreviations for the states to label the vertices: ID for Idaho, MT for Montana, WY for Wyoming, UT for Utah, and CO for Colorado. The precise placement of these vertices is not important.

Whenever two states share a common border, we connect the respective vertices with an edge. For example, Idaho shares a common border with Montana, with Wyoming, and with Utah. Continuing in this manner, we obtain the following graph.

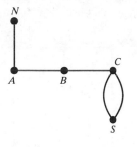

Check Point 4.

We use the letters in Figure 15.13 to label each vertex. Only one door connects the outside, E, with room B, so we draw one edge from vertex E to vertex B. Two doors connect the outside, E, to room D, so we draw two edges from E to D. Counting doors between the rooms, we complete the following graph.

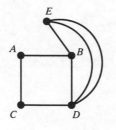

Check Point 5.

We label each of the corners and intersections with an upper-case letter and use points to represent the corners and street intersections. Now we are ready to draw the edges that represent the streets the security guard has to walk. Each street only needs to be walked once, so we draw one edge to represent each street. This results in the following graph.

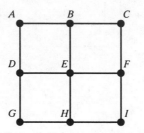

Check Point 6.

We systematically list which pairs of vertices are adjacent, working alphabetically. Thus, the adjacent vertices are A and B, A and C, A and D, A and E, B and C, and E and E.

Exercise Set 15.1

1. There are six edges attached to the Pittsburgh vertex, so Pittsburgh plays six games during the week. One edge connects the Pittsburgh vertex to the St. Louis vertex, so one game is against St. Louis. One edge connects the Pittsburgh vertex to Chicago, so one game is against Chicago. Two edges connect the Pittsburgh vertex to the Philadelphia vertex, so two games are against Philadelphia. Two edges connect the Pittsburgh vertex to the Montreal vertex, so two games are against Montreal.

3. No. Montreal is farther north than New York but is drawn lower on the graph. However, the graph is not drawn incorrectly. Only the games between teams are important, and these are represented by the edges. Geographic position is not relevant.

5. Possible answers:

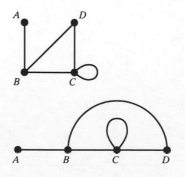

7. Both graphs have vertices *A*, *B*, *C*, and *D* and edges *AB*, *AC*, *AD*, and *BD*. The two graphs have the same number of vertices connected in the same way, so they are the same.

Possible answer:

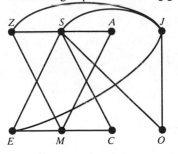

9. We label each student's vertex with the first letter of his or her name. An edge connecting two vertices represents a friendship prior to forming the homework group. The following graph results.

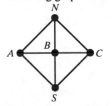

11. Label one vertex *N*, for North Gothamville. Label another *S*, for South Gothamville. Label the islands, from left to right, *A*, *B*, and *C*. Label three vertices accordingly. Use edges to represent bridges. The following graph results.

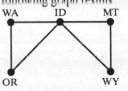

13. We use the abbreviations WA, OR, ID, MT, and WY to label the vertices representing Washington, Oregon, Idaho, Montana, and Wyoming. The following graph results.

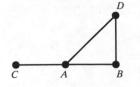

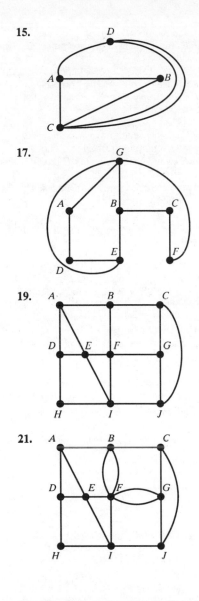

15.

17.

19.

21.

23. The degree of a vertex is the number of edges at that vertex. Thus, vertex *A* has degree 2, vertex *B* has degree 2, vertex *C* has degree 3, vertex *D* has degree 3, vertex *E* has degree 3, and vertex *F* has degree 1. (The loop at *E* counts for 2.)

25. Vertices *B* and *C* each have an edge connecting to *A*, so *B* and *C* are adjacent to *A*.

27. Starting at vertex *A*, we proceed to vertex *C*, then vertex *D*. This is one path from *A* to *D*. For a second path, start at vertex *A*, then proceed to vertex *B*, then *C*, then *D*.

29. The edges not included are the edge connecting *A* to *C*, and the edge connecting *D* to *F*.

321

31. While edge *CD* is included, the graph is connected. If we remove *CD*, the graph will be disconnected. Thus, *CD* is a bridge.

33. Edge *DF* is also a bridge. With it, the graph is connected. If *DF* is removed, vertex *F* stands alone, so the graph is disconnected.

35. Vertices *A, B, G, H,* and *I* each have two attached edges, which is an even number of edges. Thus *A, B, G, H,* and *I* are even vertices. Vertex *C* has five attached edges, vertex *E* has one, and vertices *D* and *F* have three. These are odd numbers of edges. Thus *C, E, D,* and *F* are odd vertices.

37. Vertex *F* has edges connecting to vertices *D, G,* and *I*. Thus *D, G,* and *I* are adjacent to *F*.

39. Begin at vertex *B*. Proceed to vertex *C*, then vertex *D*, then vertex *F*. This is one path from *B* to *F*. For a second path, begin at *B*, then proceed to *A*, then *C*, then *D*, then *F*.

41. Begin at vertex *G*. proceed to vertex *F*, then vertex *I*, then vertex *H*, then vertex *G*. This is a circuit. (The counterclockwise order also works.)

43. Begin at vertex *A*. Proceed to vertex *B*, then vertex *C*, then around the loop to *C* again, then vertex *D*, then vertex *F*, then vertex *G*, then vertex *H*, then vertex *I*.

45. *G, F, D, E, D* requires that edge *DE* be traversed twice. This is not allowed within a path.

47. *H, I, F, E* is not a path because no edge connects vertices *F* and *E*.

49. Possible answer:

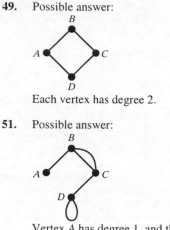

Each vertex has degree 2.

51. Possible answer:

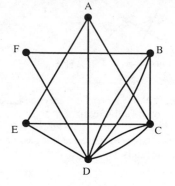

Vertex *A* has degree 1, and the rest have degree 3.

53–65. Answers will vary.

67. Use vertices to represent the six members.

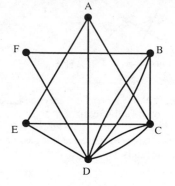

15.2   Euler Paths and Euler Circuits

Check Point 1.

We use trial and error to find one such path. The following figure shows a result.

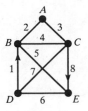

Using vertex letters to name the path, we write $D, B, A, C, B, E, D, C, E$.

Check Point 2.

We use trial and error to find an Euler circuit that starts at G. The following figure shows a result.

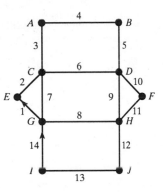

Using vertex letters to name the circuit, we write $G, E, C, A, B, D, C, G, H, D, F, H, J, I, G$.

Check Point 3.

  a. A walk through every room and the outside, using each door exactly once, means that we are looking for an Euler path or Euler circuit on the graph in Figure 15.34(b). This graph has exactly two odd vertices, namely B and E. By Euler's theorem, the graph has at least one Euler path, but no Euler circuit. It is possible to walk through every room and the outside, using each door exactly once. It is not possible to begin and end the walk in the same place.

  b. Euler's theorem tells us that a possible Euler path must start at one of the odd vertices and end at the other. We use trial and error to find such a path, starting at vertex B (room B in the floor plan), and ending at vertex E (outside in the floor plan). Possible paths follow.

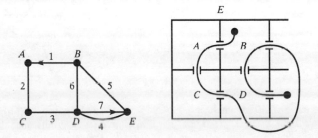

Check Point 4.

The graph has no odd vertices, so we can begin at any vertex. We choose vertex C as the starting point. From C we can travel to A, B, or D. We choose to travel to D.

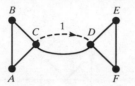

Now the remaining edge CD is a bridge, so we must travel to either E or F. We choose F.

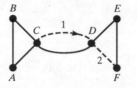

We have no choices for our next three steps, which are bridges. We must travel to E, then D, then C.

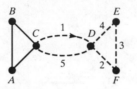

From C, we may travel to either A or B. We choose B. Then we must travel to A, then back to C.

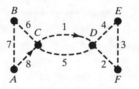

The above figure shows the completed Euler circuit. Written using the letters of the vertices, the path is C, D, F, E, D, C, B, A, C.

Exercise Set 15.2

1.

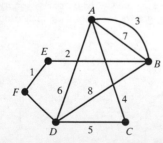

This path does not include edge FD, so it is neither an Euler path nor an Euler circuit.

3.

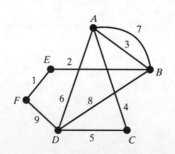

This path travels through each edge of the graph once, and only once. It begins and ends at F. Therefore, it is an Euler circuit.

5.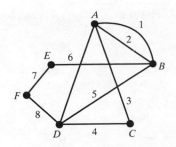

This path does not include edge *AD*, so it is neither an Euler path nor an Euler circuit.

7. a. There are exactly two odd vertices, namely *A* and *B*, so by Euler's theorem there is at least one Euler path.

b.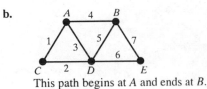

This path begins at *A* and ends at *B*.

9. a. There are no odd vertices, so by Euler's theorem, there is at least one Euler circuit.

b.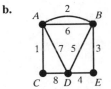

This circuit begins and ends at *C*.

11. There are more than two odd vertices, namely *B*, *D*, *G*, and *K*. Therefore by Euler's theorem, there are no Euler paths and no Euler circuits.

13. Since the graph has no odd vertices, it must have an Euler circuit, by Euler's theorem.

15. Since the graph has exactly two odd vertices, it has an Euler path, but no Euler circuit, by Euler's theorem.

17. Since the graph has more than two odd vertices, it has neither an Euler path nor an Euler circuit, by Euler's theorem.

19. a. All vertices are even, so there must be an Euler circuit.

b.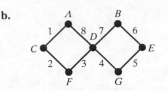

21. a. There are exactly two odd vertices, so there must be an Euler path.

b.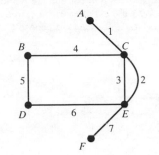

23. a. There are more than two odd vertices, so there is neither an Euler path nor an Euler circuit.

25. a. There are exactly two odd vertices, so there is an Euler path.

b.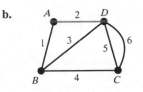

27. a. There are no odd vertices, so there is an Euler circuit.

b.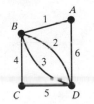

29. a. There are more than two odd vertices, so there is neither an Euler path nor an Euler circuit.

31. a. There are exactly two odd vertices, so there is an Euler path.

b.

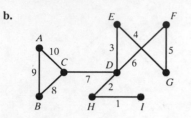

33. The two odd vertices in the graph are *A* and *C*. We start with *A*, so we must progress next to *B*. From *B*, we may travel to *C*, *D*, or *E*. We choose *C*.

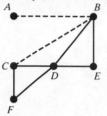

Next we travel to *F*, then *D*, then *E*.

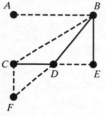

Finally we travel to *B*, then *D*, and last, to *C*. We label each step taken.

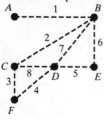

35. The two odd vertices in the graph are *A* and *C*. We start with *A*, then travel to *B*, *C*, and *E*.

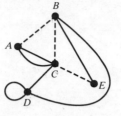

We continue on to *B*, *D*, *D*, *C*, *A*, and *C*. We label each step taken.

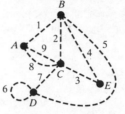

37. We begin with *A*, and travel to *D*, *H*, *G*, *F*, *E*, *B*, and *C*.

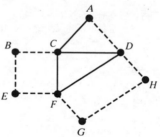

We continue on to *F*, *D*, *C*, and back to *A*. We label each step.

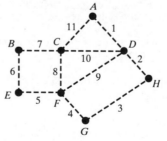

39. We begin with A, and travel to C, G, K, H, I, L, J, F, B, E, and D.

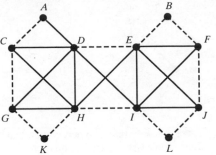

We continue on to C, H, G, D, H, E, F, I, J, E, I, D, and back to A. We label each step.

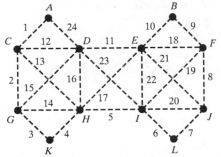

41. The graph that models the neighborhood has no odd vertices, so an Euler circuit exists with any vertex, including B, as the starting point.

43.

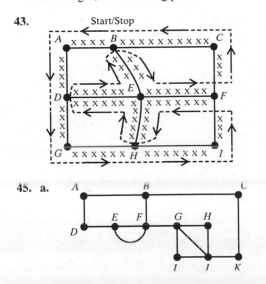

45. a.

b. There are exactly two odd vertices, namely E and B. Therefore the guard should begin at one of these vertices and end at the other.

47. a. Label the vertices N for North Bank, S for South Bank, and A and B for the two islands. Draw edges to represent bridges.

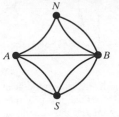

b. The graph has exactly two odd vertices, N and B, so residents can walk across all the bridges without crossing the same bridge twice.

c.

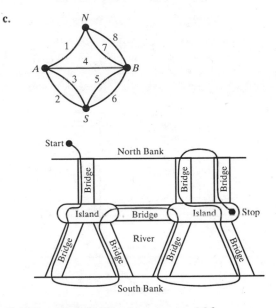

49. Use NJ to label the New Jersey vertex, M for Manhattan, SI for Staten Island, and LI for Long Island. Each edge represents a bridge.

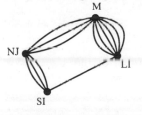

There are exactly two odd vertices, M and LI, so the graph has an Euler path. Therefore it is possible to visit each location, using each bridge or tunnel exactly once.

51. a.

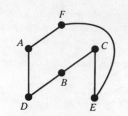

b. There are no odd vertices, so the graph has an Euler circuit. Therefore, it is possible to walk through each room and the outside, using each door exactly once.

c.

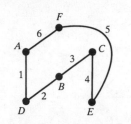

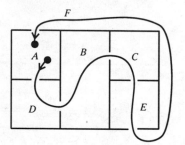

53.

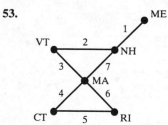

55. a.

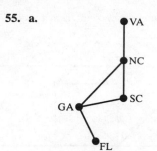

b. There are more than two odd vertices, so no Euler path exists. Therefore it is not possible to travel through these states, crossing each border exactly once.

d. For the same reason as in (b), this is not possible.

57. Answers will vary.

59. Answers will vary.

61. Answers will vary.

63. Answers will vary.

65. Answers will vary.

67. Answers will vary.

15.3   Hamilton Paths and Hamilton Circuits

Check Point 1.

   a. A Hamilton path must pass through each vertex exactly once. The graph has many Hamilton paths. An example of such a path is *E, C, D, G, B, A, F.*

   b. A Hamilton circuit must pass through every vertex exactly once and begin and end at the same vertex. The graph has many Hamilton circuits. An example of such a circuit is *E, C, D, G, B, F, A, E.*

Check Point 2.

In each case, we use the expression $(n-1)!$. For three vertices, substitute 3 for n in the expression. For six and ten vertices, substitute 6 and 10, respectively, for n.

   a.   A complete graph with three vertices has $(3-1)! = 2! = 2 \cdot 1 = 2$ Hamilton circuits.

   b.   A complete graph with six vertices has $(6-1)! = 5! = 5 \cdot 4 \cdot 3 \cdot 2 \cdot 1 = 120$ Hamilton circuits.

   c.   A complete graph with ten vertices has $(10-1)! = 9! = 9 \cdot 8 \cdot 7 \cdot 6 \cdot 5 \cdot 4 \cdot 3 \cdot 2 \cdot 1 = 362{,}880$ Hamilton circuits.

Check Point 3.

The trip described by the Hamilton circuit *A, C, B, D, A* involves the sum of four costs:

$124 + $126 + $155 + $157 = $562.

Here, $124 is the cost of the trip from *A* to *C*; $126 is the cost from *C* to *B*; $155 is the cost from *B* to *D*; and $157 is the cost from *D* to *A*. The total cost of the trip is $562.

Check Point 4.

The graph has four vertices. Thus, using $(n-1)!$, there are $(4-1)! = 3! = 6$ possible Hamilton circuits. The 6 possible Hamilton circuits and their costs are shown.

Hamilton Circuit	Sum of the Weights of the Edges	=	Total Cost
A, B, C, D, A	20 + 15 + 50 + 30	=	$115
A, B, D, C, A	20 + 10 + 50 + 70	=	$150
A, C, B, D, A	70 + 15 + 10 + 30	=	$125
A, C, D, B, A	70 + 50 + 10 + 20	=	$150
A, D, B, C, A	30 + 10 + 15 + 70	=	$125
A, D, C, B, A	30 + 50 + 15 + 20	=	$115

The two Hamilton circuits having the lowest cost of $115 are *A, B, C, D, A* and *A, D, C, B, A.*

Check Point 5.

The Nearest Neighbor method is carried out as follows:

• Start at *A*.

• Choose the edge with the smallest weight: 13. Move along this edge to *B*.

• From *B*, choose the edge with the smallest weight that does not lead to *A*: 5. Move along this edge to *C*.

• From *C*, choose the edge with the smallest weight that does not lead to a city already visited: 12. Move along this edge to *D*.

• From *D*, the only choice is to fly to *E*, the only city not yet visited: 154.

• From *E*, close the circuit and return home to *A*: 14.

An approximate solution is the Hamilton circuit *A, B, C, D, E, A*. The total weight is
13 + 5 + 12 + 154 + 14 = 198.

Exercise Set 15.3

 1. One such path is *A, G, C, F, E, D, B*.

 3. One such circuit is *A, B, G, C, F, E, D, A*.

 5. One such path is *A, F, G, E, C, B, D*.

 7. One such circuit is *A, B, C, E, G, F, D, A*.

 9. **a.** This graph is not complete. For example, no edge connects *A* and *B*. Therefore it may not have Hamilton circuits.

11. **a.** This graph is complete: there is an edge between each pair of vertices. Therefore it must have Hamilton circuits.

   b. There are 6 vertices, so the number of Hamilton circuits is $(6-1)! = 5! = 120$.

13. **a.** This graph is not complete. For example, no edge connects *G* and *F*. Therefore it may not have Hamilton circuits.

15. $(3-1)! = 2! = 2$

17. $(12-1)! = 11! = 39,916,800$

19. 11

21. $9 + 8 + 11 + 6 + 2 = 36$

23. $9 + 7 + 6 + 11 + 3 = 36$

25. $40 + 24 + 10 + 14 = 88$

27. $20 + 24 + 12 + 14 = 70$

29. $14 + 12 + 24 + 20 = 70$

31. On a complete graph with four vertices, there are 6 distinct Hamilton circuits. These are listed in Exercises 25–30. We have already computed the weight of each possible Hamilton circuit, as required by the Brute Force Method. The optimal solutions have the smallest weight, 70. They are *A, C, B, D, A*, and *A, D, B, C, A*.

33. Starting from *B*, the edge with smallest weight is *BD*, with weight 12. Therefore, proceed to *D*. From *D*, the edge having smallest weight and not leading back to *B* is *DC*, with weight 10. From *C*, our only choice is *CA*, with weight 20. From *A*, return to *B*. Edge *AB* has weight 40. The total weight of the Hamilton circuit is
12 + 10 + 20 + 40 = 82.

35.

Hamilton Circuit	Sum of the Weights of the Edges	=	Total Weight
A, B, C, D, E, A	500 + 305 + 320 + 302 + 205	=	1632
A, B, C, E, D, A	500 + 305 + 165 + 302 + 185	=	1457
A, B, D, C, E, A	500 + 360 + 320 + 165 + 205	=	1550
A, B, D, E, C, A	500 + 360 + 302 + 165 + 200	=	1527
A, B, E, C, D, A	500 + 340 + 165 + 320 + 185	=	1510
A, B, E, D, C, A	500 + 340 + 302 + 320 + 200	=	1662
A, C, B, D, E, A	200 + 305 + 360 + 302 + 205	=	1372
A, C, B, E, D, A	200 + 305 + 340 + 302 + 185	=	1332
A, C, D, B, E, A	200 + 320 + 360 + 340 + 205	=	1425
A, C, E, B, D, A	200 + 165 + 340 + 360 + 185	=	1250
A, D, B, C, E, A	185 + 360 + 305 + 165 + 205	=	1220
A, D, C, B, E, A	185 + 320 + 305 + 340 + 205	=	1355

Using the Brute Force method, we compute the sum of the weights of the edges for each possible Hamilton circuit, as in the table above. The smallest weight sum is 1220, representing a total cost of $1220 for airfare. This results from the Hamilton circuit A, D, B, C, E, A. Thus, the sales director should fly to the cities in this order.

37.

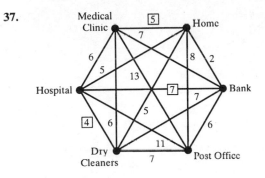

39. $2 + 6 + 7 + 4 + 6 + 5 = 30$

41. Label the vertices H for Home, B for Bank, P for Post Office, and M for Market.

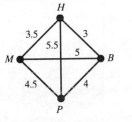

43. From Home, the closest errand is the Bank, 3 miles away. From the Bank, the closest remaining errand is the Post Office, 4 miles away. From the Post Office, the last remaining errand is the Market, 4.5 miles away. From the Market, Home is 3.5 miles away. The total distance for this Hamilton circuit is 3 + 4 + 4.5 + 3.5 = 15 miles. This is the same route found in Exercise 42.

45. Answers will vary.

47. Answers will vary.

49. Answers will vary.

51. Answers will vary.

53. Answers will vary.

55. Answers will vary.

57. Answers will vary.

15.4   Trees

Check Point 1.

The graph in Figure 15.51(c) is a tree. It is connected and has no circuits. There is only one path joining any two vertices. Every edge is a bridge; if removed, each edge would create a disconnected graph. Finally, the graph has 7 vertices and 7 – 1, or 6, edges.

The graph in Figure 15.51(a) is not a tree because it is disconnected. There are 7 vertices and only 5 edges, not the 6 edges required for a tree.

The graph in Figure 15.51(b) is not a tree because it has a circuit, namely *A, B, C, D, A*. There are 7 vertices and 7 edges, not the 6 edges required for a tree.

Check Point 2.

A spanning tree must contain all six vertices shown in the connected graph in Figure 15.55. The spanning tree must have one edge less than it has vertices, so it must have five edges. The graph in Figure 15.55 has eight edges, so we must remove three edges. We elect to remove the edges of the circuit *C, D, E, C*. This leaves us the following spanning tree.

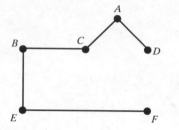

Check Point 3.

Step 1. Find the edge with the smallest weight. This is edge *DE*; mark it.

Step 2. Find the next-smallest edge in the graph. This is edge *DC*; mark it.

Step 3. Find the next-smallest edge in the graph that does not create a circuit. This is edge *DA*; mark it.

Step 4. Find the next-smallest edge in the graph that does not create a circuit. This is *AB*; mark it.

The resulting minimum spanning tree is complete. It contains all 5 vertices of the graph, and has 5 – 1, or 4, edges. Its total weight is 12 + 14 + 21 + 22 = 69. It is shown below.

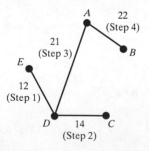

Exercise Set 15.4

1. Yes, this graph is a tree. It has 3 edges on 4 vertices, is connected, and has no circuits. Every edge is a bridge.

3. No, this graph is not a tree. It is disconnected.

5. Yes, this graph is a tree. It has 3 edges on 4 vertices, is connected, and has no circuits. Every edge is a bridge.

7. No, this graph is not a tree. It has a circuit.

9. Yes, this graph is a tree. It has 6 edges on 7 vertices, is connected, and has no circuits. Every edge is a bridge.

11. i.; If the graph contained any circuits, some points would have more than one path joining them.

13. ii.; a tree with n vertices must have $n - 1$ edges.

15. ii.; a tree has no circuits.

17. iii.

19.

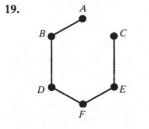

21.

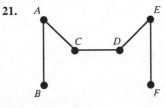

23.

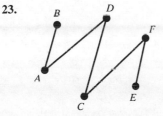

25. Kruskal's algorithm results in the following figure.

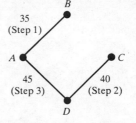

This minimum spanning tree has weight
$35 + 40 + 45 = 120$.

27. Kruskal's algorithm results in the following figure.

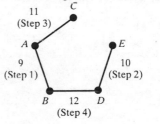

This minimum spanning tree has weight
$9 + 10 + 11 + 12 = 42$.

29. Kruskal's algorithm results in the following figure.

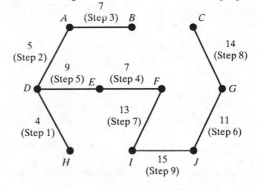

This minimum spanning tree has weight
$4 + 5 + 7 + 7 + 9 + 11 + 13 + 14 + 15 = 85$.

31. Kruskal's algorithm results in the following figure.

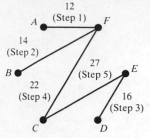

This minimum spanning tree has weight
12 + 14 + 16 + 22 + 27 = 91.

33.

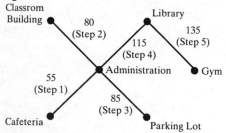

35. a.

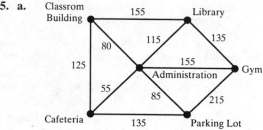

b. Kruskal's algorithm is shown in the following figure.

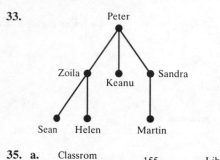

The total length of the sidewalks that need to be sheltered by awnings is
55 + 80 + 85 + 115 + 135 = 470 feet.

37. Kruskal's method is shown in the figure.

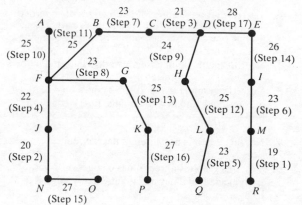

The smallest number of feet of underground pipes is
19 + 20 + 21 + 22 + 23 + 23 + 23 + 23 + 24 + 25 + 25 + 25 + 25 + 26 + 27 + 27 + 28 = 406 feet.

39. Answers will vary.

41. Answers will vary.

43. Answers will vary.

45. Answers will vary.

47. Answers will vary.

49. The graph has 6 vertices and 6 edges. A spanning tree on 6 vertices has 5 edges. Only three edges are able to be removed without a disconnected graph resulting. These are *BC*, *BF*, and *CF*. The result of removing any one of these edges is a spanning tree.

Review Exercises

1. Each graph has 5 vertices, *A, B, C, D,* and *E*. Each has one edge connecting *A* and *B*, one connecting *A* and *C*, one connecting *A* and *D*, one connecting *A* and *E*, and one connecting *B* and *C*. Both graphs have the same number of vertices, and these vertices are connected in the same ways. A third way to draw the same graph is

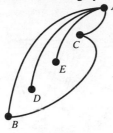

2. *A*: 5 (A loop add degree 2.); *B*: 4; *C*: 5; *D*: 4; *E*: 2

3. Even: *B, D, E*; odd: *A, C*

4. *B, C,* and *E*

5. Possible answer: *E, D, B, A* and *E, C, A*

6. Possible answer: *E, D, C, E*

7. Yes. A path can be found from any vertex to any other vertex.

8. No. There is no edge which can be removed to leave a disconnected graph.

9. *AD, DE,* and *DF*

10.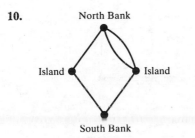

11. Use the states' abbreviations to label the vertices representing them.

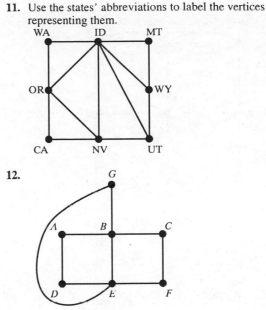

12.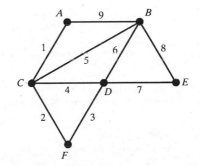

13. **a.** Neither. There are more than two odd vertices.

14. **a.** Euler circuit: there are no odd vertices.

b.

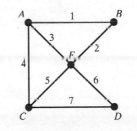

15. **a.** Euler path: there are exactly two odd vertices.

b.

16. There are exactly two odd vertices, *G* and *I*. We start at *G* and continue to *D*, *A*, *B*, *C*, *F*, *I*, *H*, and *E*.

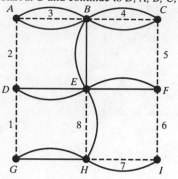

We continue erasing edges as we go, till we have completed an Euler path ending at *I*.

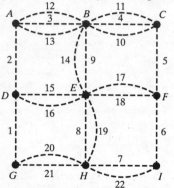

17. We may begin anywhere, since there are no odd vertices. We erase edges as we go, till we have the Euler circuit. We begin at *A*.

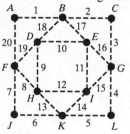

18. **a.** Yes, they would. The graph has exactly two odd vertices, so there is an Euler path.

b.

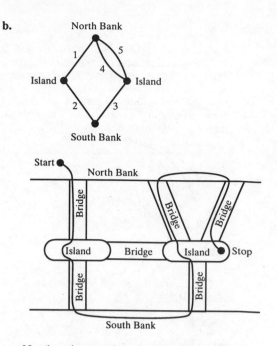

c. No; there is no such path. Since the graph has odd vertices, it does not have an Euler circuit.

19. Yes, it is possible. There are exactly two odd vertices, and therefore there is an Euler path (but no Euler circuit).

20. **a.** Yes it is possible. There are no odd vertices, so there is an Euler circuit.

b.

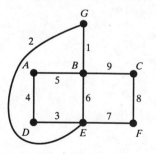

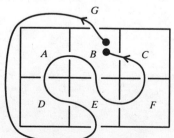

21. a.

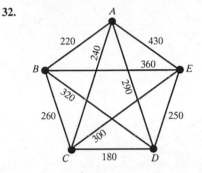

b. Yes. There are exactly two odd vertices C and F, so there is an Euler path.

c. The guard should begin at C and end at F, or vice versa.

22. A, E, C, B, D, A

23. D, B, A, E, C, D

24. a. No, because this is not a complete graph. It may not have Hamilton circuits.

25. a. Yes, because this is a complete graph.

b. $(4-1)! = 3! = 6$

26. a. No, because this is not a complete graph. It may not have Hamilton circuits.

27. a. Yes, because this is a complete graph.

b. $(5-1)! = 4! = 24$

28.

A, B, C, D, A:	$4 + 6 + 5 + 4$	$= 19$
A, B, D, C, A:	$4 + 7 + 5 + 2$	$= 18$
A, C, B, D, A:	$2 + 6 + 7 + 4$	$= 19$
A, C, D, B, A:	$2 + 5 + 7 + 4$	$= 18$
A, D, B, C, A:	$4 + 7 + 6 + 2$	$= 19$
A, D, C, B, A:	$4 + 5 + 6 + 4$	$= 19$

29. These are the only possible Hamilton circuits on a graph with 4 vertices. The lowest weight, 18, occurs on the circuits A, B, D, C, A and A, C, D, B, A. These are the optimal solutions.

30. Start with A. Then edge AC has the smallest weight, 2, of all edges starting at A. Proceed to C. From C, edge CD has the smallest weight, 5, of edges not returning to A. From D, we must travel DB, with weight 7, to B. We return to A along BA, with weight 4. The total weight of this Hamilton circuit is $2 + 5 + 7 + 4 = 18$.

31. Start with A. Of all paths leading from A, the path with smallest weight is AB, with weight 4. Proceed to B. The path with smallest weight leading from B, but not to A, is BE, with weight 6. The path with smallest weight leading from E, but not to A or B, is ED, with weight 4. From D, we proceed along DC, with weight 3, to C, the only remaining vertex. We then return to A along CA, with weight 7. The total weight is $4 + 6 + 4 + 3 + 7 = 24$.

32.

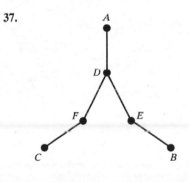

33. Start at A. The lowest cost from A, $220, is on edge AB. From B, the lowest cost other than returning to A is $260, on edge BC. From C, the lowest cost to a new city is $180, on edge CD. From D, the salesman must fly to E for $250, then return to A for $430. The total cost of this circuit is $220 + 260 + 180 + 250 + 430 = \$1340$.

34. Yes. It is connected, has no circuits, has 6 edges on 7 vertices, and each edge is a bridge.

35. No. It has a circuit.

36. No. It is disconnected.

37.

A graph with vertices A at top, connected down to D, which branches to F and E; F connects to C and E connects to B.

38.

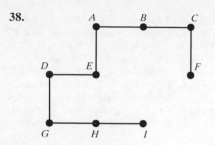

39. Kruskal's algorithm is demonstrated in the figure.

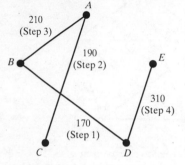

The total weight is
170 + 190 + 210 + 310 = 880.

40. Kruskal's algorithm is demonstrated in the figure.

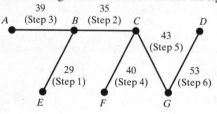

The total weight is
29 + 35 + 39 + 40 + 43 + 53 = 239.

41. The figure demonstrates Kruskal's algorithm and the layout of the cable system.

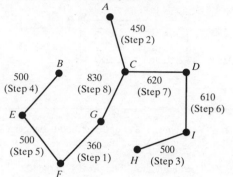

The smallest length of cable needed is
360 + 450 + 500 + 500 + 500 + 610 + 620 + 830 = 4370 miles.

Chapter Test

1. *A*: 2; *B*: 2; *C*: 4; *D*: 3; *E*: 2; *F*: 1

2. *A, D, E* and *A, B, C, E*

3. *B, A, D, E, C, B*

4. *CF*

5.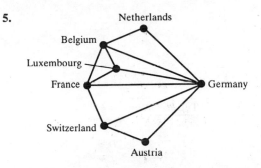

6. **a.** Euler path: there are exactly two odd vertices.

   b.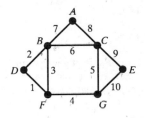

7. **a.** Neither: there are more than two odd vertices.

   b. N/A

8. **a.** Euler circuit: there are no odd vertices.

   b.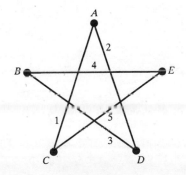

9. We begin at *A*, then proceed to *E, I, H*, and so on, erasing edges once they have been crossed. The result is shown in the figure.

   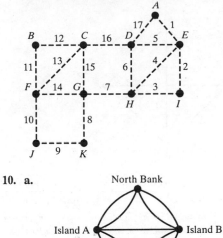

10. **a.**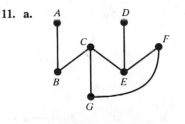

    b. Yes: there are exactly two odd vertices.

    c. It should begin at one of the islands, and end at the other island.

11. **a.**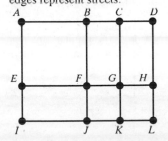

    b. No: there are more than two odd vertices.

12. **a.** Let vertices represent intersections, and let edges represent streets.

    b. No: there are more than two odd vertices.

13. A, B, C, D, G, F, E, A and A, F, G, D, C, B, E, A.

14. $(5 - 1)! = 4! = 24$

15.

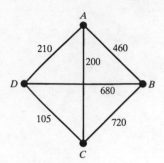

16.

Hamilton Circuit	Sum of the Weights of the Edges	=	Total Cost
A, B, C, D, A	460 + 720 + 105 + 210	=	$1495
A, B, D, C, A	460 + 680 + 105 + 200	=	$1445
A, C, B, D, A	200 + 720 + 680 + 210	=	$1810
A, C, D, B, A	200 + 105 + 680 + 460	=	$1445
A, D, B, C, A	210 + 680 + 720 + 200	=	$1810
A, D, C, B, A	210 + 105 + 720 + 460	=	$1495

The optimal route is *A, B, D, C, A* or *A, C, D, B, A*. The total cost for this route is $1445.

17. Starting from *A*, the edge with smallest weight is *AE*, with weight 5. Proceed to *E*. From *E*, the edge with smallest weight, and not leading back to *A*, is *ED*, with weight 8. From *D*, the edge with smallest weight, and to a new vertex, is *DC*, with weight 4. From *C*, only *B* remains. Edge *CB* has weight 5. Return to *A* by edge *BA*, with weight 11. The total weight of this Hamilton circuit is
5 + 8 + 4 + 5 + 11 = 33.

18. No; it has a circuit, namely *C, D, E, C*.

19.

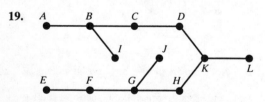

20. Kruskal's algorithm is shown in the figure.

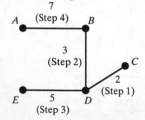

The total weight of the minimum spanning tree is 2 + 3 + 5 + 7 = 17.